Air Pollution Control Engineering for Environmental Engineers

Air Pollution Control Engineering for Environmental Engineers is a concise yet comprehensive book that can be read and used from cover to cover, presenting topics that are fundamental for environmental engineering students, engineers, and professionals in the fields of air pollution control engineering and management. It covers topics including regulatory approaches to managing air pollution, emissions calculations, and control technologies for various air pollutants. It also presents practical and contemporary issues, such as fugitive emission and leak detection and repair (LDAR). Subjects in the specifications of Fundamentals of Engineering (FE) and Professional Engineering (PE) exams are also covered. With real-world engineering design and calculation examples, the reader's understanding and common sense needed for air pollution control and management will be enhanced.

- Updated throughout.
- Provides well-digested practical information for both engineering students and engineering professionals in the fields of air pollution control engineering and management.
- Covers subjects in the specifications of FE and PE exams relevant to air pollution control.
- Includes practical and meaningful engineering design and calculation examples.

Fundamentals of Environmental Engineering
Series Editor: Jih Fen (Jeff) Kuo

Air Pollution Control Engineering for Environmental Engineers (First Edition)
Jeff Kuo

Site Assessment and Remediation for Environmental Engineers
Cristiane Q. Surbeck and Jeff Kuo

Air Pollution Control Engineering for Environmental Engineers (Second Edition)
Jih Fen (Jeff) Kuo

For more information about this series, please visit: https://www.routledge.com/Fundamentals-of-Environmental-Engineering/book-series/CRCFUNOFENV

Air Pollution Control Engineering for Environmental Engineers

Second Edition

Jih Fen (Jeff) Kuo

CRC Press
Taylor & Francis Group
Boca Raton London New York

CRC Press is an imprint of the
Taylor & Francis Group, an **informa** business

Designed cover image: Shutterstock

Second edition published 2026
by CRC Press
2385 NW Executive Center Drive, Suite 320, Boca Raton FL 33431

and by CRC Press
4 Park Square, Milton Park, Abingdon, Oxon, OX14 4RN

CRC Press is an imprint of Taylor & Francis Group, LLC

First edition published by CRC Press 2023

ISBN: 978-1-032-87451-7 (hbk)
ISBN: 978-1-032-87454-8 (pbk)
ISBN: 978-1-003-53270-5 (ebk)

DOI: 10.1201/9781003532705

Typeset in Times
by codeMantra

Contents

Preface

I worked for engineering consulting companies and the Los Angeles County Sanitation Districts before starting my teaching career at California State University, Fullerton (CSUF) in 1995. My industrial experience has given me an in-depth understanding of what an environmental engineering student and/or a junior environmental professional should know about air quality and pollution control. In addition to teaching a course on air pollution control engineering, I have also worked on several externally funded projects related to air quality recently and in the remote past at CSUF.

I believe that it is essential for an engineer to obtain professional licensure. This book is not a preparation book for the Fundamentals of Engineering (FE) and/or Professional Engineering (PE) exams. However, I considered these two exams when I wrote this book. The specifications of the FE exam reflect what an engineering undergraduate should have learned from school, while those of the PE exam are what a practicing engineer should know.

A student should have learned the basics and common sense of engineering as much as possible from their collegiate courses before starting a professional engineering career. Although they may be able to do engineering calculations, they often lack a good understanding of the meanings of the results and are unable to justify the values obtained. Throughout the years, the USEPA has published technical articles that are available to the state, tribal, and local regulatory agency personnel as well as the public. Although the courses of USEPA's Air Pollution Training Institute (APTI) (now the Air Knowledge, https://airknowledge.gov/) are mostly restricted to state and local environmental agency personnel, the training materials are available to the public for educational purposes. These technical documents and course materials cover various topics and are intended for different types of personnel (e.g., engineers, scientists, inspectors, and policymakers). This book extracts useful information relevant to environmental engineers from them. In addition, many examples are given in this book, and a discussion is provided at the end of most examples to clarify the concept and to give the practical meanings of the derived results.

There are many textbooks on air pollution control engineering available in the market. Although all the information in those books is valuable and practical, it would be nearly impossible for a lecturer to cover all the content in one semester. In addition, students may not be able to grasp and retain the vital information that the authors intend to convey. This book is shorter but still serves as a sufficient introductory book for air quality control engineering in a three-semester-unit course. In a succinct format, this practical engineering book provides well-digested information for engineering students and junior engineering professionals. My hope and the objective of this book are for readers to read it from cover to cover and gain the common sense needed to be well equipped and successful in their engineering profession.

The U.S. customary system units and SI are commonly used in engineering practice. This book will use both systems of units. The primary purpose is to help users of this book develop a good sense of the numerical values they might encounter in both systems of units.

Author Biography

Dr. Jih Fen (Jeff) Kuo worked in the environmental engineering industries for over 10 years before joining the Department of Civil and Environmental Engineering at California State University, Fullerton, in 1995. His areas of research in environmental engineering include dechlorination of halogenated aromatics by ultrasound, fines/bacteria migration through porous media, biodegradability of heavy hydrocarbons, surface properties of composite mineral oxides, kinetics of activated carbon adsorption, wastewater filtration, THM formation potential of ion exchange resins, UV disinfection, sequential chlorination, nitrification/denitrification, removal of target compounds using nanoparticles, persulfate oxidation of persistent chemicals, microwave oxidation for wastewater treatment, landfill gas recovery and utilization, greenhouse gases control technologies, fugitive methane emissions from the gas industry, and stormwater runoff treatment. He received a B.S. degree in chemical engineering from National Taiwan University, an M.S. in chemical engineering from the University of Wyoming, an M.S. in petroleum engineering, and an M.S. and a Ph.D. in environmental engineering from the University of Southern California. He is a professional civil, mechanical, and chemical engineer registered in California.

1 Air Pollutants and Their Adverse Effects

1.1 CHARACTERISTICS OF COMMON AIR POLLUTANTS

Air pollutants can be in the form of gas, liquid, or solid. They can be organic, inorganic, or microbial species.

1.1.1 Gas, Vapor, Particulate Matter, Dust, Soot, Aerosol, Smoke, Fog, and Smog

Many terms are being used to describe the state of an air pollutant. Air is a mixture of gases. *Gas* is a state of matter that has no fixed shape or volume. *Vapor* refers to a gas phase where the same substance exists in its liquid state (or in the solid state) under those conditions. For example, we say that our ambient air contains nitrogen gas and water vapor. No liquid or solid nitrogen coexists with nitrogen gas, while water or ice coexists with water vapor under ambient conditions.

In addition to gases and vapors, air also contains *particulate matter* (PM), which includes suspended solids and liquids such as dust, soot, smoke, and liquid droplets, as well as living issues such as pollen, bacteria, spores, mold, and fungus. *Dust* is a loose term applied to solid particles capable of temporary suspension in air or other gases. Dust does not diffuse, but settles under the influence of gravity. Dust particles can comprise minerals, organic matter, and various other substances. *Soot* primarily consists of particles of amorphous carbon and tar generated from the incomplete combustion of hydrocarbons, and it may contain a mix of chemicals such as heavy metals and organic compounds.

Aerosols are PMs (liquid or solid) suspended in the air, and they can be natural (e.g., fog, sea spray, tiny dust, and biological particles such as pollen and bacteria) or from anthropogenic (human-made) sources (e.g., soot from incomplete combustion, sulfates from fossil fuel combustion). *Fog* is a meteorological phenomenon consisting of tiny water droplets in the air near the ground. *Smoke* is an aerosol, a visible mixture of gases and fine particles generated by combustion. The particles are present in sufficient quantity to be observable regardless of the presence of other solids. *Smog* is a term derived from smoke and fog, and it was first used around the 1950s to describe the combination of smoke and fog in London. *Photochemical smog* is air contamination caused by photochemical reactions among nitrogen oxides and hydrocarbons activated by sunlight. Ozone (O_3), along with other "secondary" pollutants (defined later), is a compound of concern (COC) in photochemical smog.

DOI: 10.1201/9781003532705-1

1.1.2 Inorganics, Organics, or Microbial

Air pollutants, such as nitrogen dioxide (NO_2) and benzene (C_6H_6), can be pure inorganic or organic molecules. It can also be inorganic minerals, such as asbestos, or living matter, such as bacteria. It should be mentioned that an air pollutant can be a combination of two or more pollutants. For example, toxic heavy metals such as lead (Pb) can get a<u>d</u>sorbed onto PM, and sulfur dioxide (SO_2) can get a<u>b</u>sorbed into tiny water droplets, making both PM and water droplets more toxic and unacceptable.

1.1.3 Volatile Organic Compounds (VOCs)

Volatile organic compounds (VOCs) are of great environmental concern because of their mobility and toxicity. They are organic compounds with high vapor pressures that can be emitted from solids or liquids under ambient conditions. With regard to air pollution, as specifically defined in 40CFR Part 51.100(s), VOCs are any compound of carbon, excluding carbon monoxide (CO), carbon dioxide (CO_2), carbonic acid (H_2CO_3), metallic carbides or carbonates, and ammonium carbonate ($(NH_4)_2CO_3$), which participates in atmospheric photochemical reactions. Many organic compounds are volatile, but they are not considered VOCs by the U.S. Environmental Protection Agency (USEPA) in air pollution management because they have negligible photochemical reactivity. In addition to chlorofluorocarbons (CFCs), hydrofluorocarbons (HFCs), and hydrochlorofluorocarbons (HCFCs), the other notable exceptions include acetone (CH_3-CO-CH_3), dimethyl carbonate ($OC(OCH_3)_2$), ethane (C_2H_6), methane (CH_4), methyl acetate (CH_3COOCH_3), 1,1,1-trichloroethane (1,1,1-TCA; $C_2H_3Cl_3$), methyl formate ($HCOOCH_3$), methylene chloride (dichloromethane, CH_2Cl_2), siloxanes, and tetrachloroethylene (perchloroethylene, PCE, C_2Cl_4). Many of these exceptions (e.g., 1,1,1-TCA, PCE) are of concern to water or wastewater pollution as VOCs; however, they are not considered as VOCs in air regulations.

1.1.4 Primary vs. Secondary Air Pollutants

The USEPA imposes primary standards for drinking water to protect public health and instills secondary standards to regulate substances that may cause aesthetic effects. The USEPA also imposes primary and secondary air quality <u>standards</u> for the ambient air. However, "primary" and "secondary" have different meanings for <u>types</u> of air pollutants.

A *primary air pollutant* is emitted directly from a source, such as an industrial stack or a car exhaust pipe; examples include CO and nitric oxide (NO). In the atmosphere, it has the same physical and chemical forms as when it was emitted. A *secondary air pollutant* is not directly emitted from a source but is formed from chemical or photochemical reactions among the primary air pollutants in the atmosphere. Ozone is an excellent example of a secondary pollutant formed from photochemical reactions among sunlight, nitrogen dioxide (NO_2), and hydrocarbons (HCs). Acid rain is another example because it is formed by reactions between water droplets, nitrogen oxides (NO_x), and sulfur oxides (SO_x). NO_x refers to a group of reactive gases composed of nitrogen and oxygen. The two most significant ones, in terms of environmental and health impacts, are NO and NO_2. SO_x refers to a group of gases composed of sulfur and oxygen. The two most significant ones, in terms of environmental and health impacts, are sulfur dioxide (SO_2) and sulfur trioxide (SO_3). Some air pollutants can be both primary and secondary. NO_2 is mainly a secondary air pollutant formed from NO oxidation in the ambient air, but it can also be a primary air pollutant emitted directly from combustion processes. VOCs can be primary or secondary air pollutants. Larger PMs in the ambient air (e.g., dust, volcano ash) are typically primary air pollutants. Finer PMs are generally secondary air pollutants generated from physical (e.g., condensation), chemical, or photochemical reactions in the atmosphere.

1.2 GROUPS OF AIR POLLUTANTS

The USEPA categorizes the air pollutants into two groups: outdoor and indoor. The outdoor pollutants are further classified into (i) criteria air pollutants, (ii) hazardous air pollutants (HAPs), and (iii) greenhouse gases (GHGs).

1.2.1 Criteria Air Pollutants

The Clean Air Act (CAA), the primary federal law of the U.S. for controlling air quality, requires the USEPA to set National Ambient Air Quality Standards (NAAQS) for six *criteria pollutants* in outdoor air to protect human health, properties, ecosystems, and the environment. They are (i) CO, (ii) NO_2, (iii) SO_2, (iv) ground-level O_3, (v) PM, and (vi) lead (Pb). They are present in the ambient air all over the U.S. The term "criteria pollutants" derives from the requirement that the USEPA describe these pollutants' characteristics and potential health and welfare effects. These standards are set or revised based on these criteria.

1.2.2 Hazardous Air Pollutants

HAPs are air pollutants that have been listed under Section 112(b) of the Clean Air Act Amendments (CAAA) of 1990. They are categorized under National Emission Standards for Hazardous Air Pollutants (NESHAPs) or HAPs regulations. They are also known as toxic air pollutants or air toxics. *Air toxics* are known or suspected to cause cancer or other serious health effects, such as reproductive and childbirth issues, as well as adverse environmental effects. The original list included 189 pollutants; since 1990, the USEPA has modified the list through rulemaking to 187 HAPs. The majority of the 187 HAPs are volatile organic compounds. In addition to the commonly mentioned toxic organics (e.g., acetaldehyde (CH_3CHO), benzene (C_6H_6), chlorine (Cl_2), chlorobenzene (C_6H_5Cl), chloroform ($CHCl_3$), 1,4-dioxane ($C_4H_8O_2$), formaldehyde (HCHO), methylene chloride, phenol (C_6H_5OH), styrene ($C_6H_5CH{=}CH_2$), tetrachloroethylene, toluene ($C_6H_5CH_3$), and vinyl chloride ($H_2C{=}CHCl$)), pesticides (e.g., chlordane and lindane), heavy metals (e.g., mercury (Hg), cadmium (Cd), lead, and chromium (Cr) compounds), acids (e.g., hydrochloric acid (HCl) and hydrofluoric acid (HF)), cyanide compounds, and asbestos, the list also includes compounds that are often considered not that toxic such as methanol

(CH_3OH) and phosphorus (P). The list can be found at https://www.epa.gov/haps/initial-list-hazardous-air-pollutants-modifications.

Lead is a criteria pollutant and an air toxic out of the six criteria pollutants. HAPs are generally localized and are associated with specific emission sources, while criteria pollutants are widespread and monitored with ambient air quality standards. Unlike the criteria pollutants, there are no set concentration limits for HAPs in the ambient air; emission standards regulate them.

1.2.3 Greenhouse Gases

Greenhouse gases (GHGs) are those gases that can trap heat in the atmosphere. Elevated concentrations of GHGs due to human activities are considered the main contributors to global warming and climate change. The GHGs of concern are carbon dioxide (CO_2), methane, nitrous oxide (N_2O), and fluorinated gases (i.e., hydrofluorocarbons (HFCs), perfluorocarbons (PFCs), sulfur hexafluoride (SF_6), and nitrogen trifluoride (NF_3)).

1.2.4 Indoor Air Pollutants

Indoor air pollutants are those found within and around buildings and structures that affect the health and comfort of the building occupants. These pollutants may originate from outdoor infiltration, indoor sources, or human activities. They can pose significant health risks because people spend a lot of time indoors. Common indoor air pollutants include PM, CO, NO_2, O_3, VOCs, secondhand smoke, Pb, asbestos, biological contaminants (e.g., mold, dust mites, pet dander, bacteria, and viruses), and radon (Rn).

1.3 SOURCES OF AIR POLLUTANTS

Emissions are the total substances discharged. *Air emission source*s include anything that releases pollutants into the air. There are many sources of air emissions, and many terms are used to describe the sources. For example, >75% of emissions of criteria pollutants, except ozone, originate from combustion. Some commonly used terms for air pollution sources are described in this section.

1.3.1 Natural, Biogenic, and Anthropogenic Sources

Examples of natural sources of air pollutants include wildfires, wind-blown dust, volcanoes, and vegetation. *Biogenic emissions* are emissions from natural sources, such as plants and trees. These sources emit HAPs (e.g., formaldehyde and acetaldehyde) and large quantities of non-HAP VOCs. *Anthropogenic emissions* result from human activities, and they are of particular interest because they can be reduced through regulatory and voluntary actions, leading to air quality improvements. Examples of anthropogenic sources include combustion, industrial processes, agriculture, and transportation.

1.3.2 Stationary vs. Mobile Sources

Sources of air pollutant emissions can also be grouped into two categories: stationary sources and mobile sources. A *stationary source* is any building, structure, facility, or installation that emits or may emit any air pollutant. *Mobile sources* include on-road vehicles and non-road (or off-road) vehicles and equipment (i.e., agricultural equipment, construction equipment, recreational vehicles, marine vessels, aircraft, and rail locomotives).

1.3.3 Point, Area, Line, and Volume Sources

Stationary sources can be divided into two main subcategories: point and area. *Point sources* consist of a single emission source with an identified location at a facility. Facilities could have multiple point sources located on-site. Point sources, such as boilers, spray booths, or degreasers, are usually associated with manufacturing and industrial processes. An *area source* encompasses a variety of small emission sources dispersed within a specific area that cumulatively contribute to substantial emissions; examples include residential heating (e.g., wood stoves, fireplaces, water heaters), small businesses (e.g., dry cleaners, gas stations), and consumer products (e.g., barbecue lighter fluid and hair spray).

A *line source* is a source of emissions that emanates from a defined linear (one-dimensional) geometry. A highway can be a line source of air pollution. A *volume source* is a three-dimensional emission source (i.e., emissions are distributed over a specific area and height). Dust emissions from wind erosion of coal piles and fugitive VOC emissions from equipment components (e.g., flanges and valves) within an oil refinery are examples of volume sources (see Section 1.3.5 for a definition of fugitive emissions).

1.3.4 Major vs. Area Sources of HAPs

Major sources of HAPs are defined explicitly as facilities having air toxics emissions at or above the *significant source threshold*, which is 10 tons/year of a single HAP or 25 tons/year of any combination of HAPs, as defined by the CAA. Smaller stationary sources that emit below the significant source threshold are *area sources of HAPs.*

1.3.5 Fugitive Emissions

Fugitive emissions refer to pollutants released into the ambient air unintentionally or without passing through a controlled discharge point (e.g., vent, stack). These emissions often escape from industrial processes, pressurized equipment, evaporation, or operational activities.

1.4 ADVERSE EFFECTS OF AIR POLLUTION

Air pollution would adversely impact human health, aesthetics (e.g., visibility and smells), properties, ecosystems, and our global environment. A *receptor of air pollution* is an entity exposed to air pollutants. It can be any human, human body organs, ecosystem (e.g., plants, animals, wetlands, forests), or built environment (structures, paints, metals, and construction materials) affected by air pollution. Human body organ receptors include the skin, eyes, nose, and respiratory system. *Sensitive receptors* are people (e.g., children, the elderly, pregnant women, and patients) with increased sensitivity to air pollution or environmental contaminants. *Sensitive receptor locations* include schools, parks and playgrounds, daycare centers, nursing homes, and hospitals. This section briefly describes the adverse effects of air pollutants.

1.4.1 Human Health

Air pollutants can enter the human body by inhalation, ingestion, and dermal contact. They may remain in the lungs or the bloodstream, which goes through all body parts. They may undergo chemical changes in the body, especially in the liver, before being exhaled or excreted. The effects of air pollution on human health vary from one person to another. However, the elderly, infants, pregnant women, and patients with chronic lung or heart diseases are considered the most sensitive receptors to the impacts of air pollution. The adverse effects can be acute or chronic.

Lead (Pb). Once inside the human body, Pb is distributed in the blood throughout the body and accumulates. The most encountered adverse impacts are neurological effects in children and cardiovascular effects in adults. Infants and young children are susceptible to lead exposures, even at low levels, which may contribute to their behavioral problems, learning deficits, and lower IQs.

Carbon monoxide (CO). A high CO concentration in the inhaled air will reduce the amount of oxygen transported in the bloodstream to the vital organs (i.e., heart and brain). At very high levels, CO can cause dizziness, confusion, unconsciousness, and even death. It poses extraordinary threats to people with some types of heart disease.

Ground-level O_3. Ozone can cause muscles in the airways of the respiratory system to constrict, trapping air in the alveoli, which leads to wheezing and shortness of breath. Long-term exposure to ozone may be linked to aggravation of asthma and permanent lung disease. Children, older adults, outdoor workers, and people with reduced intake of specific nutrients (e.g., vitamins C and E), with specific genetic characteristics, or with asthma are at significant risk from ozone exposure.

Nitrogen dioxide (NO_2). Breathing air with high NO_2 concentrations can irritate the airways of the respiratory system. Short-term exposures to elevated levels of NO_x can aggravate respiratory diseases, particularly asthma. Longer-term exposures may contribute to the development of asthma and potentially increase susceptibility to respiratory effects. Children, the elderly, and people with asthma are generally at greater risk from NO_2 exposure. NO_2 can react with other atmospheric chemicals to form PM and ozone, the other two criteria pollutants.

Sulfur dioxide (SO_2). Short-term exposures to elevated levels of SO_2 can harm the respiratory system and make breathing difficult. Children, the elderly, and people with asthma are generally at greater risk from SO_2 exposure. SO_x can react with other atmospheric chemicals to form PM, another criteria pollutant.

Particulate matter (PM). PM can accumulate in the respiratory system and cause numerous adverse health effects, especially in the lungs and heart. Particles with diameters larger than 10 micrometers (μm, or micron) would be retained in the upper part of the respiratory system, where physiological processes can quickly expel them. However, the smaller ones (e.g., $PM_{2.5}$ and PM_{10}) can enter deeper into the system. $PM_{2.5}$ and PM_{10} consist of airborne particles with diameters of ≤ 2.5 and $\leq 10\,\mu m$, respectively. The liquid and soluble solid particles may be absorbed into body tissues and pose additional health risks, including premature death in people with heart or lung disease, nonfatal heart attacks, irregular heartbeat, aggravated asthma, decreased lung function, and increased respiratory problems. Children, the elderly, and people with heart or lung diseases are generally at greater risk from PM exposure.

Hazardous air pollutants (HAPs). People exposed to HAPs at sufficient concentrations may have an increased chance of getting cancer or experiencing other serious health effects, including damage to the immune system, as well as neurological, reproductive, developmental, respiratory, and other health problems.

1.4.2 Ecosystems

Lead. Lead can enter soil and sediments from the air through deposition, and it is persistent in the environment. Elevated lead levels in the environment can result in decreased growth and reproductive rates in animals and plants, as well as neurological effects in vertebrates.

Ground-level O_3. Ozone can adversely affect sensitive vegetation by reducing photosynthesis, slowing plants' growth, and increasing their risks of disease, damage from insects, effects of other pollutants, and harm from severe weather. These impacts on sensitive plants can then negatively impact ecosystems, including loss of species diversity, changes in habitat quality, water, and nutrient cycles, and the specific assortment of plants in the forest.

Nitrogen dioxide (NO_2). NO_2 can react with other atmospheric chemicals to form acid rain, harming sensitive ecosystems such as lakes and forests. The nitrate particles originating from NO_x make the air hazy and reduce visibility. NO_x in the atmosphere can also contribute to nutrient pollution in coastal waters.

Sulfur dioxide (SO_2). Gaseous SO_x can harm vegetation at high concentrations by damaging foliage and decreasing growth. SO_x can also react with other atmospheric chemicals to form acid rain, harming sensitive ecosystems such as lakes and forests. The particles originating from SO_x make the air hazy and reduce visibility.

Particulate matter (PM). Fine particles are the leading cause of haze and reduced visibility in parts of the U.S. PM can be carried long distances and then settle on the ground or surface water. The adverse effects of this deposition could (i) make surface water acidic and change its nutrient balance, (ii) deplete nutrients in the soil, (iii) harm sensitive forests and farm crops, (iv) affect the diversity of ecosystems, and (v) contribute to acid rain effects.

Hazardous air pollutants (*HAPs*). Like humans, animals may experience health problems if taking in sufficient doses of air toxics. In addition, some toxic air pollutants (e.g., mercury) can be deposited into soil or surface water, where vegetation can be extracted and ingested by animals, eventually entering the food chain.

1.4.3 Built Environment

Deposition of particles and acid rain can stain and damage stone and other materials, including facets of buildings and culturally significant objects such as statues and monuments. Paints, metals, plastics, and construction materials degrade faster when exposed to certain air pollutants.

1.4.4 Regional and Global Environment

Acid rain can cause problems beyond the national scale. Air pollution can also be global, such as ozone layer depletion, global warming, and climate change.

1.4.5 Useful Data Sources

USEPA's Report on the Environment (ROE) shows how the U.S. environment and human health conditions change over time (https://cfpub.epa.gov/roe/). The ROE presents indicators of national trends in five theme areas: air, water, land, human exposure and health, and ecological conditions. The ROE air indicators address three fundamental questions: (i) What are the trends in outdoor air quality, and how do they affect human health and the environment? (ii) What are the trends in GHG emissions and concentrations, and how do they affect human health and the environment? (iii) What are the trends in indoor air quality, and how do they affect human health? Most of the trends are presented graphically and are easy to understand.

National-Scale Air Toxics Assessment (NATA) is the USEPA's ongoing comprehensive evaluation of air toxics in the U.S. This assessment aims to understand the potential health risks associated with exposure to HAPs and to guide regulatory and policy decisions to improve air quality and public health. The NATA contains recent emissions data and uses models to make broad estimates of health risks over geographic areas of the country (https://www.epa.gov/national-air-toxics-assessment). The USEPA has succeeded NATA with the Air Toxics Screening Assessment, or AirToxScreen, starting with the 2017 AirToxScreen. The EPA updates AirToxScreen annually (https://www.epa.gov/AirToxScreen). The AirToxScreen analysis includes four main steps: (i) compile a National Emissions Inventory (NEI), (ii) estimate ambient concentrations of air toxics across the U.S., (iii) estimate population exposures, and (iv) characterize potential health risks from inhalation.

EXERCISE QUESTIONS

1. About anthropogenic mercury emissions in the U.S,
   a. What are the primary source categories?
   b. Which sector is the most prominent emission source?
   c. What is the total emission of the most recent year available on the USEPA's Report on the Environment website (https://cfpub.epa.gov/roe/)?
2. a. What are the median and average ambient 8-h ozone concentrations in the U.S. in the most recent year available on the USEPA's Report on the Environment website (https://cfpub.epa.gov/roe/)?
   b. How much was the reduction (in percentage) in median concentrations from 1978 to the most recent year?
3. a. What are "current asthma prevalence" and "asthma attack prevalence"?
   b. What is the age-adjusted asthma attack prevalence in U.S. adults (18+ years) in the most recent year available on the USEPA's Report on the Environment website (https://cfpub.epa.gov/roe/)?
   c. How much reduction (in percentage) from 2002 to the most recent year?

2 Properties of Air and Compounds of Concern

2.1 OUR ATMOSPHERE

The Earth's atmosphere is a layer of gases surrounding the planet held by gravitational force. It is vital to life by providing essential gases, protecting against harmful solar radiation, and regulating temperatures. Significant components of dry ambient air include nitrogen (78.08% by volume or by mole), oxygen (20.95%), and argon (0.93%). Carbon dioxide (CO_2), the primary greenhouse gas (GHG), accounts for most of the balance (0.04%) with other gases in trace amounts. Water vapor concentrations vary by location. Our atmosphere also contains particulate matter (PM).

Atmospheric pressure decreases as the altitude increases due to the decreasing mass of air above. The Earth's atmosphere has five primary layers: (i) troposphere, (ii) stratosphere, (iii) mesosphere, (iv) thermosphere, and (v) exosphere. Figure 2.1 illustrates the temperature profile of the atmosphere.

The *troposphere* is the lowest layer of the atmosphere, extending from the Earth's surface to about 8–15 km (5–9 miles) in altitude, thinner at the poles and thicker at the equator. It is part of the atmosphere that we live in. Its thickness is less than 3% of the entire atmosphere; however, it contains about 75% of all the atmospheric mass and almost all its water vapor. The troposphere's temperature decreases with altitude, averaging about 6.5°C/km (3.6°F/1,000 ft). The air in the troposphere will move horizontally and vertically and redistribute heat, moisture, and other constituents, including air pollutants, along with its movement. The *tropopause* is the upper limit of the troposphere, which marks the boundary between the troposphere and the stratosphere; it helps regulate the flow of energy and moisture in the atmosphere, impacting climate dynamics and weather. Temperatures are relatively constant at the tropopause, often around −60°C (−76°F). The airlines often fly around the tropopause. The lowest part of the troposphere is called the *planetary/atmospheric boundary layer,* and this layer interacts with the Earth's surface through processes such as surface friction, solar heating and surface cooling, and evaporation and condensation of water. It is characterized by turbulent air flow and rapid temperature, humidity, and wind speed changes.

The *stratosphere* starts above the troposphere and ends about 50 km (31 mi) above the Earth's surface, containing much of the atmosphere's ozone. Regarding air pollution, the troposphere is the layer in which we have most of our daily activities and almost all the weather, while the *ozone layer*, which absorbs harmful ultraviolet (UV) radiation from the Sun, is within the stratosphere. The temperature in this layer increases with altitude because of the absorption of solar UV radiation, and the temperature rise creates a stable layer that limits vertical mixing.

DOI: 10.1201/9781003532705-2

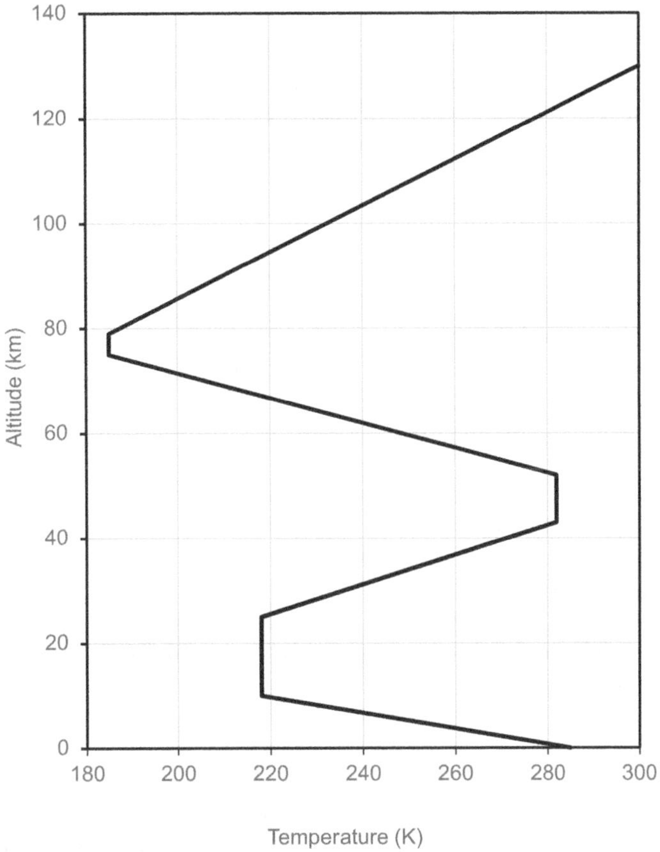

FIGURE 2.1 Temperature profile of the atmosphere.

The *mesosphere* starts above the stratosphere and ends about 80 km (50 mi) above the Earth's surface. The mesosphere contains gases similar to those in the lower layers, but at much lower densities. Its temperature decreases with height due to a lack of significant ozone and other heat-absorbing gases, reaching a minimum temperature of about −90°C (−130°F) at the *mesopause* from the Earth's surface. Most meteors burn up in this layer upon entering the Earth's atmosphere.

The *thermosphere* is the fourth layer of the Earth's atmosphere, extending from approximately 80 km (50 mi)–700 km (440 mi) above the Earth's surface, whose lowest part contains the *ionosphere* since the energetic solar radiation knocks electrons off molecules and atoms in this region. Its temperature increases with altitude because of solar UV and X-ray radiation absorption.

The *exosphere* is the outermost layer of the Earth's atmosphere, extending from the top of the thermosphere to approximately 10,000 km (6,200 mi) above the Earth's surface and merging with the solar wind at its top. Molecules found here are extremely low density, so this layer doesn't behave like a gas, and the molecules can escape into space. Most Earth satellites orbit in the thermosphere or the exosphere due to the low air resistance resulting from low concentrations of gases.

While there's no clear boundary between the Earth's atmosphere and outer space, most scientists use a delineation known as the *Karman line*, located 100 km (62 mi) above the Earth's surface, to denote this transition point since 99.99997% of the Earth's atmosphere lies beneath this point (NASA, 2024).

2.2 FACTORS AFFECTING PROPERTIES OF AIR

This section discusses temperature and pressure (two important parameters affecting air properties), the Ideal Gas Law, and the standard conditions.

2.2.1 Temperature

Temperature affects the density and pressure of air. Warmer air is less dense and tends to rise, while colder air is denser and tends to sink. The difference in air density drives atmospheric circulation and weather patterns and, consequently, affects the transport of air pollutants.

In our daily activities, we express temperatures (T) in degrees Fahrenheit (°F) or degrees Celsius (°C). A temperature can be converted from degrees Celsius to degrees Fahrenheit (or vice versa) by:

$$T(\text{in }^\circ\text{F}) = 1.8 \times T(\text{in}^\circ\text{C}) + 32 \tag{2.1}$$

However, when temperature is part of an equation describing a physical or chemical relationship (e.g., the Ideal Gas Law), the temperature should be the absolute temperature. The *absolute temperature* is a temperature scale that starts with zero at the *absolute zero temperature* (i.e., the temperature at which kinetic energy is negligible). For the International System (SI) of units, degree Kelvin (K) expresses an absolute temperature. The following equation converts a temperature in degrees Celsius to degrees Kelvin:

$$T\ (\text{in K}) = T(\text{in }^\circ\text{C}) + 273.15 \approx T(\text{in }^\circ\text{C}) + 273 \tag{2.2}$$

For the U.S. customary units, absolute temperatures are expressed in degrees Rankine (°R). The following equation converts a temperature from degrees Fahrenheit to degrees Rankine:

$$T(\text{in }^\circ\text{R}) = T(\text{in }^\circ\text{F}) + 459.67 \approx T(\text{in }^\circ\text{F}) + 460 \tag{2.3}$$

Therefore, water boils at 100°C, 212°F, 373 K, or 672°R when pressure (P) = 1 atm. A temperature change in 1°C or 1 K equals a change of 1.8°F or 1.8°R. The conversion between temperatures in degrees Rankine and Kelvin is simple:

$$T(\text{in }^\circ\text{R}) = 1.8 \times T(\text{in K}) \tag{2.4}$$

In chemistry, STP stands for *Standard Temperature and Pressure*, a set of conditions for experimental measurements to allow comparison between different datasets. The *standard temperature* in science is 0°C (32°F). The USEPA has different standard temperatures: 25°C (77°F) for air monitoring and 20°C (68°F) for most other applications (USEPA, 2012a). The standard temperatures used by various groups are listed in Table 2.1.

2.2.2 Pressure

Atmospheric pressure decreases with altitude. Variations in pressure affect wind patterns and weather systems. High-pressure areas are typically associated with clear skies, while low-pressure regions can lead to stormy weather.

A stationary object may be subjected to tensile and compressive forces. It will also be subjected to shear/friction force in motion. *Pressure/stress* is defined as force per unit area. Pressure is also an important parameter affecting the properties of air and gas/vapor. *Pressure drop* (*head loss*) is a matter of concern for air movement in treatment units (e.g., absorbers) and transport systems (e.g., ductwork and blowers). A larger head loss means we must provide more energy to overcome it, which will incur a higher operating cost.

The *standard barometric/atmospheric pressure* is the average atmospheric pressure at sea level, 45° north latitude at 35°F (~275 K), and it is equal to 1 atmosphere (atm). This is often used as a baseline for comparing atmospheric conditions. The barometric pressure at a location would depend on its altitude, latitude, and weather conditions. The most used SI and U.S. customary units for pressure are Newton/m^2 (N/m^2 or Pascal (Pa)) and lb/in^2 (psi), respectively. However, many different units are also being used, and an engineer often needs to convert a pressure value from one unit to another. Common units of pressure and the relationships among them are given below. It would be a good idea to memorize or have them readily available.

TABLE 2.1
Standard Temperatures

	Standard Temperature	
	(°C)	(°F)
Science (chemistry)	0	32
USEPA (general)	20	68
USEPA (air monitoring)	25	77
Combustion	15.6	60
Industrial hygiene	21.1	70

Source: USEPA (2012a).

1 atmosphere (atm)

$$= 1.01325 \times 10^5 \text{ N/m}^2 = 1.01325 \times 10^5 \text{ Pa} = 101.325 \text{ kN/m}^2 \approx 101.3 \text{ kN/m}^2$$
$$= 1.013 \text{ bar} = 1{,}013 \text{ milli-bar (mbar)}$$
$$= 14.696 \text{ lb/in}^2 \text{ (psi)} \approx 14.7 \text{ psi}$$
$$= 760 \text{ millimeters of mercury (mm-Hg)} = 760 \text{ Torr} = 29.92 \text{ in-Hg}$$
$$= 33.92 \text{ ft-H}_2\text{O} = 407 \text{ in-H}_2\text{O} = 10.33 \text{ m-H}_2\text{O}$$

A pressure gauge reading tells the difference in pressure between the system and its surroundings. The *gauge pressure* is positive when the system is higher than its surroundings. The gauge pressure will be negative when the system is under a vacuum. The *absolute pressure* (P_{abs}) is the sum of the barometric/atmospheric pressure (P_b) and the gauge pressure (P_g):

$$P_{abs} = P_b + P_g \tag{2.5}$$

2.2.3 Properties of Ideal Gases

An *ideal gas* is a "theoretical" gas that follows the Ideal Gas Law under all conditions. For ideal gases, there are no intermolecular forces between gas molecules, and these molecules occupy no volume. There is no such thing as an ideal gas in reality. However, the behaviors of real gases and air deviate insignificantly from those of an ideal gas under ambient P and T.

Boyle's Law states that an ideal gas's volume (V) is inversely proportional to the system pressure (P) when the system temperature (T) is held constant. *Charles' Law* states that P is proportional to T when V is constant. Combining these two laws into the *Ideal Gas Law*:

$$PV = \left(\frac{m}{\text{MW}}\right) RT = nRT \tag{2.6}$$

where m is the mass, MW is the molecular weight, n is the number of moles of the gas, R is the universal gas constant, and T is the absolute temperature of the system. The following are the values of R in some commonly used units:

$$R = 0.082 \frac{\text{L} \cdot \text{atm}}{\text{g-mole} \cdot \text{K}} = 8.314 \frac{m^3 \cdot Pa}{\text{g-mole} \cdot \text{K}} = 0.73 \frac{\text{ft}^3 \cdot \text{atm}}{\text{lb-mole} \cdot °\text{R}} = 10.731 \frac{\text{ft}^3 \cdot \text{psi}}{\text{lb-mole} \cdot °\text{R}}$$

Example 2.1: Molar Volume of an Ideal Gas

Find the volume of one g-mole and one lb-mole of an ideal gas at $T = 0°\text{C}$ and $P = 1$ atm.

Solution:

a. For one g-mole of an ideal gas,

$$V = \frac{nRT}{P} = \left[(1 \text{ g-mole})\left(0.082\ \frac{\text{L}\cdot\text{atm}}{\text{g-mole}\cdot\text{K}}\right)(273 \text{ K})\right] \div 1 \text{ atm} = 22.4 \text{ L}$$

b. For 1 lb-mole of an ideal gas,

$$V = \left[(1\text{lb-mole})\left(10.731\frac{\text{ft}^3\cdot\text{psi}}{\text{lb-mole}\cdot{}^\circ\text{R}}\right)(492^\circ\text{R})\right] \div 14.7 \text{ psi} = 359 \text{ ft}^3$$

Discussion: Using similar approaches, one can readily find that volumes of 1 g-mole ideal gas @T=20°C and 25°C (P=1 atm) are 24.05 and 24.46 L, respectively, while those of 1 lb-mole ideal gas would be 385 and 392 ft^3 @T=68°F and 77°F, respectively. Having these values readily available or memorized may not be a bad idea.

In chemistry, the STP is defined by the International Union of Pure and Applied Chemistry (IUPAC) as T=273.15 K and P=100 kPa. However, the standard conditions used in practice are often different. For example, the USEPA generally has 760 mm-Hg (or 1 atm) as the standard P. However, the standard temperature for stack sampling is 20°C (68°F), while that for ambient air sampling is 25°C (77°F) (also see Table 2.1). The South Coast Air Quality Management District (SCAQMD) in southern California has 60°F or 68°F as the standard temperature in their regulations. It is always a good practice to state the conditions of STP in data reporting. This book uses P=1 atm and T=20°C as the standard conditions unless otherwise specified.

2.3 BASIC PROPERTIES OF AIR

This section discusses basic air properties, including molecular weight, density, viscosity, specific heat capacity, and Reynolds number.

2.3.1 Molecular Weight

Since air is a mixture of many gases, the "average" molecular weight (MW) of air can be estimated from its composition. For most air pollution calculations, it is often satisfactory to consider that the composition of dry air is 21% O_2 and 79% N_2 by volume (also by mole). Thus,

$$\text{MW}_{\text{dry air}} \approx [(32)(21\%)+(28)(79\%)] \approx 29 \text{ g/g-mole (lb/lb-mole)}$$

$$\text{Molar ratio of } N_2 : O_2 \text{ in air} \approx (79\%) \div (21\%) \approx 3.76$$

$$\text{Mass ratio of } N_2 : O_2 \text{ in air} \approx [(28)(79\%)] \div [(32)(21\%)] \approx 3.29$$

The apparent *MW* for <u>wet</u> air can be found by including the volume (or molar) fraction of water (f_{water}). Thus,

$$\mathrm{MW}_{wet\ air} = (1 - f_{water})(\mathrm{MW}_{dry\ air}) + (f_{water})(\mathrm{MW}_{water}) \tag{2.7}$$

Since the MW of water (= 18) is smaller than the MW of dry air (29), the MW of wet air is less than the MW of dry air (see Eq. 2.7).

2.3.2 Density

The *density* (ρ) of a material is the ratio of its mass (m) and the volume (V) it occupies. For an ideal gas:

$$\rho = \left(\frac{m}{V}\right) = \frac{P \times \mathrm{MW}}{R \times T} \tag{2.8}$$

Since MW of water (= 18) is smaller than that of dry air (≈29), the MW and density of wet air are smaller than those of dry air (see Eqs. 2.7 and 2.8).

Example 2.2: Density of Air

Determine the density of dry air (a) under the standard conditions (T=20°C and P=1 atm), (b) T=200°C and P=1 atm, and (c) T=200°C and P=2 atm.

Solution:

a. Assuming MW of dry air=29, then at T=20°C and P=1 atm:

$$\rho = \frac{(1\ \text{atm})\left(29\,\frac{\text{g}}{\text{g-mole}}\right)}{\left(0.082\,\frac{\text{L}\cdot\text{atm}}{\text{g-mole}\cdot\text{K}}\right)(293\ K)} = 1.21\frac{\text{g}}{\text{L}} = 1.21\frac{\text{kg}}{\text{m}^3}$$

or

$$\rho = \frac{(14.7\ \text{psi})\left(29\,\frac{\text{lb}}{\text{lb-mole}}\right)}{\left(10.731\,\frac{\text{ft}^3\cdot\text{psi}}{\text{lb-mole}\cdot{}^\circ\text{R}}\right)(528^\circ\text{R})} = 0.075\ \text{lb/ft}^3$$

b. At T=200°C and P=1 atm:

$$\rho = \frac{(1)(29)}{(0.082)(473\ \text{K})} = 0.75\ \text{g/L} = 0.75\frac{\text{kg}}{\text{m}^3}$$

c. At $T=200°C$ and $P=2$ atm:

$$\rho = \frac{(2)(29)}{(0.082)(473\,\text{K})} = 1.50\text{ g/L} = 1.50\frac{\text{kg}}{\text{m}^3}$$

Discussion:

1. Since the molar volume of an ideal gas is 24.05 L @ $T=20°C$ and $P=1$ atm (see Example 2.1), its density can be readily estimated as (29)/(24.05) = 1.21 g/L.
2. Under these conditions, the density of dry air (1.21 g/L) is ~800 times smaller than that of water, which is 1 kg/L (i.e., 1,000 g/L or 1,000 kg/m^3).
3. Since the volume of an ideal gas increases with temperature, its density will decrease with temperature. The ratio of densities at two different temperatures can be directly obtained from the inverse ratio of the two absolute temperatures.
4. The density of an ideal gas is directly proportional to the system pressure.

2.3.3 Viscosity

The *viscosity* of a fluid (i.e., air or liquid) measures its internal resistance to flow. This resistance is due to intermolecular friction between fluid layers when they slide against each other. The *shearing stress* (τ_{shear}) between layers of a fluid is related to the *velocity gradient (du/dy)* between the liquid layers (see the following equation) [Note: a *gradient* is a change in the value of a quantity (e.g., concentration) with respect to a change in a given variable (usually distance). It can be considered a driving force that causes movement from one location to another. In environmental engineering applications, temperature, pressure, concentration, and hydraulic gradients are commonly encountered].

$$\tau_{shear} = \mu\left(\frac{du}{dy}\right) \tag{2.9}$$

where μ is the *dynamic/absolute viscosity* (or just *viscosity*). Common viscosity units in SI are N·s/m^2 (Pa·s) or Poise, while the U.S. customary system is lb_f-s/ft^2. Conversions among commonly used units are [Note: 1 N = 1 kg·m/s^2; 1 slug = 32.2 lb_m; 1 lb_f = 32.2 lb_m·ft/s^2; 1 lb_f = 4.448 N]:

- 1 N·s/m^2 = 1 Pa·s = 1 kg/(m·s) = 10 Poise = 0.0209 lb_f-s/ft^2 = 0.0209 slug/(ft-s) = 0.672 lb_m/ft-s
- 1 Poise = 100 centipoise (cp) = 0.0672 lb_m/ft-s = 0.00209 lb_f-s/ ft^2

In engineering applications, *kinematic viscosity* (ν) is also used, which is the ratio of the dynamic/absolute viscosity (μ) and the mass density (ρ) of the fluid (i.e., $\nu=\mu/\rho$). The standard units of SI for kinematic viscosity are m^2/s, Stokes (St), or centi-Stokes (cSt), while those of the U.S. customary system are ft^2/s.

- 1 St = 100 cSt = 1 cm^2/s = 10^{-4} m^2/s = 1.08×10^{-3} ft^2/s

Dynamic viscosity is used (i) to calculate the force required to move a fluid and (ii) to understand the flow characteristics in various systems. In contrast, kinematic viscosity is used for fluid flow under the influence of gravity, such as in the sedimentation process.

Gas molecules are farther apart than water molecules, resulting in smaller intermolecular cohesion. The absolute/dynamic viscosity values of air and water at 293 K are 1.81×10^{-5} and 1.01×10^{-3} N·s/m^2, respectively (Table 2.2), which means that water is ~55 times more viscous than air under ambient conditions. However, the kinematic viscosity of air at the STP is larger than that of water (1.50×10^{-5} vs. 1.01×10^{-6} m^2/s), mainly because the air density is ~800 times smaller.

The viscosity of a fluid is relatively independent of pressure but depends significantly on temperature. Liquid viscosity decreases when the temperature increases; the gas has the opposite trend because the increasing temperature would cause more collisions among the gas molecules. The viscosity of air at any temperature can be estimated using the viscosity value at a reference absolute temperature (μ_{ref}) and the equation below (USEPA, 2012a). The viscosity value of 1.81×10^{-5} N·s/m^2 in Table 2.2 can be used as the reference value @ $T = 293$ K.

$$\frac{\mu}{\mu_{ref}} = \left(\frac{T}{T_{ref}}\right)^{0.768} \tag{2.10}$$

Example 2.3: Viscosity of Air

Use the value of air viscosity @ $T = 20°C$, given in Table 2.2, to estimate the viscosity of air at $T = 200°C$ and $P = 1$ atm.

Solution:

$$\frac{\mu}{\mu_{ref}} = \left(\frac{T}{T_{ref}}\right)^{0.768} \rightarrow \frac{\mu}{1.81 \times 10^{-5}} = \left(\frac{473}{293}\right)^{0.768}$$

$$\mu_{473K} = 2.61 \times 10^{-5} \text{kg/(m} \cdot \text{s)} = 5.46 \times 10^{-7} \text{lb}_f\text{-s/ft}^2 = 1.76 \times 10^{-5} \text{lb}_m\text{/ft-s)}$$

Discussion:

1. As expected, the viscosity of air increases with temperature.
2. Using the conversion factors given earlier, one can readily convert a viscosity value between different units.

TABLE 2.2
Viscosities of Air and Water ($T = 293$ K and $P = 1$ atm)

	Viscosity (N·s/m^2)	Kinematic Viscosity (m^2/s)
Air	1.81×10^{-5}	1.50×10^{-5}
Water	1.01×10^{-3}	1.01×10^{-6}

2.3.4 Specific Heat Capacity

Commonly used energy units are Joule (J), calorie, and British thermal unit (BTU). A *Joule* is the unit of work/energy in the SI, and it is the work done on an object when the force of one Newton (N) acts on that object in the direction of its motion for one meter (= 1 N·m). One *calorie* is the energy needed to raise the temperature of 1 g of water by 1°C, while one *BTU* is the energy needed to raise the temperature of 1 pound of water by 1°F at the temperature when the water has its greatest density (~39°F or ~4°C). One Joule is about one-quarter of one calorie and is about one-thousandth of one BTU. Conversions among these common units are:

- 1 J = 1 N·m = 9.45×10^{-4} BTU = 0.737 ft-lb_f
- 1 BTU = 1,055 J = 252 cal
- 1 J = 0.239 calorie (1 calorie = 4.184 J)

Heat capacity is the amount of heat required to raise the temperature of the entire object by one degree. In comparison, *specific heat capacity* (or *specific heat*) is the amount of heat needed to raise the temperature of a unit mass of the object by one degree. Standard units of specific heat are J/kg·K, BTU/lb-°F, or calories/g·°C. The conversions among these units are:

- 1 BTU/lb-°F = 1 kcal/kg·°C = 4.184 kJ/kg·K
- 1 kJ/kg·K = 0.239 BTU/lb-°F = 0.239 kcal/kg·°C

Heat can be added to a gas mass, while the volume or pressure of the system is kept constant, so there are *specific heat at constant volume* (C_v) and *specific heat at constant pressure* (C_p). For an ideal gas,

$$C_p = C_v + R \tag{2.11}$$

where R is the universal gas constant. At room temperatures, $C_v = (3/2)R$ and $C_p = (5/2)R$ for monatomic gases (e.g., He, Ne, and Ar); and $C_v = (5/2)R$ and $C_p = (7/2)R$ for diatomic gases (e.g., O_2 and N_2). Since air is mostly diatomic O_2 and N_2, it is typically treated as diatomic.

From the definitions of calorie and BTU, we should be able to tell that the heat capacity of water under room temperatures is equal to or about 1 BTU/lb-°F, 1 cal/g·°C, 1 kcal/kg·°C, or 4.184 kJ/kg·K. The specific heat of a gas stream depends on its composition and temperature. Table 2.3 tabulates the specific heat of air, CO_2, and water vapor at several temperatures. For a temperature not listed in this table, an estimate can be made by interpolating between two known values. As shown, the heat capacity increases with temperature. Water vapor's heat capacity values are approximately twice those of air and CO_2. Consequently, the heat capacity of a gas stream containing a large amount of water vapor would be significantly higher than that of drier air. In addition, it would take more heat to raise the temperature of a gas stream when wet.

Most energy-related calculations for gas combustion deal with flue gases at a temperature $\leq$ 1,500°F (815°C). For those cases, the heat capacity of dry gas is about 0.25

TABLE 2.3
Specific Heat Capacity of Air, CO_2, and Water Vapor in BTU/lb-°F or kcal/kg°C

Temperature		Specific Heat Capacity		
(°F)	(°C)	Air	CO_2	Water Vapor
68	20	0.242	0.200	0.445
212	100	0.244	0.218	0.452
500	260	0.249	0.245	0.470
1,100	593	0.260	0.285	0.526
2,200	1,204	0.278	0.315	0.622
3,000	1,649	0.297	0.325	0.673
3,800	2,093	0.303	0.330	0.709

Source: USEPA (2012b).

BTU/lb-°F (1.05 kJ/kg·K), while water vapor is about 0.50 BTU/lb-°F (2.1 kJ/kg·K). These values can be used for most calculations without introducing significant errors (USEPA, 2012b).

Example 2.4: Specific Heat Capacity of a Gas Stream

Estimate the specific heat capacity of an exhaust gas stream with 15% water vapor, with a stack temperature of 700°F (371°C).

Solution:

a. Interpolating the values at 500°F and 1,100°F (Table 2.3), the specific heat capacities of dry air and water vapor at 700°F can be estimated as 0.253 and 0.489 BTU/lb-°F, respectively.
b. Heat capacity of the gas stream = (0.253)(1 − 15%) + (0.489)(15%)

$$= 0.289 \text{ BTU/lb-°F} \left(= 1.21 \text{ kJ/kg·K}\right).$$

2.3.5 Fluid Reynolds Number

Fluid Reynolds number (*Re*) is a dimensionless quantity used in fluid mechanics to predict flow patterns in different fluid flow situations. It is the ratio of the inertial force to the viscous force of a flowing fluid, as follows:

$$Re = \frac{vL\rho}{\mu} \left(= \frac{vD\rho}{\mu} \text{ for a circular pipe} \right) \tag{2.12}$$

where L is the characteristic system dimension, and v, ρ, and μ are the fluid's velocity, density, and dynamic viscosity, respectively. L is four times the hydraulic radius (R_H).

The *hydraulic radius* is the cross-sectional area of the flow divided by the wetted perimeter. $R_H = \pi r^2/2\pi r = r/2 = D/4$ (r = radius and D = diameter) for a circular pipe with a full flow. Consequently, its characteristic system dimension equals its diameter (i.e., $L = D$) for a circular pipe with a full flow.

The fluid Reynolds number is used in fluid mechanics to predict the flow regime of a fluid. When $Re \leq 2{,}000$, viscous forces dominate and the flow is *laminar* (i.e., the layers of fluid move in parallel, and there is no mixing of the molecules between the layers; the flow is smooth and orderly). When $Re \geq 4{,}000$, inertial forces dominate and the fluid moves in a turbulent manner (i.e., the flow is chaotic and mixed). The *transition region* is between the laminar and turbulent regions (i.e., $2{,}000 < Re < 4{,}000$). The Reynolds number is frequently used in engineering applications, as it helps to determine flow patterns, frictional forces, and energy losses in fluid conveyance systems.

The formula for the *Reynolds number of particles* (Re_p) for particles traveling in air differs from that described in Eq. 2.12. It will be presented in Chapter 6.

2.4 PROPERTIES OF COMPOUNDS OF CONCERN IN AIR

2.4.1 Vapor Pressure

A compound's *vapor pressure* (P^{vap}) is the equilibrium pressure of a vapor with its liquid/solid in a closed container at a given temperature. It indicates the tendency of molecules to escape from a liquid/solid and enter the gaseous phase. As the temperature increases, the vapor pressure increases. Intermolecular forces among molecules in a compound determine its vapor pressure. Therefore, some compounds are more volatile than others. The *boiling point* of a liquid is the temperature at which its vapor pressure equals the atmospheric pressure and the liquid begins to boil. Thus, when $P = 1$ atm = 760 mm-Hg, P^{vap} of water equals 760 mm-Hg @T = 100°C (i.e., the boiling point of water), while it is only 18 mm-Hg @T = 20°C.

The *Antoine equation* is an empirical equation, commonly used to estimate P^{vap} at a specific temperature as follows:

$$\ln\left(P^{vap}\right) = A - \frac{B}{C + T} \tag{2.13}$$

where A, B, and C are the Antoine constants; their values can be readily found on the Internet or in chemistry handbooks. It should be noted that these Antonie constants are empirical constants, and, for a specific compound, their values may differ for different temperature ranges. In addition, the reported values of these constants can be different in different research. One should also pay attention to the units of T and P^{vap} that go with these values [Note: this suggestion applies to all empirical equations].

The *Clausius–Clapeyron equation* correlates P^{vap} and the absolute temperature. It assumes that the enthalpy of vaporization (ΔH^{vap}) is independent of temperature:

$$\ln\frac{P_1^{vap}}{P_2^{vap}} = -\frac{\Delta H^{vap}}{R}\left[\frac{1}{T_1} - \frac{1}{T_2}\right]. \tag{2.14}$$

where R is the universal gas constant and T is the absolute temperature. Values of ΔH^{vap} can be readily found on the Internet or in chemistry handbooks. The unit of R should match those of T and ΔH^{vap}.

For an ideal solution, the vapor–liquid equilibrium follows *Raoult's law*, which relates the vapor pressure of each component in an ideal solution and its mole fraction in the liquid phase as follows:

$$P_A = P^{vap} \times x_A \tag{2.15}$$

where P_A is the partial pressure of compound A in the vapor phase and x_A is the mole fraction of compound A in the solution. Raoult's law holds only for *ideal solutions* in which the intermolecular forces between different components are similar to those within each pure component [Note: Henry's law is considered more applicable for dilute aqueous solutions commonly encountered in environmental engineering applications – described later].

The *partial pressure* is the pressure that a single component of a gas mixture would exert if it occupied the entire volume at the same temperature. The partial pressure of compound A (P_A) is equivalent to its mole fraction in the gas phase (y) multiplied by the total pressure (P_{total}), assuming its behavior is independent of other gases:

$$P_A = P_{total} \times y_A \tag{2.16}$$

The total pressure is the sum of all partial pressures (i.e., *Dalton's Law of Partial Pressure*). For example, the ambient air contains 21% oxygen by mole (or by volume). Thus, the mole fraction of oxygen in the air is 0.21, and its partial pressure is 0.21 atm (= 0.21 × 1 atm). It should be noted here that 1% is $10{,}000 \times 10^{-6}$; that is, 10,000 parts per million by volume (ppmV) [Note: the ppmV, or ppm (v/v), is determined by comparing the volume of one constituent with the total volume of the substance]. Gas concentrations are always expressed in ppm (v/v, or volume/volume) instead of the ppm (w/w, mass/mass, or weight/weight) format often used for liquids or solids. This book will use ppm as an abbreviation of ppmV or ppm (v/v). Thus, the oxygen concentration in our ambient air is about 210,000 ppm. If the CO_2 concentration at a location is 400 ppm, then its mole fraction is 0.0004. The corresponding CO_2 concentration in percentage is 0.04% by mole (or by volume). Its partial pressure would be 4.0×10^{-4} atm or 0.304 mm-Hg (= 760 mm-Hg × (4.0×10^{-4})).

Example 2.5: Estimate Partial Pressures of a Mixed Solvent

An above-ground storage tank contains 50% (by wt.) of toluene ($C_6H_5CH_3$; MW = 92) and 50% of ethylbenzene ($C_6H_5C_2H_5$; MW = 102). Estimate the maximum toluene and ethylbenzene concentrations (in ppm) in the void space of the tank. The ambient temperature is 20°C. Note: At 20°C, the vapor pressure of toluene = 22 mm-Hg and that of ethylbenzene = 7 mm-Hg.

Solution:

a. Using a total mass of 100 g as the basis for calculations, for 50% by weight of toluene, its mole fraction = (moles of toluene) ÷ [(moles of toluene) + (moles of ethyl benzene)]

$$= (50/92)/\left[(50/92)+(50/106)\right] = 0.535 = 53.5\%$$

b. Using Eq. 2.15, the partial pressure of toluene in the void

$$= (22)(0.535) = 11.78 \text{ mm-Hg} = 0.0155 \text{ atm} = 15{,}500 \text{ ppm}$$

c. The partial pressure of ethylbenzene in the void

$$= (7)(1-0.535) = 3.25 \text{ mm-Hg} = 4{,}300 \text{ ppm}$$

Discussion: The vapor concentrations are those in equilibrium with the solvent. An equilibrium between the liquid and vapor phases can occur in a closed container or a stagnant space.

Example 2.6: Using the Antoine Equation to Estimate Vapor Pressure

The empirical constants for the Antoine equation for benzene are $A = 15.9008$, $B = 2788.51$, and $C = -52.36$ for P^{vap} in mm-Hg and T in K (Reid et al., 1987). Estimate the vapor pressure of benzene at $T = 20°C$ and 25°C and $P = 1$ atm using the Antoine equation.

Solution:

a. Using Eq. 2.13, at 20°C (293 K):

$$\ln\left(P^{vap}\right) = A - \frac{B}{T+C} = 15.9008 - \frac{2788.51}{(293-52.36)} = 4.322$$

$$P^{vap} = \underline{75.3 \text{ mm-Hg}}$$

b. Using Eq. 2.13, at 25°C (298 K):

$$\ln\left(P^{vap}\right) = 15.9008 - \frac{2788.51}{(298-52.36)} = 4.557$$

$$P^{vap} = \underline{95.3 \text{ mm-Hg}}$$

Discussion: As expected, the vapor pressure of benzene at 25°C (95.3 mm-Hg) is larger than that at 20 °C (75.3 mm-Hg).

Example 2.7: Using the Clausius–Clapeyron Equation to Estimate Vapor Pressure

Use the vapor pressure value of benzene at 20°C (obtained from Example 2.6) and the Clausius–Clapeyron Equation to estimate its vapor pressure at 25°C. Note: Enthalpy of vaporization of benzene = 33,830 J/g-mole (Lide, 1992).

Solution:

Using Eq. 2.14,

$$\ln\frac{75.3}{P_2^{vap}} = -\frac{33{,}840}{8.314}\left[\frac{1}{293}-\frac{1}{298}\right]$$

P^{vap} of benzene at 25°C = 95.1 mm-Hg

Discussion:

1. The value (95.1 mm-Hg) derived in this example using the Clausius–Clapeyron Equation is essentially the same as the value (95.3 mm-Hg) in Example 2.6 using the Antoine equation.
2. The value of R used = 8.314 m³·Pa/g-mole·K = 8.314 J/g-mole·K [Note: 1 J = 1 N·m and 1 Pa = 1 N/m²].

2.4.2 Humidity

Humidity refers to the amount of water vapor in the air, often expressed in several ways. *Absolute humidity* (*AH*) is the mass of water vapor ($m_{water\ vapor}$) in a given air volume (V_{air}), often in g/m³; it is a function of T and P. This is a direct measure of the water vapor content in the air.

$$\text{Absolute Humidity (AH)} = \left(m_{water\ vapor}/V_{air}\right) \quad (2.17)$$

Specific humidity (*SH*, or *humidity ratio*) is the ratio of the mass of water vapor to the total air mass, including water vapor, often expressed in g/kg or kg/kg.

$$\text{Speific Humidity (SH)} = m_{water\ vapor}\Big/\left(m_{water\ vapor} + m_{dry\ air}\right) \quad (2.18)$$

Relative humidity (*RH*) measures the absolute humidity relative to the *saturation humidity* (i.e., RH = 100%) at that temperature (i.e., the percentage of water vapor compared to the maximum amount of water vapor the air can hold). When RH ≥ 100%, condensation can occur, resulting in dew or precipitation. It can be defined as the ratio of the partial pressure of water vapor (P_{water}) to its vapor pressure in equilibrium with pure water under the same conditions:

$$\text{Relative Humidity (RH)} = \left(P_{water}/P_{water}^{vap}\right) \times 100\% \quad (2.19)$$

Dew-point, dry-bulb, and wet-bulb temperatures are essential parameters to describe the status of air regarding humidity. The *dew-point temperature* (DPT) is the temperature at which the air becomes saturated (i.e., RH = 100%) and water vapor starts to condense out of the air. The higher the dew point, the more the moisture in the air (i.e., more humid). Since the saturation vapor pressure increases with temperature, warm air can hold more water vapor, even if RH stays the same (e.g., the "hot and humid" subtropical weather).

The *dry-bulb temperature* (DBT) is measured with a standard thermometer, which is the ambient air temperature. The *wet-bulb temperature* (WBT) can be measured by wrapping the thermometer bulb in a wet fabric. With the cooling of the bulb from adiabatic evaporation of water, the WBT is lower than the corresponding DBT and higher than the DPT [Note: "*adiabatic*" denotes a process or condition in which heat does not leave or enter the system concerned. In other words, there is no heat exchange between the system and its surroundings]. When the humidity of the air is high, the difference between DBT and WBT is less because of a lower tendency for water to evaporate from the wet fabric. DBT and WBT are equal when RH = 100%. In addition, a smaller difference between DBT and DPT indicates that the RH of the air is high.

In summary, DBT is the actual air temperature measured by a regular thermometer and is not affected by the moisture content of the air. WBT is the temperature that air would cool to by evaporating water into the air, and it is affected by both temperature and humidity. DPT is the temperature at which air becomes saturated and water vapor condenses into liquid; regardless of the current temperature, it is a fixed value for a given moisture content. The DBP, WBP, and DPT are all equal at 100% RH. The general relationship among them is:

$$\text{DPT} \leq \text{WBP} \leq \text{DBP} \tag{2.20}$$

The relationships among them and some thermal properties of moist air can be retrieved from a psychrometric chart.

Example 2.8: Calculation of Dew-Point Temperature

The flue gas from a combustion chamber contains 10% water by volume. Estimate its DPT ($P = 1$ atm) using the Antoine equation below (for water vapor between 1°C and 100°C):

$$\log_{10} P(\text{in mm-Hg}) = 8.07131 - \frac{1730.63}{T(\text{in °C}) + 233.426}$$

Solution:

a. $P_{water} = y_{water} \times P = (10\%) \times (760) = 76.0$ mm-Hg

b. $\log_{10}(76.0) = 8.07131 - \dfrac{1730.63}{T(\text{in °C}) + 233.426}$

$T = 46.1°C$

Discussion:

1. Antonine constants are readily available on the Internet. However, we need to pay attention to the units associated with the parameters since they are empirical constants. For example, the equation above has "$\log_{10}$", not the "natural log", and T is in °C, not Kelvin.
2. Water vapor pressure at 46°C is 75.7 mm-Hg (from the Internet), which is essentially the same as 76.0 mm-Hg calculated in part (a).

2.4.3 Psychrometric Chart

A *psychrometric chart* is a graphical representation of the physical and thermal properties of moist air, including (i) DBT, (ii) humidity ratio or specific humidity of air, (iii) WBT, (iv) RH, (v) DPT, (vi) specific volume of air, and (vii) total heat content of air. The NCEES PE/Environmental Reference Book (NCEES, 2025) includes two psychrometric charts (one in SI and the other in U.S. customary units for sea level, temperatures ranging from 0°C to 50°C or 32°F to 122°F), and $P = 1$ atmosphere from the American Society of Heating, Refrigerating and Air-Conditioning Engineers, Inc. (ASHRAE). They are also included in this book as Figures 2.2 and 2.3 (Permission granted by ASHRAE on 02/04/2025).

In these two charts, DBTs are shown by straight vertical lines (scaling from 0°C to 50°C or 35°F to 120°F at the bottom horizontal axis), and the humidity ratios (scaling from 0.000 to 0.030 on the right-most vertical axis) are shown by the horizontal lines. The RH lines are curved and extend from lower left to upper right (scaling from 0% to 100%); the 0% RH curve is the bottom horizontal axis, and 100% RH is the topmost curve.

Wet-bulb temperatures are shown by straight lines sloping from upper left to lower right (scaling from 5°C to 30°C or 35°F to 85°F with these values shown next to the 70% RH curve); the specific volumes are indicated by straight lines sloping down from top to bottom (ranging from 0.78 to 0.94 m^3/kg of dry air or 12.5 to 15.0 ft^3/lb of dry air). The enthalpies are also shown by straight lines sloping from upper left to lower right (ranging from 10 kJ/kg of dry air in the bottom left corner to 120 kJ/kg of dry air in the top right corner of the chart, or 10 to 60 BTU/lb of dry air). The DPT can be read on the WBT scale (the topmost curve) along a horizontal line of constant humidity ratio [Note: the numerical values in this paragraph are those shown on the figures to facilitate understanding of these two figures].

One can locate a specific point on a psychrometric chart with two of the seven properties known. One can get the values of the other five properties from its location. For example, the DBT on the horizontal axis can be identified and moved vertically to meet the specific humidity level or a WBT line if one has the measurement from a hygrometer. The other properties, such as enthalpy, DPT, or RH, can be obtained consequently.

Example 2.9: Use of Psychrometric Chart (SI)

At a site where the DBT = 20°C and RH = 50%, find the (a) humidity ratio of dry air, (b) WBT, (c) DPT, (d) specific volume of dry air, and (e) heat content of dry air.
Solution*:*

Locate the intersection point of the T = 20°C vertical line and the RH = 50% curve in Figure 2.2.

a. Draw a horizontal line from the intersection point to the right vertical axis; the humidity ratio of the dry air = ~0.0073 kg moisture/kg dry air.
b. The WBT 13.5°C at that location.
c. Draw a horizontal line from the intersection point to the left-most curve, where the DPT = ~9°C.
d. At that location, the specific volume of the air = ~0.84 m^3/kg of air.
e. At that location, the total heat content of the air = ~39 kJ/kg of air.

Discussion:

1. The values were obtained from reading the figure; they are approximate values, but good enough for most practical uses.
2. Assuming the MW of air = 29, the molar volume of the air can be estimated as (0.84 L/g)(29 g) = 24.36 L, which is relatively close to the molar volume of 24.05 L using the Ideal Gas Law.

Example 2.10: Use of Psychrometric Chart (U.S. Customary Units)

At a site where the DBT = 68°F and RH = 50%, find the (a) humidity ratio of dry air, (b) WBT, (c) DPT, (d) specific volume of dry air, and (e) heat content of dry air.
Solution*:*

Locate the intersection point of the T = 68°F vertical line and the RH curve = 50% in Figure 2.3.

a. Draw a horizontal line from the intersection point to the right vertical axis, the humidity ratio of the dry air = ~0.0073 lb moisture/lb dry air.
b. The WBT = ~57°F at that location.
c. Draw a horizontal line from the intersection point to the left-most curve, the DPT = ~48°F.
d. At that location, the specific volume of the dry air = ~13.4 ft^3/lb of dry air.
e. At that location, the total heat content of the dry air = ~17 BTU/lb of dry air.

Discussion:

1. The values obtained are essentially the same as the corresponding ones in Example 2.9.
2. Note: (a) 1 kg moist/kg dry air = 1 lb moist/lb dry air; (b) 1 m^3/kg of dry air = (35.3/2.2) ft^3/lb of dry air = 16.0 lb^3/ft of dry air ; and (c) 1 BTU/lb of air = (1.055 × 2.2) kJ/kg of dry air = 2.324 kJ/kg of dry air (also see the next section).

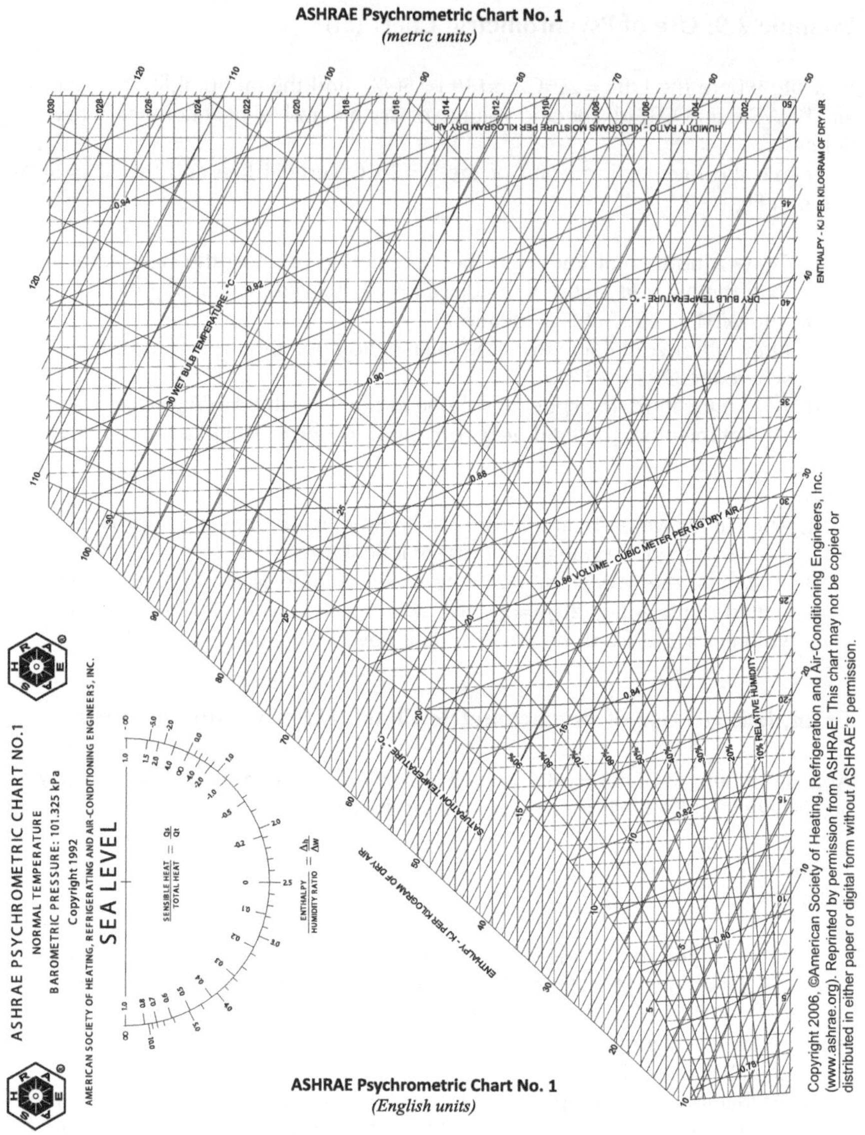

FIGURE 2.2 ASHRAE psychrometric chart no. 1 (SI).

2.4.4 Heating Value

Combustion is a chemical reaction in which a substance with energy content (i.e., fuel) reacts with an oxidizing agent (typically oxygen) that releases heat and light. This exothermic reaction is the basis for many energy generation processes. It is also the primary source of air pollutants and greenhouse gases. The combined influent

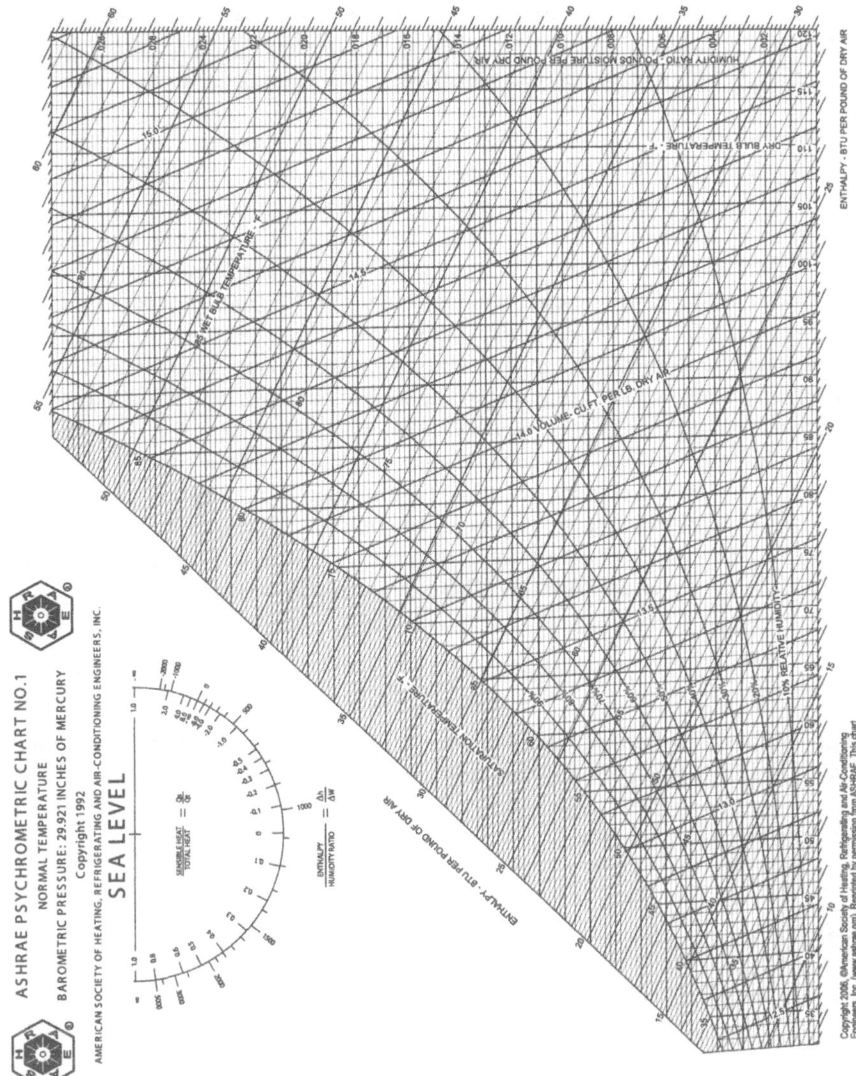

FIGURE 2.3 ASHRAE psychrometric chart no. 2 (U.S. customary units).

to the combustion chamber should possess sufficient energy to sustain burning. Here are some terms commonly used regarding energy/heat (more in Chapter 7).

Sensible heat (H_s) is the amount of energy needed to cause a change in temperature of a unit mass of a substance without any phase change. It is the product of specific heat capacity (C_p) and the temperature change (ΔT), as follows:

$$H_s = C_p \times \Delta T \quad (2.21)$$

Latent heat is the energy absorbed or released during a phase change while the temperature remains constant during the transition. *Heat of vaporization* (H_v) is the energy required for a unit mass of a substance to transform itself from a liquid state to a vapor state at its boiling point. For example, water needs to absorb heat to evaporate. The *latent heat of water vaporization* (or *latent heat of water vapor condensation*) at its boiling point (i.e., 100°C @ $P=1$ atm) is 970.3 BTU/lb or 2,255 kJ/kg.

Enthalpy is a fundamental concept in thermodynamics, and it represents the total heat content of a system, accounting for both internal energy and the energy associated with pressure and volume. It reflects the energy in a system relative to that at reference conditions. The standard reference conditions are $T=20$°C and $P=1$ atm (typ.).

Higher heating value (HHV), also called *gross heating value*, is the total amount of heat released from the complete combustion of fuel at 15.56°C (60°F), and its combustion products, including water vapor, cooled down to 15.56°C (60°F). The *lower heating value* (LHV), or the *net heating value* (NHV), is the HHV minus the latent heat of water formed from combustion. HHV and LHV are two measures used to quantify the energy content of fuels. HHV is used in applications where the full energy potential of the fuel is considered. In contrast, LHV is often used in applications where the condensation of water vapor is not achievable, such as in internal combustion engines or gas turbines. The relationship between them is:

$$\text{HHV} = \text{LHV} + \left(\text{Mass}_{\text{water formed}} \times \text{Latent heat of water vaporization}\right) \tag{2.22}$$

Heating values are often expressed in energy per unit mass or volume. The standard units for heating value per unit mass are:

- 1 kJ/kg = 0.43 BTU/lb
- 1 BTU/lb = 2.324 kJ/kg

The standard units for heating value per unit volume are:

- 1 kJ/m^3 = 0.02685 BTU/ft^3
- 1 BTU/ft^3 = 37.24 kJ/m^3

Suppose the heat content per unit mass of a gas stream is given; the corresponding heating value per unit volume can be readily calculated by multiplying the given value by the density of the gas stream.

Organic compounds generally contain large heating values and can serve as energy sources for combustion. The higher the organic concentration in a gas stream, the higher the heat content; consequently, the lesser the auxiliary fuel requirement. *Dulong's formula* has been commonly used to estimate heating values of fuels or organic substances based on the elemental composition of the fuel or the compounds. If the heating value of a COC is not available, Dulong's formula can be used:

$$\text{Heating Value}\left(\text{in } \frac{\text{Btu}}{\text{lb}}\right) = 145.4\ \text{C} + 620\left(\text{H} - \frac{\text{O}}{8}\right) + 41\text{S} \tag{2.23}$$

$$\text{Heating Value}\left(\text{in }\frac{\text{kJ}}{\text{kg}}\right) = 337.9\ \text{C} + 1{,}441\left(\text{H} - \frac{\text{O}}{8}\right) + 95\text{S} \tag{2.24}$$

where C, H, O, and S are the elements' compound percentages (by weight).

According to the U.S. Energy Information Administration (EIA), the average heat content of natural gas in 2023 was 1,036 BTU/ft^3 (38,600 kJ/m^3) (http://www.eia.gov). It should be noted that "*therm*" is a standard energy unit for natural gas, and one therm = 100,000 BTU.

Example 2.11: Estimate the Heating Value of Methane

Use Dulong's formula to estimate the heating value of methane (CH_4) gas.

Solution:

a. MW of $CH_4 = 12 \times 1 + 1 \times 4 = 16$

Weight percentage of C in methane = 12 ÷ 16 = 75%

Weight percentage of H in methane = 4 ÷ 16 = 25%

$$\text{Heating Value}\left(\text{in }\frac{\text{kJ}}{\text{kg}}\right) = 337.9(75) + 1{,}441(25) = 61{,}325\frac{\text{kJ}}{\text{kg}}$$

$$\text{Heating Value}\left(\text{in }\frac{\text{BTU}}{\text{lb}}\right) = 145.4(75) + 620(25) = 26{,}405\frac{\text{BTU}}{\text{lb}}$$

b. Using Eq. 2.9 to find the density of methane at $T = 20°\text{C}$ (68°F):

$$\rho = \left(\frac{m}{V}\right) = \frac{P \times \text{MW}}{R \times T} = \left(\frac{16}{24.05}\right) = 0.665\,\text{kg/m}^3$$

$$\rho = \left(\frac{m}{V}\right) = \left(\frac{16}{383}\right) = 0.418\ \text{lb/ft}$$

c. Heating value per unit volume = (61,325 kJ/kg)(0.665 kg/m^3) = 41,080 kJ/m^3

$$= (26{,}405\ \text{BTU/lb}) \times (0.418\ \text{lb/ft}^3) = 1{,}103\ \text{BTU/ft}^3$$

Discussion:

1. Methane is the principal constituent of natural gas, representing about 95% of the mixture (Kuo et al., 2015). The heating value of methane calculated from Dulong's formula, 1,103 BTU/ft^3 or 41,080 kJ/m^3, is very close to the reported average heating value of the natural gas, 1,036 BTU/ft^3 (38,600 kJ/m^3).

2. The weight percentage of C in methane is 75%. A value of 75, not 75%, should be used in Dulong's formula shown in Eq. 2.23 or 2.24.
3. Methane gas is lighter than air. Its density can also be found by multiplying the air density by the molecular weight ratio of methane and air (i.e., 16/29).

2.5 INTERACTION AMONG GAS, LIQUID, AND SOLID

2.5.1 Diffusion

Molecular diffusion of molecules is caused by a concentration gradient (i.e., from a higher-concentration location to a lower-concentration location) due to their random thermal motion. Due to molecular diffusion, a COC in the air could travel in directions different from the airflow, and it can also migrate in stagnant air due to diffusion. The equation below is the one-dimensional form of the popular *Fick's law of diffusion*, which describes how concentration changes over time due to diffusion:

$$\frac{\partial G}{\partial t} = D\frac{\partial^2 G}{\partial^2 x} \tag{2.25}$$

where G is the COC concentration, D is the diffusion coefficient (length^2/time), t is the time, and x is the distance. It would be harder for a compound of a larger MW to diffuse in a fluid, especially when it is more viscous. Generally, the diffusion coefficient is inversely proportional to the square root of the molecular weight (MW) and inversely proportional to the viscosity (μ) of the air, as follows:

$$\frac{D_1}{D_2} = \sqrt{\frac{\text{MW}_2}{\text{MW}_1}} \tag{2.26}$$

$$\frac{D_1}{D_2} = \frac{\mu_{\text{air},2}}{\mu_{\text{air},1}} \tag{2.27}$$

Typical molecular diffusion coefficients of gaseous compounds in air range from 10^{-6} to 10^{-5} m^2/s, while those of the soluble compounds in water are about 10,000 times smaller (i.e., in the order of magnitude of 10^{-9} m^2/s). For example, the diffusion coefficients of benzene (C_6H_6) @T=298 K & P=1 atm are 0.92×10^{-5} and 1.02×10^{-9} m^2/s in air and water, respectively.

Eddy diffusion is driven by turbulent flow and the movement of eddies/swirls in a fluid. It is a macroscopic process that results from the chaotic mixing caused by turbulence. At the same time, molecular diffusion is a microscopic process that can occur even if the fluid is stagnant. The values of eddy diffusion coefficients are typically larger than the corresponding molecular diffusion coefficients when the fluid flow is turbulent. The diffusivities of particles in air will be discussed in Chapter 6.

2.5.2 Vapor–Liquid Equilibrium

In addition to gaseous components, air contains moisture and some liquid droplets. Gaseous components tend to cross the interfacial boundaries between air and liquid droplets to get absorbed/dissolved. On the other hand, dissolved species also tend to leave the liquid droplets and move into the air. *Absorption* is the process in which one substance (e.g., a gas or vapor) is taken up by a liquid. Absorption is a standard air pollution control process that uses liquid to remove gaseous COCs (more discussion in Chapter 7). Consequently, it is essential to understand the interactions between vapor and liquid.

As mentioned earlier, Raoult's law describes the vapor–liquid equilibrium (see Eq. 2.15). Henry's law has been found more applicable to aqueous solutions of lower COC concentrations, which are commonly encountered in environmental engineering applications [Note: Henry's law assumes no chemical reaction between gas and liquid, and it does not work well for highly reactive gases such as NH_3 and SO_2]. *Henry's law* states that, at equilibrium, the partial pressure of a compound A (P_A) in the gas phase above a liquid is proportional to the concentration of that compound in the liquid (C_A), as follows:

$$P_A = H_A \times C \tag{2.28}$$

where H_A is *Henry's constant* (or *Henry's law constant*) of compound A. This equation shows a linear relationship between the liquid and gas concentrations. The higher the liquid concentration, the higher the gas concentration, or vice versa. A compound with a large Henry's constant prefers to stay in the air. In some air pollution books or literature, Henry's law is written oppositely as $C_A = H_A \times P_A$; Henry's constant in this format is the inverse of the one shown in Eq. 2.28.

Henry's law can also be expressed in the following form:

$$G = H \times C \tag{2.29}$$

where C is the COC concentration in the liquid phase, and G is the corresponding concentration in the gas phase. Henry's law has been widely used in various disciplines to describe the distribution of a compound between the gas and liquid phases. The units of Henry's constant reported in the literature vary considerably. The units commonly encountered include atm/mole fraction, atm/M, M/atm, atm/(mg/L), and dimensionless. When inserting a value of Henry's constant into one of the two equations above, it is essential to check if its units match the two concentrations. Engineers typically use the units they are familiar with and often have difficulties performing the necessary unit conversions. The use of Henry's constant in dimensionless form has increased because it is easier (see the design calculations of absorbers in Chapter 7 later). Note that it is not a "(mole fraction)/(mole fraction)" dimensionless unit. The actual meaning of Henry's constant in this dimensionless format is (concentration in the gas phase)/(concentration in the liquid phase), which can be [(mg/L)/(mg/L)]. To be more precise, it has a unit of (unit volume of liquid)/(unit volume of air). Table 2.4 shows conversions among commonly used units for Henry's constant.

The Henry's constant of a compound is practically the ratio of its vapor pressure and solubility if both are measured at the same temperature:

$$H = (\text{Vapor Pressure}) \div (\text{Solubility}) \tag{2.30}$$

This equation implies that the higher the vapor pressure, the larger the Henry's constant. It also indicates that the lower the solubility, the larger the Henry's constant. For most organic compounds, the vapor pressure increases with temperature, while the solubility decreases. Consequently, as defined in Eq. 2.30, Henry's constant should increase with temperature.

Example 2.12: Unit Conversions for Henry's Constant

The Henry's constant of benzene (C_6H_6) in water at 25°C is 5.55 atm/M. Convert this value to (a) dimensionless units and (b) units of atm.

Solution:

a. From Table 2.4,

$$H = H^*RT = 5.55 = H^*(0.082)(273 + 25)$$

$$H^* = \underline{0.227}\,(\text{dimensionless})$$

b. Also from Table 2.4,

$$H = (H^*RT)[1{,}000\gamma/W]$$
$$= [(0.227)(0.082)(273 + 25)][(1{,}000)(1)/(18)] = \underline{308.3\text{ atm}}$$

TABLE 2.4
Unit Conversions for Henry's Constant

Desired Unit for Henry's Constant (*H*)	Conversion Equation
atm/M, or atm·L/mole	$H = H^*RT$
atm·m^3/mole	$H = H^*RT/1{,}000$
M/atm	$H = 1/(H^*RT)$
atm/(mole fraction in liquid), or atm	$H = (H^*RT)[1{,}000\gamma/W]$
(mole frac. in vapor)/(mole frac. in liquid)	$H = (H^*RT)[1{,}000\gamma/W]/P$

H^*, Henry's constant in the dimensionless form; γ, specific gravity of the solution (1 for a dilute solution); W, equivalent MW of solution (18 for dilute aqueous solution); R, 0.082 atm/(K)(M); T, system temperature in Kelvin; P, system pressure in atm (usually =1 atm); M, solution molarity in (g-mole/L).

Source: Kuo and Cordery (1988).

Discussion:

1. Benzene is a VOC of concern and appears in most, if not all, databases of Henry's constant values. It may not be a bad idea to memorize benzene's Henry's constant=0.23 under ambient conditions.
2. To convert the Henry's constant of another COC in the database, multiply the ratio of Henry's constants (in any units) of COC and benzene by 0.23. For example, to convert the Henry's constant of methylene chloride (CH_2Cl_2), 2.03 atm/M to dimensionless, multiply the ratio by 0.23: [(2.03)/(5.55)] × (0.23) = 0.084 (dimensionless).

Example 2.13: Estimate Henry's Constant from Solubility and Vapor Pressure

The vapor pressure of benzene (C_6H_6) and its solubility in water at 298 K are 95.2 mm-Hg and 1,780 mg/L, respectively. Estimate Henry's constant of benzene from the given information.

Solution:

a. P^{vap} = 95.2/760 = 0.125 atm
b. C = 1,780 mg/L = 1.78 g/L
   = (1.78 g/L) ÷ (78 g/g-mole) = 0.0228 mole/L = 0.0228 M
c. H = (0.125 atm) ÷ (0.0228 M) = 5.48 atm/M

Discussion: The calculated value, 5.48 atm/M, is essentially the same as the value, 5.55 atm/M, given in Example 2.12.

Example 2.14: Use Henry's Law to Estimate Equilibrium Concentrations

An air sample was taken from the headspace of a landfill leachate collection tank. The concentration of tetrachloroethylene (PCE, C_2Cl_4) in the air sample was 1,250 ppm. Estimate the PCE concentration in the leachate. Note: T = 298 K, Henry's constant of PCE = 25.9 atm, and MW of PCE = 165.8.

Solution:

a. 1,250 ppmV = 1,250 × 10^{-6} atm = 1.25 × 10^{-3} atm = P_A
b. From Eq. 2.28,

$$P_A = 1.25 \times 10^{-3}\,\text{atm} = H_A \times C_A = (25.9\ \text{atm/M}) \times (C_A)$$

$$C_A = 4.82 \times 10^{-5}\,\text{M} = (4.82 \times 10^{-5}\,\text{mole/L}) \times (165.8\ \text{g/mole})$$

$$= 8.0 \times 10^{-3}\,\text{g/L} = \underline{8.0\ \text{mg/L}} = \underline{8.0\ \text{ppm}}$$

We can use the dimensionless Henry's constant to solve this problem.

a. $H = H^*RT = 25.9 = H^*(0.082)(273 + 20)$

$$H^* = 1.08\,(\text{dimensionless})$$

b. Use Eq. 4.17 to convert ppm to mg/m³ (see Chapter 4):

$$1,250 \text{ ppm} = (1,250)[(165.8/24.05)] \text{ mg/m}^3$$
$$= 8,620 \text{ mg/m}^3 = 8.62 \text{ mg/L}$$

c. Use Eq. 2.29:

$$G = 8.62 \text{ mg/L} = H \times C = 1.08 \times C$$

So, $C = \underline{8.0 \text{ mg/L}} = \underline{8.0 \text{ ppm}}$

Discussion:

1. These two approaches yield identical results.
2. The Henry's constant of PCE is five times larger than benzene's (1.08 vs. 0.227).
3. A concentration of 8.0 mg/L of PCE in the leachate is in equilibrium with a gas concentration of 1,250 ppmV. The numeric value of the gas concentration (1,250 ppm) is much higher than that of the corresponding liquid concentration (8.0 ppm) – more discussion on this can be found in Kuo (2014).

2.5.3 VAPOR–SOLID EQUILIBRIUM

Adsorption is the process by which molecules from a gas move across the bulk onto the surface of a solid or liquid. In addition to gaseous components, an air stream contains particulates, including liquid droplets. Gaseous components may attach to the surface of liquids/solids through adsorption. Here, let us focus on the gas and solid interaction. Interactions among three distinct components cause adsorption: (i) *adsorbent*, the solid onto which the target compound is adsorbed/attached; (ii) *adsorbate*, the target compound that is being adsorbed onto the surface of the solid; and (iii) *carrier gas*, the inert part of the gas stream, typically air, from which the target compounds are to be removed. In adsorption, the adsorbate is removed from the carrier gas and taken by the adsorbent. Adsorption is a standard air pollution control process that uses solids (e.g., activated carbon, silica gel, zeolites, alumina, biochar, clay minerals, and carbon nanotubes) to remove gaseous contaminants. Consequently, it is essential to understand the interactions between gas and solid.

An *adsorption isotherm* describes the equilibrium relationship between the gas and solid phases at a constant temperature in a system where solid and gas phases coexist. The most popular isotherms are the Langmuir isotherm and the Freundlich isotherm; both were derived in the early 1900s. The *Langmuir isotherm* has a theoretical basis that assumes a monolayer coverage of the adsorbent surface by the adsorbate, while the *Freundlich isotherm* has a semi-empirical relationship. For a Langmuir isotherm, the concentration on the solid increases with increasing concentration in the gas phase until a maximum concentration on the solid is reached. The Langmuir isotherm can be expressed as follows:

$$q = q_{\max} \frac{K_L G}{1 + K_L G} \tag{2.31}$$

where q is the adsorbed concentration on the solid surface, G is the concentration in the gas phase, K_L is an empirical constant, and q_{max} is the maximum adsorbed concentration.

The Freundlich isotherm is often expressed in the following form:

$$q = K_F G^{1/n} \tag{2.32}$$

Both K_F and $1/n$ are empirical constants. The empirical constants in Eqs. 2.31 and 2.32 differ for different adsorbates, adsorbents, and carrier gases. The values also differ for various temperatures and concentration ranges for a compound. When using an isotherm, we should ensure consistent units among the parameters and empirical constants. Both isotherms describe the equilibrium relationships between the concentrations of COCs in the gas phase and those adsorbed on the solid. The higher the vapor concentration, the higher the concentration on the solid surface would be.

Both absorption and adsorption processes are used in air pollution control to remove contaminants. A<u>d</u>sorption is a surface phenomenon in which gaseous compounds are removed by adhering to the surface of an adsorbent (e.g., activated carbon). A<u>b</u>sorption is a bulk phenomenon in which the gaseous compounds or particulates cross the air–liquid interface and enter the liquid (e.g., water).

EXERCISE QUESTIONS

1. At what temperature does the numerical value on the Celsius scale match the Fahrenheit scale?
2. Between 100 pounds and 500 N, which is a larger force?
3. A diver is 200 ft (61 m) under the sea; what would be the total pressure exerted on the diver (in atm)?
4. Estimate the molecular weight of an air sample containing 2% moisture by volume. Does it have a higher or lower density than dry air?
5. Find the volume of 1 g-mole and 1 lb-mole of an ideal gas at T=20°C (= 68°F) and P=1 atm.
6. Determine the density of air at (a) T=77°F and P=1 atm, (b) T=600°F and P=1 atm, and (c) T=600°F and P=0.9 atm.
7. Use the value of air viscosity @ T=20°C, given in Table 2.2, to estimate the viscosity of air at T=100°C and P=1 atm. What would be the corresponding density and kinematic viscosity of air?
8. The gas stream contains 10% water vapor and the stack temperature is 300°C (572°F). Estimate the specific heat capacity of flue gas from an emission stack.
9. An above-ground storage tank contains 40% (by wt.) of toluene and 60% of ethylbenzene. Estimate the maximum toluene and ethylbenzene concentrations (in ppm) in the void space of the tank. The ambient temperature is 20°C. Note: At 20°C, the vapor pressure of toluene = 22 mm-Hg and that of ethylbenzene = 7 mm-Hg.

10. The empirical constants of the Antoine equation for benzene (C_6H_6) are $A = 15.9008$, $B = 2788.51$, and $C = -52.36$ for P^{vap} in mm-Hg and T in K.
    a. Estimate the vapor pressure of benzene at 30°C (= 303 K).
    b. Use this value and benzene's enthalpy of vaporization (33,830 J/g-mole) to estimate its vapor pressure at 15°C.
11. The flue gas from a combustion chamber contains 15% of water by volume. Estimate its DPT ($P = 1$ atm) using the Antoine equation below (for water vapor between 1°C and 100°C):

$$\log_{10} P(\text{in mmHg}) = 8.07131 - \frac{1730.63}{T(\text{in}°\text{C}) + 233.426}$$

12. At a site where the DBT = 30°C and RH = 40%, find the (a) humidity ratio of dry air, (b) WBT, (c) DPT, (d) specific volume of dry air, and (e) heat content of dry air.
13. At a site where the DBT = 86°F and RH = 40%, find the (a) humidity ratio of dry air, (b) WBT, (c) DPT, (d) specific volume of dry air, and (e) heat content of dry air. Compare your results with those of Question #12.
14. a. Use Dulong's formula to estimate the heating value of propane (C_3H_8).
    b. Compare this heating value with methane (Example 2.11).
15. Henry's constant of toluene in water at 25°C is 6.7 atm/M. Convert this value to (a) dimensionless units and (b) units of atm.
16. The vapor pressure of toluene and its solubility in water at 298 K are 22.0 mm-Hg and 515 mg/L, respectively.
    a. Estimate its Henry's constant from the given information.
    b. Compare results with those of Question #16 above.
17. An air sample was taken from the headspace of a landfill leachate collection tank. The methylene chloride concentration in the air sample was 1,000 ppm. Estimate the methylene chloride concentration in the leachate [Note: $T = 298$ K and Henry's constant of methylene chloride = 2.03 atm/M].

BIBLIOGRAPHY

Kuo, J. (2014). *Practical Design Calculations for Groundwater and Soil Remediation*, 2nd edition, CRC Press, Boca Raton, FL.

Kuo, J.F.; Cordery, S.A. (1988). "*Discussion of Monograph for Air Stripping of VOC from Water*," *J. Environ. Eng.* 114(5), 1248–50.

Kuo, J.; Hicks, T.C.; Drake, B.; Chan, T.F. (2015). "*Estimation of Methane Emission for California Natural Gas Industry*," *J. Air Waste Manag. Assoc.* 65(7), 844–55.

Lide, D.R. (1992). *Handbook of Chemistry and Physics*, 73rd edition, CRC Press, Boca Raton, FL.

NASA. (2024). *Earth's Atmosphere: A Multi-layer Cake*, edited by NASA Editorial Team, https://science.nasa.gov/earth/earth-atmosphere/earths-atmosphere-a-multi-layered-cake/, last updated on 10/22/2024.

NCEES. (2025). *PE/Environmental Reference Handbook (Version 1.6)*, National Council of Examiners for Engineering and Surveying (NCEES), Greenville, SC.

Reid, R.C., Prausnitz, J.M. and Poling, B.F. (1987). *The Properties of Liquids and Gases*, 4th edition, McGraw-Hill, Inc., New York, NY.

USEPA. (2012a). *APTI 413: Control of Particulate Matter Emissions - Student Manual*, 5th edition, Prepared by Tidewater Operations Center of C^2 Technologies for Air Pollution Training Institute, United States Environmental Protection Agency, Research Triangle Park, NC.

USEPA. (2012b). *APTI 427: Combustion Source Evaluation - Student Manual*, 3rd Edition, Prepared by ICES Ltd. for Air Pollution Training Institute, United States Environmental Protection Agency, Research Triangle Park, NC.

3 Regulatory Approaches to Solve Air Pollution Problems

3.1 REGULATORY FRAMEWORK

The U.S. Congress passes laws that govern the U.S. For the statutes related to protecting the environment and public health, Congress has authorized the Environmental Protection Agency (USEPA), as a regulatory agency, to create and enforce regulations to implement these laws. In addition, several Presidential Executive Orders (EOs) play a central role in the USEPA's activities.

A proposed regulation will first be published for comments in the Federal Register (FR), the official daily publication for rules, proposed rules, and notices of federal agencies and organizations, as well as EOs and other presidential documents. Once a final decision is issued, the final regulation is codified and incorporated into the U.S. Code of Federal Regulations (CFR). Title 40 (Protection of Environment) of the CFR deals with the USEPA's mission of protecting human health and the environment. These regulations are mandatory requirements that can apply to individuals, businesses, state or local governments, non-profit institutions, or others. A summary of the U.S. environmental laws and EOs can be found at https://www.epa.gov/laws-regulations/laws-and-executive-orders.

3.2 CLEAN AIR ACT

The Clean Air Act (CAA) of 1970 is a comprehensive federal law of the U.S. that regulates air emissions from stationary and mobile sources to ensure that air quality meets health-based standards and protects public health and the environment from the harmful effects of air pollution. It was initially passed in 1963. One of its early goals was to set and achieve *National Ambient Air Quality Standards* (NAAQS) in every state by 1975 to address the public health and welfare risks posed by certain widespread air pollutants and to regulate emissions of hazardous air pollutants (HAPs). The setting of these standards was coupled with directing the states to develop state implementation plans (SIPs) applicable to appropriate industrial sources in the states to achieve these standards. To set new goals and dates for achieving the attainment of NAAQS, the CAA was amended in 1977 and 1990.

Eleven titles are associated with the Clean Air Act Amendment of 1990 (CAAA of 1990), as shown in Table 3.1.

DOI: 10.1201/9781003532705-3

TABLE 3.1
Titles of CAAA of 1990

I	National Ambient Air Quality Standards
II	Mobile Sources
III	Air Toxics
IV	Acid Deposition Control
V	Permits
VI	Stratospheric Ozone and Global Climate Protection
VII	Enforcement
VIII	Miscellaneous Provisions
IX	Clean Air Research
X	Disadvantaged Business Concerns
XI	Clean Air Employment Transition Assistance

TABLE 3.2
NAAQS of CO

Pollutant	Primary/Secondary	Averaging Time	Level	Form
CO	Primary	8 hours	9 ppm	Not to be exceeded more than once per year
		1 hour	35 ppm	

3.2.1 National Ambient Air Quality Standards (NAAQS)

The NAAQS aims to protect public health, including protecting the health of "sensitive" populations such as asthmatics, children, and older people (i.e., the *primary standards*), as well as public welfare, including protection against decreased visibility and damage to animals, crops, vegetation, and buildings (i.e., the *secondary standards*). The USEPA periodically reviews the NAAQS primary and secondary standards to promulgate necessary revisions. These standards establish the maximum allowable pollutant concentration levels in ambient air. The current primary and secondary standards of the six criteria pollutants (i.e., CO, lead, NO_2, ozone, SO_2, and particulate matter (PM)) are shown in Tables 3.2 to 3.7 (USEPA, 2024a).

It is interesting to note that CO is the only criteria pollutant that does not have a secondary NAAQS.

The USEPA issued standards for PM in 1971; PM_{10} replaced total suspended particulate (TSP) in 1987; and $PM_{2.5}$ standards were added in 1997.

3.2.2 Air Quality Implementation Plans

The NAAQS are enforced by designating attainment and non-attainment areas with the standards. An *attainment area* is a geographic region that meets or exceeds the air quality standard for a criteria pollutant. In contrast, a *non-attainment area* is a geographic region that does not meet one or more of the NAAQS criteria for pollutants.

TABLE 3.3
NAAQS of Pb

Pollutant	Primary/Secondary	Averaging Time	Level	Form
Lead	Primary & secondary	Rolling 3-month avg.	0.15 μg/m³ [a]	Not to be exceeded

[a] In areas designated as non-attainment for Pb standard before the promulgation of the current (2008) standards and for which implementation to attain or maintain the current (2008) standards has not been submitted and approved, the previous standards (1.5 μg/m³ as a calendar quarter average) also remain in effect.

TABLE 3.4
NAAQS of NO_2

Pollutant	Primary/Secondary	Averaging Time	Level	Form
NO_2	Primary	1 hour	100 ppb	98th percentile of 1-hour daily maximum concentrations, averaged over 3 years
	Primary & secondary	1 year	53 ppb [a]	Annual mean

[a] The level of the annual NO_2 standard is 0.053 ppm. It is shown in terms of ppb for a more precise comparison to the 1-hour standard level.

TABLE 3.5
NAAQS of O_3

Pollutant	Primary/Secondary	Averaging Time	Level	Form
Ozone	Primary & secondary	8 hours	0.070 ppm [a]	Annual fourth-highest daily maximum 8-hour concentration, averaged over 3 years

[a] The final rule was signed on October 1, 2015, and became effective on December 28, 2015. The previous (2008) O_3 standards have not been revoked and remain in effect in some areas. Additionally, some areas may have certain continuing implementation obligations under the prior revoked 1-hour (1979) and 8-hour (1997) O_3 standards.

The states need to revise their SIPs to include measures to bring the non-attainment areas into attainment.

A *SIP* is a collection of regulations and documents used by a state, territory, or local air district to reduce air pollution in areas that do not meet NAAQS. Contents of a typical SIP should include (i) state-adopted control measures, which consist of either regulations or source-specific requirements, (ii) state-submitted "non-regulatory" components, and (iii) additional requirements promulgated by USEPA.

TABLE 3.6
NAAQS of SO_2

Pollutant	Primary/ Secondary	Averaging Time	Level	Form
SO_2	Primary	1 hour	75 ppb	99th percentile of 1-hour daily maximum concentrations, averaged over 3 years
	Secondary	3 hours	0.5 ppm	Not to be exceeded more than once per year

[a] The previous SO_2 standards (0.14 ppm 24-hour and 0.03 ppm annual) will additionally remain in effect in specific areas: (i) any area for which it is not yet 1 year since the effective date of designation under the current (2010) standards, and (ii) any area for which an implementation plan providing for attainment of the current (2010) standard has not been submitted and approved and which is designated non-attainment under the previous SO_2 standards or is not meeting the requirements of a SIP call under the previous SO_2 standards (40 CFR 50.4(3)). A SIP call is a USEPA action requiring a state to resubmit all or part of its State Implementation Plan to demonstrate attainment of the required NAAQS.

TABLE 3.7
NAAQS of PM

Pollutant		Primary/ Secondary	Averaging Time	Level (μg/m³)	Form
PM	$PM_{2.5}$	Primary	1 year	9.0	Annual mean, averaged over 3 years
		Secondary	1 year	15.0	Annual mean, averaged over 3 years
		Primary and secondary	24 hours	35	98th percentile, averaged over 3 years
	PM_{10}	Primary and secondary	24 hours	150	Not to be exceeded more than once per year on average over 3 years

The USEPA must develop a *Federal Implementation Plan* (FIP) if a state fails to submit a SIP or the plan does not fully comply with the NAAQS.

The USEPA provides a living document, the *Menu of Control Measures* (MCM), which contains existing emission reduction measures as well as relevant information concerning the efficiency and cost-effectiveness of these measures to help reduce air pollutant emissions to comply with the NAAQS (see https://www.epa.gov/air-quality-implementation-plans/menu-control-measures-naaqs-implementation).

3.2.3 Emission Standards for New and Modified Sources

Emission limitations depend on whether a source is in an attainment or non-attainment area, and they can be different for new and existing sources. *Existing sources* are facilities or emission sources already in operation before implementing the *New Source Performance Standards* (NSPS), which can be found in 40 CFR Part 60; they are distinguished from "*new sources*", which are facilities or emission units constructed or modified after the implementation of the NSPS. The NSPS sets technology-based standards for new and modified stationary sources and sets emission limitations for industrial source categories. A *modified source* means there is a physical change in the operation of the source that will cause an increase in emissions. The standards for existing sources may be less stringent than those for new sources, reflecting the technical and economic feasibility of retrofitting older facilities.

New or modified sources located in attainment areas must comply with any applicable NSPS or, in general, apply *Best Available Control Technology* (BACT). BACT is the maximum degree of emission reduction available, considering economic, energy, and environmental factors. For example, for a proposed coal-fired power plant, the BACT for SO_2 might be a flue gas desulfurization (FGD) system. These sources must also comply with the *Prevention of Significant Deterioration* (PSD) provision. This provision aims to prevent air quality deterioration in areas that meet or exceed the NAAQS.

New or modified sources located in non-attainment areas must meet the most stringent, technology-based level of control, called the *Lowest Achievable Emission Rate* (LAER). BACT and LAER are determined case-by-case, usually by state or local permitting agencies. In determining LAER, the reviewing authority must consider the most stringent emission limitation in any SIP and the lowest emission achieved by the same or similar source type. There is no consideration of economic, energy, and environmental factors when selecting LAER. In other words, LAER is stricter than BACT.

3.2.4 Emission Standards for Existing Sources

There are no federal policies for regulating existing stationary sources because it is the primary responsibility of the states. *Reasonable Available Control Technologies* (RACT) are imposed on non-attainment areas on existing stationary sources. Unlike LAER, RACT considers the economic feasibility of implementing control technologies. To provide the states guidance on setting RACT emission limits for specific source categories, the USEPA has published a series of documents referred to as *Control Technique Guidelines* and *Alternative Control Techniques* (https://www3.epa.gov/ttn/atw/ctg_act.html). The USEPA also established the *RACT/BACT/LAER Clearinghouse* (*RBLC*) to provide a central database of air pollution technology information for criteria pollutants, HAPs, and GHGs (https://cfpub.epa.gov/RBLC/).

Table 3.8 presents a brief comparison of RACT, BACT, and LAER.

TABLE 3.8
A Comparison of RACT, BACT, and LAER

	RACT	BACT	LAER
Applicability	Existing sources	New/modified sources	
	Non-attainment areas	Attainment areas	Non-attainment areas
Cost feasibility	Considered	Considered	Not considered
Stringency	Moderate	Higher	Most stringent

3.2.5 Emission Standards for HAPs

Section 112 of the 1990 CAAA addresses emissions of HAPs and requires issuing technology-based standards for primary sources and specific area sources. National emission standards for hazardous air pollutants (NESHAPs) define *major sources* as a stationary source or group of stationary sources that emit or have the potential of emitting ≥10 tons/yr of a HAP or ≥25 tons/yr of a combination of HAPs. "*Area sources*" are smaller-scale facilities that emit HAPs but at levels below the thresholds for major sources. However, these sources contribute to air pollution in the region. The USEPA established the *Maximum Achievable Control Technology* (MACT) standards for major sources by evaluating best-performing technologies and practices within each industry sector. The USEPA periodically reviews the standards to determine if any residual risk exists for that source category and, if necessary, revises the standards to address such risk.

The MACT standards apply to new and existing primary sources. For new sources, the standards must be as stringent as the best-performing source. In contrast, for the existing sources, the standards are based on the best-performing 12% of sources within their specific industry.

3.2.6 Permits

Under the USEPA's *New Source Review* (NSR) program, a company that plans to have a new or modified source must obtain an NSR permit. A Prevention of Significant Deterioration (PSD) permit applies to the sources in attainment areas, and it requires the use of BACT to minimize emissions. A non-attainment NSR permit applies to sources in non-attainment areas, and it requires the use of LAER technology.

The NSR's *Permit to Construct* (*PTC*) is a construction permit that requires the company to minimize air pollution emissions by changing the process to prevent air pollution and installing air pollution control equipment. The USEPA requires facilities that emit pollution into the air to obtain a *Permit to Operate* (*PTO*). This permit (or an "*operating permit*") contains information about how the facility will comply with established emission standards and guidelines set forth by the USEPA. An operating permit is required for any primary stationary source. Potential-to-emit limits are determined by the air quality of the geographical region where the facility is located. PTOs are often subject to periodic reviews to ensure ongoing compliance and update permit conditions based on new regulations or changes of facility operations.

3.2.7 Other Air Pollution Control Programs

The CAA also addresses emissions from mobile sources, such as vehicles and airplanes, by establishing standards for reducing vehicle emissions and improving fuel efficiency. Other air pollution control programs established by the USEPA under the 1990 CAAA include programs on acid rain, sulfur dioxide emission reduction, nitrogen oxides emission reduction, stratospheric ozone, visibility, and mobile sources. Details of these programs can be found on the USEPA websites.

3.3 AMBIENT AIR MONITORING

The ultimate goal of air pollution control is to protect humans and the ecosystem from excessive exposure to air pollutants. Implementing good ambient air monitoring programs, which determine types and concentrations of pollutants in ambient air, is needed to assess the pollution levels and their impacts on public health and the environment. It is also necessary to develop and evaluate air pollution control strategies.

Air quality at a given location is usually dynamic and complex. It changes spatially and temporally, mainly because of changes in emission rates, meteorology, and topography. In addition, physicochemical and photochemical reactions may also occur on these pollutants in the atmosphere. *Ambient air monitoring* is a systematic, long-term assessment of pollutant levels in the outdoor air environment to determine whether the areas meet NAAQS. Per requirements of the CAA, the states need to establish and operate ambient monitoring networks and report collected data to the USEPA. Since its primary purpose is to support human health objectives, monitoring stations are often installed in populated areas, near significant point sources, or near sensitive receptors. Technical information on ambient air monitoring programs of the U.S., including the networks of monitoring stations, monitoring methods, and QA/QC procedures, can be found at the website of the USEPA's Ambient Monitoring Technology Information Center (AMTIC), https://www.epa.gov/amtic.

The AMTIC website provides information on air monitoring methods for criteria pollutants, inorganic air toxics, and organic toxins. The USEPA has established and requires specific monitoring methods for criteria pollutants to determine compliance with the NAAQS, which can be found in 40 CFR Part 50 Appendices A through N. The designated *Federal Reference Methods* (FRMs) specify the procedures and technologies for the measurement to determine compliance with the primary and secondary NAAQS. A *Federal Equivalent Method* (FEM) may be used if they have been designated to be equivalent to an FRM. FEMs provide flexibility for agencies to use technologies that are validated as accurate but differ in their approach from traditional FRM methods.

3.3.1 Sample Collection for Criteria Pollutants

Pollutant concentrations in ambient air usually fluctuate. For example, ozone formation in ambient air is due to photochemical reactions. Its concentrations in the daytime would be higher than those at night. Consequently, the averaging times for sampled compounds of concern (COCs) are an essential parameter in ambient air sampling. Figure 3.1 illustrates the differences among the real-time, 1-hour, and 4-hour averaging times for gaseous pollutants. A longer averaging time may not truly reflect fluctuations in concentration and dampen peak concentrations. On the other hand, fluctuations in data from continuous air monitoring instruments may make data analysis challenging without comparing time-averaged concentrations. The selection of averaging times depends on the duration needed to collect meaningful data and the intended use of the collected data. For example, 24-hour averages would properly reflect long-term air quality trends, while 1-hour or 8-hour averaging

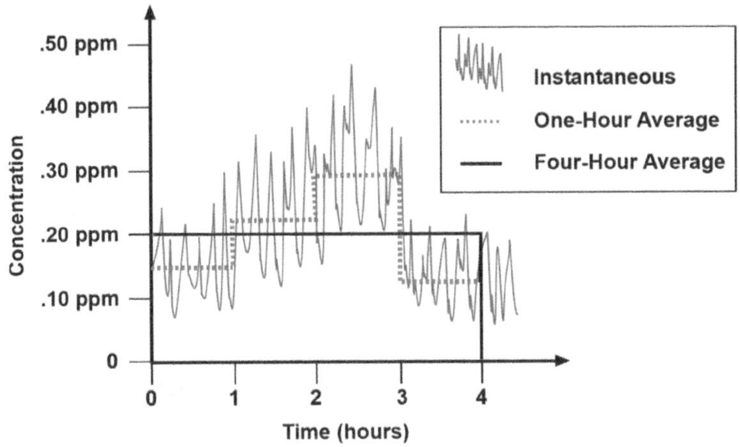

FIGURE 3.1 Averaging times for gaseous pollutants (Modified from USEPA, 2003).

time would represent pollutants that would reach peak levels within short periods (USEPA, 2003). The averaging times for NAAQS are 1-hour and 8-hour averages for CO, the rolling 3-month average for Pb, 1-hour and 1-year averages for NO_2, 8-hour for O_3, 1-hour and 3-hour averages for SO_2, and 24-hour averages for PM_{10} and $PM_{2.5}$ (and 1-year average for $PM_{2.5}$) – see Section 3.2.1 for details.

Inertial impaction and filtration are two essential removal mechanisms of particulate control devices (more details in Chapter 6), and they are most commonly used for sampling PM. *Inertial impaction* drives particles to a collection device by deflecting their original paths. Filtration can handle sizable volumetric flow rates. The particulate concentration is readily calculated using the mass collected and the air volume passed through the collection device.

Figure 3.2 illustrates a typical particle size distribution (PSD) and sources of particles in ambient air. The PSD has three peaks at about 0.01, 0.5, and 15 microns (μm or μ). Most of the larger particles are generated from natural and mechanical processes. Finer particles are usually anthropogenic (man-made) from combustion sources, or they are secondary particles. Although large particles ($\geq$15 μ) are present in the inhaled air, they would be removed in the nose and throat and do not present as much of a threat as the finer particles do because the finer ones can enter the lower respiratory system and are often toxic (e.g., containing heavy metals and polyaromatic hydrocarbons).

Figure 3.3 illustrates a typical sampling train. Air-containing pollutants are drawn into the sampling train by an air mover and pass through a sampling collection device. The device physically and/or chemically removes pollutants from the air stream and holds them for later analysis or analyzes them simultaneously with the collection. A flow meter measures the flow rate to determine the total air volume that passes through the sampling train.

Figure 3.4 illustrates a schematic of a PM_{10} sampler with an impaction inlet. Due to a relatively small mass, ambient air is drawn to a PM_{10} sampler at a constant flow rate for a relatively long duration. The typical volumetric flow rates are high (e.g., 40 ft^3/min or 1.1 m^3/min), and these samplers are often termed *high-volume samplers*. The sampler

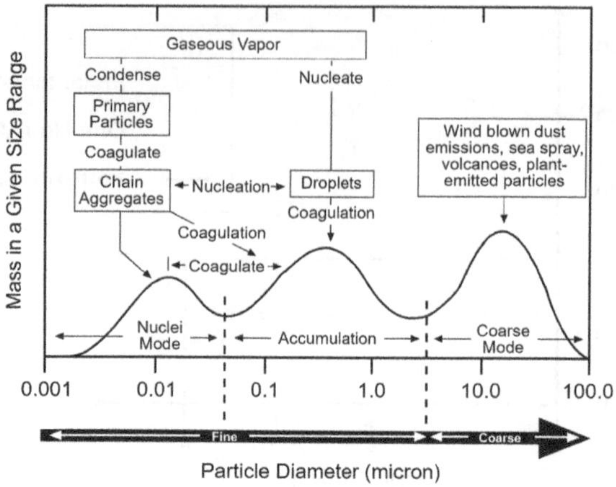

FIGURE 3.2 Size distribution and sources of atmospheric particles (USEPA, 2003).

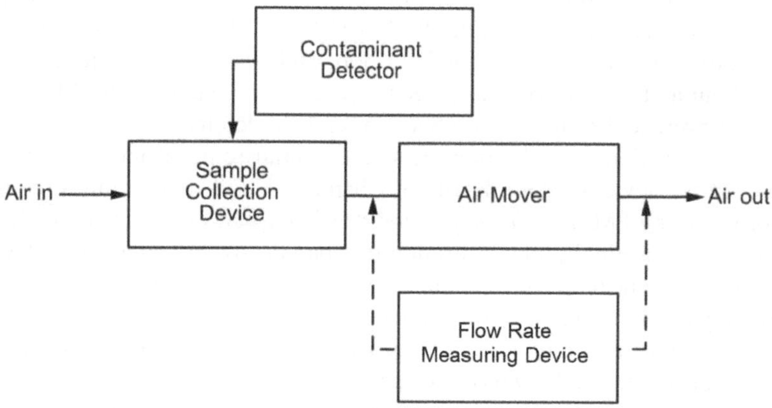

FIGURE 3.3 A typical sampling train (Modified from USEPA, 2008).

has an inlet head to remove larger particles through an impaction and gravity-settling chamber before reaching the filters. *Dichotomous cascade impactors* separate particles into distinct size fractions, typically focusing on PM_{10} and $PM_{2.5}$; they employ a series of stages, each with progressively smaller openings to separate particles based on aerodynamic diameter (see Chapter 6 for the definition of aerodynamic diameter). The typical sampling duration for PM is 24 hours (USEPA, 2003).

Absorption, adsorption, or condensation devices commonly remove gaseous pollutants from gas streams (see Chapter 7 for more details). They are also the operating principles for devices used to collect gaseous samples. The method of choice depends mainly on the physicochemical properties of the pollutants. *A<u>b</u>sorption* is the transfer of compounds from the bulk of the gas to the air–liquid interface, followed by diffusion of these compounds to the bulk of the liquid. Getting the

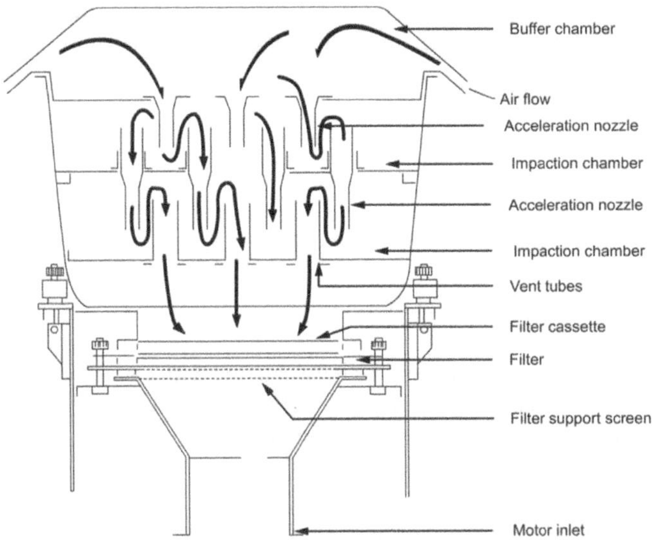

FIGURE 3.4 Schematic of a PM10 sampler with an impaction inlet (USEPA, 2008).

pollutants dissolved into a solution for subsequent analyses is essential. *Adsorption* is a process whereby gaseous compounds in the air/gas stream pass through a bed of porous material (adsorbent), and these compounds diffuse to the surface of the adsorbent and are retained. Here, the gaseous pollutants will be captured and concentrated on the surface of a solid/adsorbent (e.g., activated carbon). Following the collection, the collected pollutants are purged for subsequent analyses. For condensation or cryogenic trapping, the air stream is drawn through collection chambers with progressively lower temperatures for condensation of the pollutants. The condensate is then subjected to subsequent analyses (USEPA, 2008).

3.3.2 Analytical Methods for Criteria Pollutants

Ozone (O_3). Various analytical methods can measure O_3 concentration in the ambient air. Ozone molecules absorb UV light at a wavelength of 254 nm. In the UV photometric method, air is passed through a cell exposed to UV light, and the amount of light absorbed is proportional to the ozone concentration. In the chemiluminescence method, O_3 in the air sample reacts with ethylene (C_2H_4) or nitric oxide (NO) in a mixing zone to emit light (chemiluminescence). A photomultiplier tube then measures the light to derive the ozone concentration. The FRM for measuring ozone in ambient air is based on the UV photometric method.

Carbon monoxide (*CO*). The measurement of CO concentration is commonly done using nondispersive infrared (NDIR) photometry because CO molecules absorb IR light (particularly around 4.6 μm). The CO concentration is related to the reduction in IR intensity resulting from the IR absorption of CO. This is the FRM for measuring CO in ambient air.

Nitrogen dioxide (*NO_2*). The chemiluminescence method is widely used to measure NO_2 concentration. Nitric oxide (NO) can have a chemiluminescent reaction with ozone as:

$$NO + O_3 \rightarrow NO_2 + O_2 + h\nu \text{ (light)} \quad (3.1)$$

NO_2 is measured indirectly from the measurement of NO_x (i.e., NO and NO_2), in which NO_2 is first catalytically converted to NO:

$$NO_2 + \text{Catalytic converter} \rightarrow NO + O \quad (3.2)$$

The NO_2 concentration is the difference between one with catalytic conversion to NO and one without. This is the FRM for measuring NO_2 in the ambient air.

Sulfur dioxide (*SO_2*). Various analytical methods can measure SO_2 concentration in the ambient air. SO_2 molecules absorb UV light (typically around 210 nm) and emit fluorescence at a longer wavelength. The intensity of the emitted fluorescence is proportional to the concentration of SO_2. In the pararosaniline method (i.e., a colorimetric method), the air is bubbled through a potassium tetrachloromercurate solution with tetrachloromercurate (II) ion, $HgCl_4^{-2}$, to form a dichlorosulfitomercurate complex (see Eq. 3.3). Upon addition of pararosaniline, the complex will create a color solution, and the SO_2 concentration is then determined in a colorimeter by measuring the light absorbance of the solution (often called the West-Gaeke method).

$$SO_2 + [HgCl_4]^{-2} + H_2O \rightarrow [HgCl_2(SO_3)]^{-2} + 2\,Cl^- + 2\,H^+ \quad (3.3)$$

The FRM for measuring SO_2 in ambient air is based on UV fluorescence.

Particulate matter (*PM*). PM in ambient air is measured using various techniques that capture and quantify airborne particles based on size and concentration. The gravimetric method is the FRM, in which each filter is weighed (after moisture equilibration) before and after sampling to determine the net weight/mass gain due to the captured PM. The total volume of air sampled is determined from the measured air flow rate, corrected for the USEPA's reference conditions (25°C and 1 atm), and the sampling time.

Lead. The lead concentration in the ambient air is determined by measuring the TSP collected by a high-volume sampler. Lead is then acid-extracted from the collected PM, and the lead content of the solution is determined by atomic absorption (AA) spectrometry or inductively coupled plasma mass spectrometry (ICP-MS). This is the FRM for measuring lead concentration in ambient air.

Example 3.1: High-Volume Sampling for PM

Determine the TSP concentration (i) at actual sampling conditions and (ii) at standard conditions ($T = 25°C$ and $P = 1$ atm) using the sampling and analysis data below:

- Mass of filter before sampling = 4.4444 g
- Mass of filter after sampling = 4.7777 g

- Ambient temperature = 18°C
- Ambient pressure = 725 mm-Hg
- Starting flow rate = 1.80 m^3/min
- Ending flow rate = 1.50 m^3/min
- Sampling duration = 24 hours

Solution:

a. The average actual flow rate (Q_{act}) = (1.80 + 1.50)/2 = 1.65 m^3/min
b. The actual total sample volume (V_{act}) = (1.65)(1,440) = 2,376 m^3
c. Mass of TSP collected = 4.7777 − 4.4444 = 0.3333 g
d. TSP_{actual} = (0.3333)/(2,376) = 1.403×10^{-4} g/m^3 = 140 μg/m^3
e. Using the ideal gas law (Eq. 2.6) to convert the actual volume to volume at standard conditions:

$$V_2 = V_1\left(\frac{T_2}{T_1}\right)\left(\frac{P_1}{P_2}\right) = (2,376)\left(\frac{298}{291}\right)\left(\frac{725}{760}\right) = 2,321 \text{ m}^3$$

f. $TSP_{standard\ condition}$ = (0.3333)/(2,321) = 1.436×10^{-4} g/m^3 = 144 μg/m^3

Discussion:

1. The USEPA's standard temperature for ambient air sampling is 298 K (25°C).
2. The equation used in part (e) is based on the fact that the gas volume is proportional to the absolute temperature and inversely proportional to the pressure.

3.3.3 Air Quality Index

The *air quality index* (AQI) reports the daily ambient air quality in a specific location. It provides an easy-to-understand representation of air pollution levels and their potential health impacts. An AQI value is calculated based on ambient air concentrations of O_3, PM, CO, SO_2, and NO_2, five of the six criteria air pollutants, except lead. The range of AQI values is from 0 to 500. The AQI aims to help the public understand the measured ambient air quality and its effect on their health. The higher the AQI value, the greater the level of air pollution and the greater the health concern. An AQI value of 100 generally corresponds to the NAAQS, the level that the USEPA has set to protect public health.

The air quality conditions are considered good, moderate, unhealthy for sensitive groups, unhealthy, very unhealthy, and hazardous when an AQI value is in the range of "0–50", "51–100", "101–150", "151–200", "201–300", and "301–500", respectively. Each category has an assigned color corresponding to a different health concern level. Table 3.9 shows the ranges of AQI values, the corresponding levels of health concern, and the color codes.

The AQI for each pollutant is calculated using a linear interpolation between two breakpoints (*BPs*). The AQI is calculated for each pollutant, and the highest value becomes the overall AQI.

TABLE 3.9
Air Quality Index Chart

AQI Values	Levels of Health Concern	Color Codes
0–50	Good	Green
51–100	Moderate	Yellow
101–150	Unhealthy for sensitive groups	Orange
151–200	Unhealthy	Red
201–300	Very unhealthy	Purple
301–500	Hazardous	Maroon

$$I_p = \left(\frac{I_{Hi} - I_{Lo}}{BP_{Hi} - BP_{Lo}}\right) \times \left(C_p - BP_{Lo}\right) + I_{Lo} \tag{3.4}$$

where I_p is the index value for pollutant p, C_p is the truncated concentration of pollutant p, BP_{Hi} and BP_{Lo} are the breakpoint concentrations that bracket C_p, and I_{Hi} and I_{Lo} are the AQI values corresponding to these breakpoints. The breakpoints for the AQI are shown in Table 3.10

The following four examples (Examples 3.2–3.6) are instructions and samples for AQI calculations extracted from USEPA (2024b).

Example 3.2: Calculate a Single AQI Value

At a specific location, 8-hour O_3 = 0.07853333 ppm; determine its AQI value.

Solution:

a. Truncate the value to 0.078 ppm.
b. Refer to Table 3.1, the range that falls above and below 0.078 is (0.071–0.085), corresponding to the AQI range of 101–150.
c. Use Eq. 3.4 to calculate the AQI value:

$$I_p = \left(\frac{I_{Hi} - I_{Lo}}{BP_{Hi} - BP_{Lo}}\right) \times \left(C_p - BP_{Lo}\right) + I_{Lo} = \left(\frac{150 - 101}{0.085 - 0.071}\right) \times (0.78 - 0.071) + 101 = 126$$

Example 3.3: Calculate Multiple AQI Values

At a specific location, 8-hour O_3 = 0.078 ppm, $PM_{2.5}$ = 35.9 μg/m^3, and CO = 8.4 ppm; determine the AQI value at this location.

Solution:

a. The AQI value of 8-hour O_3 = 126 (from Example 3.2).
b. Use Eq. 3.4 to calculate the AQI value of $PM_{2.5}$:

TABLE 3.10
The Breakpoints for the AQI

These Breakpoints...							...Equal This AQI
O_3 (ppm) 8-hour	O_3 (ppm) 1-hour[a]	$PM_{2.5}$ (μg/m³) 24-hour	PM_{10} (μg/m³) 24-hour	CO (ppm) 8-hour	SO_2 (ppb) 1-hour	NO_2 (ppb) 1-hour	AQI
0.000–0.054	—	0.0–9.0	0–54	0.0–4.4	0–35	0–53	0–50
0.055–0.070	—	9.1–35.4	55–154	4.5–9.4	36–75	54–100	51–100
0.071–0.085	0.125–0.164	35.5–55.4	155–254	9.5–12.4	76–185	101–360	101–150
0.086–0.105	0.165–0.204	(55.5–125.4)[c]	255–354	12.5–15.4	186–304[c]	361–649	151–200
0.106–0.200	0.205–0.404	(125.5–225.4)[c]	355–424	15.5–304	305–604[c]	650–1249	201–300
0.201–([b])	0.405+	225.5+	425+	30.5+	605+[c]	1250+	301+[d]

Source: USEPA (2024b).

[a] Areas are generally required to report the AQI based on 8-hour O_3 values. However, there are a small number of areas where an AQI based on 1-hour O_3 values would be more precautionary. In these cases, in addition to calculating the 8-hour O_3 index value, the 1-hour O_3 value may be calculated, and the maximum of the two values reported.

[b] 8-hour O_3 values do not define higher AQI values (≥301). AQI values of 301 or higher are calculated with 1-hour O_3 concentrations.

[c] 1-hour SO_2 values do not define higher AQI values (≥200). AQI values of 200 or higher are calculated with 24-hour SO_2 concentration.

[d] AQI values between breakpoints are calculated using Eq. 3.4. For AQI values in the hazardous category, AQI values greater than 500 should be calculated using Eq. 3.4 and the concentration specified for the AQI value of 500. The AQI values of 500 are as follows: O_3 1-hour: 0.604 ppm; $PM_{2.5}$ 24-hour: 325.4 μg/m³; PM_{10} 24-hour: 604 μg/m³; CO: 50.4 ppm; SO_2 1-hour: 1,004 ppb; and NO_2 1-hour: 2,049 ppb.

$$I_p = \left(\frac{150-101}{55.4-35.5}\right)\times(35.9-35.5)+101 = 102$$

c. Use Eq. 3.4 to calculate the AQI value of CO:

$$I_p = \left(\frac{100-51}{9.4-4.5}\right)\times(8.4-4.5)+51 = 90$$

d. The AQI value at this location is 126, with O_3 as the main pollutant.

Example 3.4: Using Both 1-Hour and 8-Hour O_3 Concentrations to Find the AQI Value for a Location

At a specific location, 8-hour O_3 = 0.078 ppm, $PM_{2.5}$ = 35.9 μg/m³, and CO = 8.4 ppm (see Example 3.3 above), and 1-hour O_3 = 0.162 ppm; determine the AQI value at this location.

Solution:

a. Footnote #a of Table 3.10 – Areas are generally required to report the AQI based on 8-hour O_3 values. However, there are a small number of areas where an AQI based on 1-hour O_3 values would be more precautionary. In these cases, in addition to calculating the 8-hour O_3 index value, the 1-hour O_3 value may be calculated and the maximum of the two values reported.
b. Use Eq. 3.4 to calculate the AQI value of 1-hour O_3:

$$I_p = \left(\frac{150-101}{0.164-0.125}\right)\times(0.162-0.125)+101=148$$

c. The AQI value for 8-hour O_3 is 126 (Example 3.3). The AQI value at this location is 148 (the larger of 148 and 126), with O_3 as the main pollutant.

Example 3.5: Calculate AQI Values for SO_2

What are the USEPA's guidelines for calculating AQI values for SO_2?

Solution:

a. The USEPA strengthened the primary standard for SO_2 in 2010. Because there was not enough health information to inform changing the upper end of the AQI for SO_2, the upper end continues to use the 24-hour average SO_2 concentration. The lower end of the AQI uses the daily max 1-hour SO_2 concentration.
b. If the daily max 1-hour SO_2 concentration <305 ppb, then use the breakpoints in Table 3.10 to calculate the AQI value.
c. If a 24-hour average SO_2 concentration ≥305 ppb, then use the breakpoints in Table 3.10 to calculate the AQI value. If a 24-hour value is in this range, it will always result in a higher AQI value than a 1-hour value.
d. On rare occasions, one could have a day where the daily max 1-hour concentration is ≥305 ppb, but the 24-hour concentration is <305 ppb when trying to use the 24-hour to calculate the AQI value. If this happens, use 200 for the lower and upper AQI breakpoints (I_{Lo} and I_{Hi}) in Eq. 3.4 to calculate the AQI value based on the daily max 1-hour value. This effectively fixes the AQI value at 200 exactly, ensuring that you get the highest possible AQI value associated with your 1-hour concentration on such days (USEPA, 2024b).

Example 3.6: Calculate AQI Values for SO_2

What are the USEPA's guidelines for handling pollutant concentrations that have blank places in Table 3.10?

Solution:

a. Disregard numbers that have blank places in Table 3.10. Suppose one had a 1-hour O_3 value of 0.104 ppm and an 8-hour O_3 value of 0.078 ppm. First, disregard the 1-hour O_3 value because it is less than 0.125 ppm. Then, you calculate the index for the 8-hour O_3 value as before (USEPA, 2024b).
b. The calculated AQI value would be 126 (see Example 3.2).

3.3.4 Sampling and Analysis of Ambient HAPs

Ambient air monitoring also characterizes ambient HAP concentrations and deposition in representative areas. The data can be used to evaluate the dispersion and deposition models and the effectiveness of HAP reduction strategies (USEPA, 2003).

Of the current 187 listed HAPs, 99 are volatile organic compounds (VOCs), 49 are semi-volatile organic compounds (SVOCs), 11 are complex organic mixtures, and 28 are non-volatile inorganic compounds. Due to the variety of these HAPs and the lack of standard and documented methods, the measurements of ambient HAPs are difficult (USEPA, 2003). The USEPA has developed an updated compendium of techniques to address the need to characterize HAP concentrations in ambient air by providing procedures for sampling and analysis. Methods for inorganic toxics are essentially based on "1999 Inorganic (IO) Compendium of Methods" (EPA/625/R-96/01a). For example, the Compendium Method IO-5 guides sampling and analysis for atmospheric mercury. There are 17 compendium methods (Methods TO 1 to 17) for the determination of toxic organic (TO) compounds in ambient air (USEPA, 1999). For example, TO-15 is "Determination of VOCs in Air Collected in Specially Prepared Canisters and Analyzed by GC/MS".

A sampling train can be set up to draw air through the collection device for (i) filter-based collection to capture particulate-bound HAPs (e.g., heavy metals and less-volatile organic compounds) and (ii) sorbent tube collection (e.g., activated carbon, silica gel). In addition to a sampling chain, grab samples can be taken by evacuated containers made of glass, stainless steel, or pliable plastics (e.g., Tedlar®, Mylar®, Teflon®, aluminized PVC). A *passivated canister* means that the canister has been passivated to minimize chemical or physical interactions between the sample and the canister walls before analysis (USEPA, 2008). Passivated canisters are typically made of stainless steel, and their interior surface is coated with an inert coating. The canisters are evacuated with a vacuum before sampling. Figure 3.5 shows two examples of containers for "whole air" (grab) sampling.

Gas chromatography (GC) and GC with mass spectrometry (GC-MS) are standard methods for analyzing VOCs and SVOCs. Polyaromatic hydrocarbons (PAHs), pesticides, and certain organic compounds are often studied with high-performance liquid chromatography (HPLC). Ion chromatography (IC) is commonly used for acidic gases or when HAPs are in an ionic form. X-ray fluorescence (XRF) and inductively coupled plasma mass spectrometry (ICP-MS) are widely used for heavy metals.

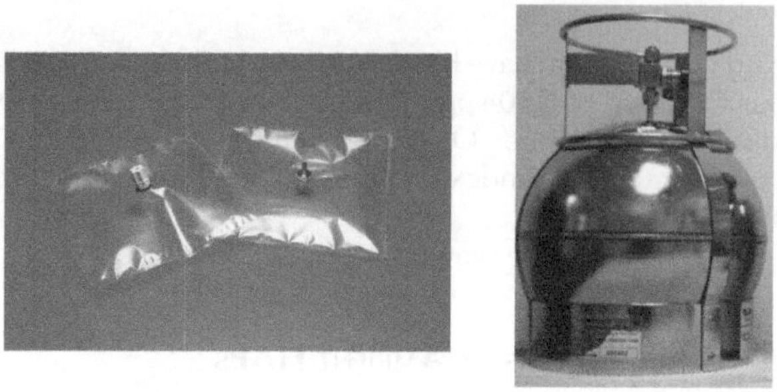

(a) (b)

FIGURE 3.5 Examples of containers for grab sampling: (a) Tedlar® bags and (b) Summa® canister.

3.4 EMISSION MONITORING AND MEASUREMENTS

Eliminating or reducing emissions of pollutants from human activities into the atmosphere is the most effective way to improve ambient air quality. Measurements of the type and quantity of these emissions serve as the basis for implementing a meaningful air pollution control program. The sampling/monitoring of particulate and gaseous emissions from a stationary source is often termed *source testing* or *source sampling*. Source testing/sampling determines the composition of the effluent and its emission rate into the atmosphere from a source (e.g., industrial stacks, vents, or exhaust systems). Typical COCs of source testing include CO, NO_x, SO_2, PM, organics, air toxics, CO_2, and H_2O.

The data/results from source testing can serve several purposes, such as (i) determining types and emission rates of COCs, (ii) assessing the performance of pollution control devices, and (iii) others, such as providing feedback to IC engines for tuning to optimize fuel efficiency and to reduce emissions (Kuo & Dow, 2017).

3.4.1 Stationary Source Emission Sampling

To obtain reliable and representative data regarding the composition of a gas stream and its emission rate, a source test should meet the following requirements:

i. The gas sampled should represent the stack effluent's total or a known portion.
ii. Samples collected for analysis should represent the sampled gas stream.
iii. The volume of the gas sample withdrawn for analysis should be accurately measured so that concentrations of COCs in the stack effluent can be accurately calculated.
iv. The flow rate of the stack effluent should be accurately determined so that emission rates of COCs can be calculated using the corresponding concentrations from (iii).

Air Emission Measurement Center (AEMC) is the USEPA's hub for information on test methods and procedures for measuring air emissions from stationary sources (https://www.epa.gov/emc). The original eight USEPA *FRMs for Source Testing* were promulgated in 1971. They were the results of the first NSPS authorized in the CAA of 1970 and focused on source categories that emitted large quantities of criteria pollutants. Subsequent FRMs reflect the emission types and characteristics of additional promulgated source categories. The original eight FRMs are (USEPA, 2013):

- FRM 1 – Sample and velocity transverse for a stationary source
- FRM 2 – Determination of stack gas velocity and volumetric flow rate
- FRM 3 – Gas analysis of CO_2, oxygen, excess air, and dry MW
- FRM 4 – Determination of moisture content in stack gas
- FRM 5 – Determination of PM emissions
- FRM 6 – Determination of SO_2 emissions
- FRM 7 – Determination of nitrogen oxides emissions
- FRM 8 – Determination of sulfuric acid mist and SO_2 emissions

The first four original FRMs provide the basis for the subsequent methods. Major elements of these methods include sample points and duration, isokinetic sampling, separation of COCs, and calculation of the results. Since the flow rate is the product of the velocity and the cross-sectional area, the sampling should start with measuring the velocity. A type "S" pitot tube is typically used, and it allows the measurement of differential pressure, which is directly related to the velocity of the air or gas stream. It comes with large sensing holes to make it less prone to particulate clogging than other pitot tubes. The gas is then extracted from sampling points distributed over the cross-section of the stack/duct, and the sample duration for each point should be exact (Figure 3.6).

For particle sampling, *isokinetic sampling* means the sample extraction rate is adjusted so that the velocity of gas into the sample nozzle (v_n) is the same as the velocity of the gas in the stack (v_s). It should be conducted so that the extracted sample will have a concentration the same as that in the stack (Figure 3.7).

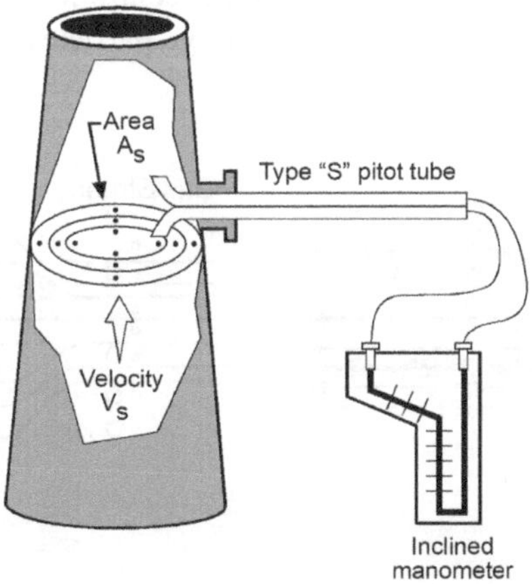

FIGURE 3.6 Velocity measurements in stack sampling (USEPA, 2003).

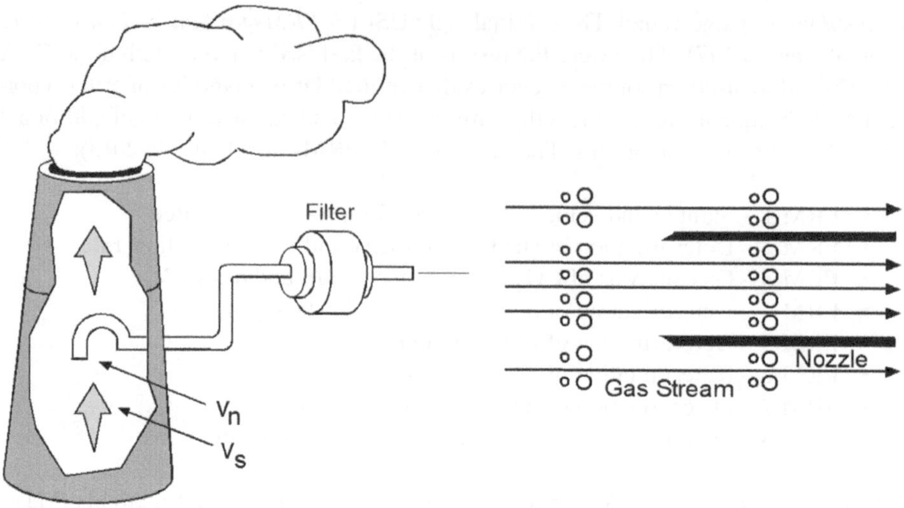

FIGURE 3.7 Isokinetic sampling of particulates from a stack (USEPA, 2003).

Over-isokinetic sampling occurs when the intake velocity of the sampling probe (v_n) is greater than the velocity of the airstream being sampled (v_s). Under this condition, gas from the regions around the nozzle will also enter the nozzle. At the same time, the large particles will not follow the streamlines into the nozzle because of their large inertia. The measured concentration will be lower than the actual concentration due to the additional volume of stack gas extracted, and the concentration is biased with small particles (Figure 3.8a). On the other hand, when $v_n < v_s$ (*under-isokinetic sampling*), the stack gas volume would be extracted at a rate lower than the isokinetic rate. However, larger particles in the streamline immediately outside the nozzle would flow into the nozzle due to their inertia. Consequently, the measured concentration will be higher than the actual concentration due to the smaller volume of stack gas extracted, which is biased with large particles (Figure 3.8b).

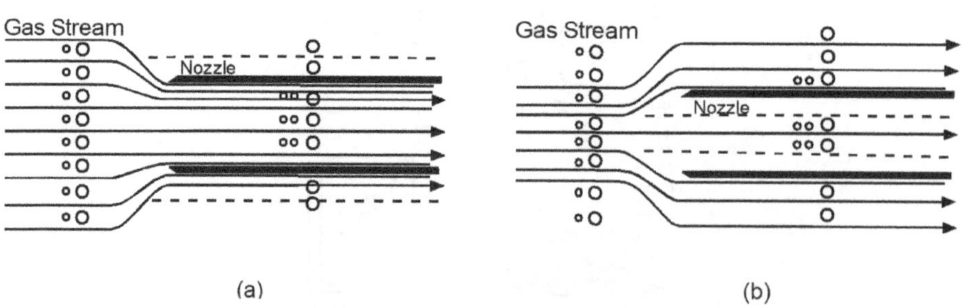

FIGURE 3.8 Non-isokinetic conditions: (a) over-isokinetic and (b) under-isokinetic (USEPA, 2003).

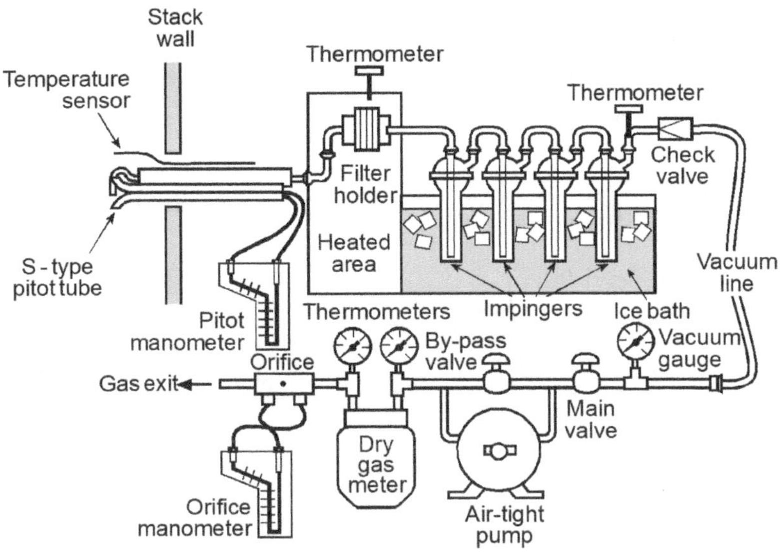

FIGURE 3.9 Schematic of a Method 5 sampling train (USEPA, 2003).

Figure 3.9 shows the sampling train for *Method 5 – Determination of Particulate Emissions from Stationary Sources*. The gas drawn from the stack is passed through a sampling train, a leak-proof series of components to capture the COCs. The first component of the train is a heated probe or a hollow glass, while a pump, the last component, moves the gas through the train. In between, solids are captured on filter paper, and water vapor and gaseous pollutants are typically captured by condensation or bubbling through chilled and sealed glass vessels (*impingers*). The gas sample volume is measured by a gas meter and then exhausted through an orifice, which adjusts the flow rate through the sampling train by a bypass valve (USEPA, 2003).

The mass of particulates collected on the filter paper and the masses of COCs in the solutions of the impingers are then measured and used to determine concentrations with the measured gas volume. These concentrations are often corrected to the standard temperature and pressure (STP). Typically, the total sampling time for each run lasts about 1 hour, and triplicate runs are conducted (shorter than the ambient air sampling, which is usually 24 hours).

3.4.2 Continuous Emission Monitoring (CEM)

The source monitoring systems can be classified into (i) extractive (e.g., the sampling train previously discussed) and (ii) in situ. The source emission testing described in the previous section is mainly done in a batch-wise mode, manually, and infrequently. It is time-consuming and labor-intensive. In addition, the operators may tend to tune up the process equipment and control devices for peak performance before the mandatory/compliance source testing.

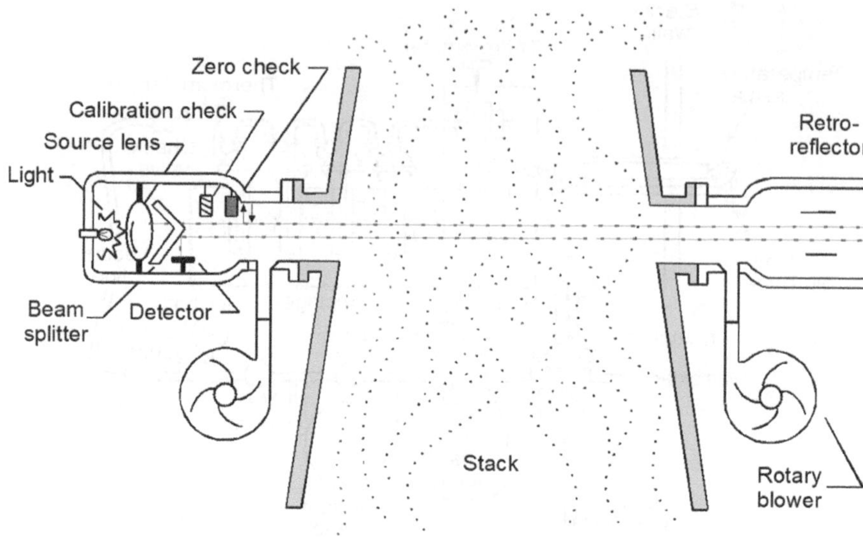

FIGURE 3.10 Schematic of a double-pass opacity monitor (USEPA, 2003).

In situ analyzers directly monitor effluent gases at their actual source-level concentrations. *Continuous emission monitoring* (CEM), without extracting a gas stream from the stack, is used to record source emissions continuously using instrumental methods. With the advances in in situ and online chemical and physical analyzers, CEM has become an integral part of source emission measurement for air pollution control and regulatory compliance for particulates, opacity, and many gaseous air pollutants. CEM is the real-time or near-real-time measurement of pollutant concentrations or emission rates from stationary sources. The key components of CEM include flow monitors, gas analyzers, and data acquisition systems. Figure 3.10 illustrates a double-pass opacity monitor. *Opacity* is a measure of the light attenuation caused by PM in the emissions, and it serves as an indirect indicator of particulate concentration. In a "double-pass" sensing configuration, a light beam is directed across the emission path, reflected, and detected after passing through the emission twice. At the same time, a "point" sensor only has measurements within a limited distance.

3.5 EMISSION INVENTORY AND EMISSION FACTORS

3.5.1 Emission Inventory

An *emission inventory* database contains the types and quantities of air pollutants discharged into the atmosphere from different sources/sectors within a specific area and a specific period. National emission inventories can be used to identify significant sources/sectors of air pollutants and use the information as the basis to target necessary regulatory actions. Emission inventories can also be an essential input to mathematical models to estimate air quality. The effect of potential regulatory actions on air quality can be predicted by applying the estimated reduction to the emission inventory data used in the air quality models. With the accumulation of data in emission inventories, emission trends over time can be readily established. Inventories can also be used to raise public awareness about sources and pollution trends in the areas where they live.

The USEPA develops and maintains three significant emission inventories. The *National Emissions Inventory* (NEI) estimates air emissions of criteria pollutants, HAPs, and greenhouse gases (GHGs) from various air emission sources, including industrial facilities, mobile sources, area sources, and natural sources (https://www.epa.gov/air-emissions-inventories). It is updated every three years and is based primarily upon information provided by state, local, and tribal air agencies for sources under their jurisdictions, as well as supplemental data developed by the USEPA.

The *Toxics Release Inventory* (TRI) program was created by Section 313 of the Emergency Planning and Community Right-to-Know Act. TRI tracks the management of certain toxic chemicals that may threaten human health and the environment. It is a resource for learning about toxic chemical releases and pollution prevention activities reported by industrial and federal facilities (https://www.epa.gov/toxics-release-inventory-tri-program). The USEPA also issues an annual report, titled *Inventory of U.S. Greenhouse Gas Emissions and Sinks* (https://www.epa.gov/ghgemissions), which tracks U.S. GHG emissions and sinks by source, economic sector, and GHG going back to 1990.

Using the data from NEI, Table 3.11 shows the primary sources and emission rates of CO, NO_x, SO_2, PM_{10}, $PM_{2.5}$, and VOCs in the U.S. in 2024 (USEPA, 2025). Lots of information can be extracted from this table. For example, CO has the most significant emission rate among these pollutants, at ~45 million tons. The primary emission source for CO and NO_2 is transportation, while stationary fuel combustion is responsible for most of the SO_2 emissions. Figure 3.11 illustrates the annual emission trends of these air pollutants. SO_2, NO_x, and CO emissions have been significantly reduced since 1970.

3.5.2 Emission Factors

Emission data are needed to build an emission inventory. Having continuous emission data from each single emission source would be impossible. Several methods are being used to estimate emissions from a specific source type. They include (i) continuous monitoring of emissions from a source (then used to estimate emissions for different operating and longer averaging times), (ii) short-term emission testing (then extrapolated to a more extended period), and (iii) the use of emission factors. The choice of method depends on the availability of data, time, and money.

An *emission factor* (EF) is a representative value that attempts to relate the quantity of a pollutant emitted with an activity level, output, or input of a specific source for the emission of that pollutant. The activity, output, or input chosen needs to be easily measured (e.g., the amount of material processed or the amount of fuel used); for example, pounds of pollutant per ton of material processed, kilograms per megawatt-hour (kg/MWh) of electricity generated, or grams per vehicle mile traveled.

TABLE 3.11
Source Categories and Emission Trends of Several Air Pollutants (in Thousand Tons)

Source Category	CO	NO_x	SO_2	PM_{10}	$PM_{2.5}$	VOC_s
Stationary fuel combustion	**4,730**	**2,263**	**897**	**899**	**838**	**629**
Electric utility	409	751	649	94	81	27
Industrial	875	967	219	285	243	116
Others	3,446	545	29	520	514	486
Industrial and other process	**3,168**	**1,150**	**621**	**1,118**	**657**	**6,950**
Chemical & allied product manufacturing	128	31	72	18	13	69
Metals processing	345	57	53	42	30	22
Petroleum & related industries	750	705	335	28	23	2,899
Other industrial processes	461	271	126	738	347	386
Solvent utilization	1	1	0	5	4	2,777
Storage & transport	6	3	1	34	13	605
Waste disposal & recycling	1,477	82	34	253	227	192
Transportation	**24,305**	**3,323**	**29**	**288**	**160**	**1,886**
Highway vehicles	12,322	1,745	10	189	67	896
Off-Highway	11,983	1,578	19	99	93	990
Miscellaneous	**12,734**	**204**	**127**	**14,114**	**3,433**	**3,317**
Wildfires	3,692	40	36	770	504	974
Others	9,042	164	91	13,344	2,929	2,343
Total	**44,937**	**6,940**	**1,674**	**16,419**	**5,088**	**12,782**

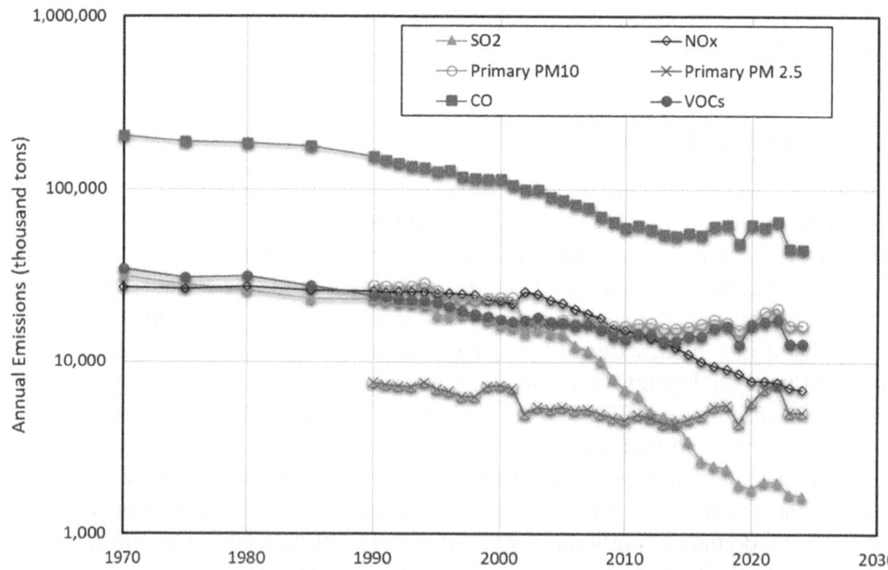

FIGURE 3.11 Trends of annual emissions of several air pollutants.

With an emission factor and a known activity rate (A), a simple multiplication yields an estimation of the emission (E):

$$E = A \times EF \times (1 - ER) \tag{3.5}$$

where ER is the destruction, removal, or capture efficiency of the control device in percentage (USEPA, 1986). More discussion on derivations of EFs is given in Chapter 11.

Emission factors are used to build emission inventories, guide air quality management decisions, and develop emission control strategies. The USEPA website on *Air Emissions Factors and Quantification* (https://www.epa.gov/air-emissions-factors-and-quantification) provides information on these tools and supports their use. USEPA AP-42, *Compilation of Air Pollutant Emission Factors*, is its primary compilation, which contains EFs and process information for more than 200 air pollution source categories. A *source category* is a specific industry sector or a group of similar emitting sources (e.g., nitric acid industry, dry cleaning, and municipal solid waste landfills). These EFs have been developed based on source test data, mass balance studies, and engineering estimates.

Example 3.7: Use of Emission Factors to Estimate the Emission Rate

Emission factors for nitrogen oxides (NO_x) from natural gas combustion for large wall-fired boilers are 140 and 100 lb/10^6 standard cubic feet (scf) for those with low-NO_x burners and for those with flue gas recirculation (FGR), respectively. Estimate the hourly and daily NO_x emissions from a large wall-fired boiler using 200,000 scf/h of natural gas continuously.

Solution:

a. NO_x emission from a 200,000 scf/h natural gas boiler with low-NO_x burners

$$= (140 \text{ lb/}10^6\text{scf}) \times (200{,}000 \text{ scf/h}) = 28 \text{ lb/h} = 672 \text{ lb/day}$$

b. NO_x emission from a 200,000 scf/h natural gas boiler with FGR

$$= (100 \text{ lb/}10^6\text{scf}) \times (200{,}000 \text{ scf/h}) = 20 \text{ lb/h} = 480 \text{ lb/day}$$

Discussion:

1. Emission factors are extracted from USEPA AP-42. The large wall-fired boilers are those with a heat input > 100 MMBTU/h. These emission factors are based on an average natural gas higher heating value (HHV) of 1,200 Btu/scf. 1 MMBTU = 1 million BTU. 1 scf = 1 standard cubic foot.
2. "Low-NO_x burners" and FGR are common practices to reduce NO_x emissions from combustion, and they are discussed in Chapter 8. Based on the emission factors, a boiler with FGR will reduce more NO_x emissions than a low-NO_x burner with the same heat input.

3.6 AIR POLLUTION MANAGEMENT

Air pollution causes adverse impacts on human health, properties, ecosystems, and the global environment. The ability to control air pollution in a coordinated manner is known as *air pollution management*. Good air pollution management typically instills effective air pollution control programs that employ strategies and tactics to reduce air pollution to achieve the goal of good air quality. These strategies can differ greatly in their purpose and function. This section first describes the air pollution control philosophies, followed by air pollution control strategies.

3.6.1 Air Pollution Control Philosophies

According to Dr. N. de Nevers, most air pollution control programs are based on one or a combination of four basic philosophies: emission standards, air quality standards, emission taxes, and cost–benefit analysis (de Nevers, 1977). These philosophies guide the development of policies, technologies, and practices aimed at minimizing emissions and mitigating the impacts of pollutants on human health and the environment.

Emission standards. The basic idea of this philosophy is that some maximum possible or practical degrees of emission control exist. Although this maximum degree of control varies among various classes of emitters, this philosophy assumes that it can be determined individually. If all emitters employ their maximum degrees of control, the cleanest possible air quality would be achieved (de Never, 1977). The NSPS and the NESHAP programs in the U.S., which employ MACT and other emission standards, are essentially based on this philosophy.

Air quality standards. This philosophy assumes that there are threshold values for all air pollutants. Below the threshold values, these pollutants will not result in damage. The standards are based on scientific research, health risk assessments, and environmental impact studies to determine safe levels of exposure. This is the basis for USEPA to establish the NAAQS using dose-response and other toxicology data (de Nevers, 1977; USEPA, 2003).

Emission taxes. This philosophy would impose a tax on each emitter based on a published scale related to the emission rates. The tax rate would be established so that most major polluters would find it more economical to install pollution control systems to reduce emissions and pay less tax (de Nevers, 1977). Some economic incentive programs exist, such as *emission trading* or *cap and trade*. The cap part is a limit on emissions backed by science. In contrast, the trade part is a market for companies to buy or sell emission allowances, incentivizing companies to save money by cutting emissions.

Cost–benefit philosophy. This philosophy assumes that neither threshold values for pollutants exist nor they are low enough (but we cannot afford to have air that clean). It is to select a cost–benefit alternative that best minimizes the damages resulting from air pollution and is most cost-effective. The cost–benefit analysis is complex because many uncertain factors and variables must be analyzed. However, simplified approaches are being used.

For example, the USEPA must take costs into account when establishing RACT and MACT (de Nevers, 1977; USEPA, 2013).

3.6.2 Air Pollution Control Strategies

Air quality management is a continuous process. Most of the information in this subsection was extracted from the USEPA website on the *Air Quality Management Process* (https://www.epa.gov/air-quality-management-process). A *control strategy* related to air quality is a set of techniques and measures identified and implemented to reduce air pollution to attain an air quality goal or standard. Air pollution control strategies should be implementable at various levels (i.e., local, regional, and national) and often involve collaboration among government agencies, industries, communities, and individuals. Environmental, engineering, and economic factors should be considered when designing an effective air quality control strategy. Environmental factors include ambient air quality conditions, meteorological conditions, and locations of the emission sources. Engineering factors include the characteristics of pollutants, gas streams, and the control system's performance. Examples of economic factors are capital cost, operation and maintenance (O&M) costs, and administrative and enforcement costs.

The old saying, "prevention is better than cure", also applies to air pollution control. Pollution prevention approaches to prevent, reduce, or eliminate pollution at its source should always be considered. Examples include using less toxic materials/fuels, employing a less-polluting industrial process, and improving the efficiency of the process. To minimize, recycle, and reuse the waste generated from air pollution control activities, it should also be part of air quality management.

EXERCISE QUESTIONS

1. Determine the TSP concentration (i) at actual sampling conditions and (ii) at standard conditions using the sampling and analysis data below:
   - Mass of filter before sampling = 3.1234 g
   - Mass of filter after sampling = 5.4321 g
   - Ambient temperature = 22°C
   - Ambient pressure = 750 mm-Hg
   - Starting flow rate = 2.00 m^3/min
   - Ending flow rate = 1.80 m^3/min
   - Sampling duration = 24 hours
2. At a specific location, CO = 13.0 ppm; determine its AQI value.
3. At a specific location, 8-hour O_3 = 0.13 ppm, $PM_{2.5}$ = 30.0 μg/m^3, and CO = 13.0 ppm; determine the AQI value at this location.
4. Emission factors for nitrogen oxides (NO_x) from natural gas combustion for small boilers are 50 and 32 lb/scf for those with low-NO_x burners and for those with FGR, respectively. Estimate the hourly, daily, and yearly NO_x emissions from a small boiler using 50,000 scf/h of natural gas continuously (i) with low-NO_x burners and (ii) with FGR.
5. The CO_2 emission of average passenger cars was approximately 400 g/mile (249 g/km). When a vehicle travels at 55 mi/h (88.5 km/h), its exhaust flow

rate is 250 ft^3/min (7.1 m^3/min), and the exhaust contains 14% by volume of CO_2 (all based on $T = 662°F$ (350°C) and $P = 1$ atm).

a. What is the molar volume of an ideal gas at $T = 662°F$ (350°C) and $P = 1$ atm?
b. Compute the CO_2 emission rate in g/mi and in g/km.
c. Is this emission rate above or below the national average?

BIBLIOGRAPHY

de Nevers, N. (1977). "*Air Pollution Control Philosophies*," *J. Air Pollut. Control Assoc.* 27(3), 197–218.

Kuo, J.; Dow, J. (2017). "*Biogas Production from Anaerobic Digestion of Food Waste and Relevant Air Quality Implications*," *J. Air Waste Manage. Assoc.* 67(9), 1000–1011. doi: 10.1080/10962247.2017.1316326.

USEPA. (1986). *Emission Factors for Equipment Leaks of VOC and HAP,* EPA/450/3-86/002, United States Environmental Protection Agency, Research Triangle Park, NC.

USEPA. (1999). *Compendium of Methods for the Determination of Toxic Organic Compounds in Ambient Air*, 2nd Edition, EPA/625/R-96/01b, United States Environmental Protection Agency, Cincinnati, OH.

USEPA. (2003). *APTI 452: Principles and Practices of Air Pollution Control - Student Manual*, 3rd Edition, Air Pollution Training Institute, United States Environmental Protection Agency, Research Triangle Park, NC.

USEPA. (2008). *APTI 435: Atmospheric Sampling Course - Student Manual,* 5th Edition, prepared by Tidewater Operations Center of C^2 Technologies for Air Pollution Training Institute, United States Environmental Protection Agency, Research Triangle Park, NC.

USEPA. (2010). *APTI 464: Analytical Methods for Air Quality Standards - Student Manual*, Air Pollution Training Institute, United States Environmental Protection Agency, Research Triangle Park, NC.

USEPA. (2012). *APTI 435: Atmospheric Sampling Course - Student Manual*, 5th Edition, prepared by Tidewater Operations Center of C^2 Technologies for Air Pollution Training Institute, United States Environmental Protection Agency, Research Triangle Park, NC.

USEPA. (2013). *APTI 450: Source Sampling for Pollutants - Student Guide*, Air Pollution Training Institute, United States Environmental Protection Agency, Research Triangle Park, NC.

USEPA. (2024a). *NAAQS Table*, https://www.epa.gov/criteria-air-pollutants/naaqs-table, last updated on 02/07/2024.

USEPA. (2024b). *Technical Assistance Document for the Reporting of Daily Air Quality - the Air Quality Index*, EPA-454/B-24-002 May 2024, Air Quality Assessment Division, United States Environmental Protection Agency, Research Triangle Park, NC.

USEPA. (2025). *Air Pollutant Emissions Trend Data*, https://www.epa.gov/air-emissions-inventories/air-pollutant-emissions-trends-data, last updated on 04/28/2025.

4 Basic Principles for Engineering Calculations

4.1 FLOW RATE AND VELOCITY

The process flow rate of a gas stream can be expressed in three different ways: (i) volumetric flow rate (volume/time), (ii) mass flow rate (mass/time), and (iii) molar flow rate (moles/time). They are related, but one may be more suitable than the others for a specific engineering calculation. For example, a volumetric flow rate (Q_v) should be used to size air ductwork or determine the gas residence time in a pollution control device. To determine the mass loading rate of a treatment unit, the mass flow rate (Q_{mass}) should be used. For calculations related to chemical reactions, a molar flow rate (Q_{molar}) would typically be more appropriate.

From the ideal gas law, a gas's volume depends on the system's temperature (T) and pressure (P). Consequently, Q_v of a gas/air stream is a function of T and P. As a gas stream goes through the unit processes/operations with different T's and P's in a treatment process train, its Q_v would change. At the same time, its mass and molar flow rates stay unchanged (assuming no chemical, biological, or thermal reactions occur).

Q_v is typically expressed in cubic meters per minute (m^3/min), cubic meters per second (m^3/s or cms), cubic feet per minute (ft^3/min or cfm), or cubic feet per second (ft^3/s or cfs). One common calculation is to convert Q_v between actual and standard conditions.

The ideal gas law can convert Q_v from one set of T and P to another. Assuming the ideal gas law is valid, conversions between $Q_{v,\text{actual}}$ and standard $Q_{v,\text{ideal}}$ of a given gas stream can be readily made by using one of the following two formulas:

$$\frac{Q_{v,\text{standard}}}{Q_{v,\text{actual}}} = \left(\frac{T_{\text{standard}}}{T_{\text{actual}}}\right)\left(\frac{P_{\text{actual}}}{P_{\text{standard}}}\right) \tag{4.1}$$

$$\frac{Q_{v,\text{actual}}}{Q_{v,\text{standard}}} = \left(\frac{T_{\text{actual}}}{T_{\text{standard}}}\right)\left(\frac{P_{\text{standard}}}{P_{\text{actual}}}\right) \tag{4.2}$$

The value of $Q_{v,\text{actual}}$ should be larger at a higher T and a lower P.

The velocity through a treatment unit or ductwork can be readily determined as the division product of the actual volumetric flow rate ($Q_{v,\text{actual}}$) and the cross-sectional area of the flow (A), as follows:

$$\text{Velocity} = Q_{v,\text{actual}}/A \tag{4.3}$$

DOI: 10.1201/9781003532705-4

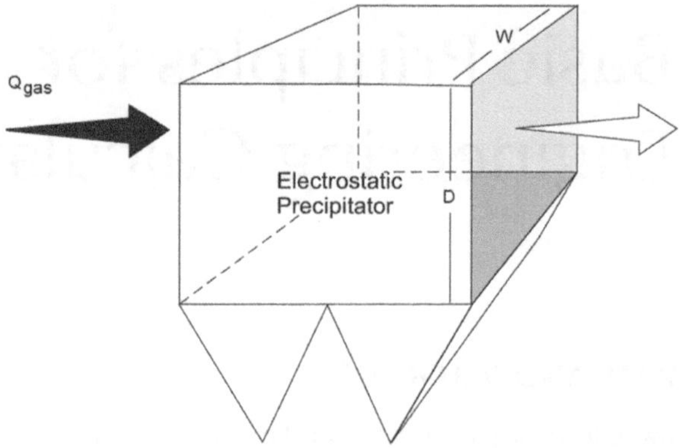

FIGURE 4.1 Example of flow rate and through velocity (Modified from USEPA, 2000).

For example, to calculate the gas flow velocity through an electrostatic precipitator for the removal of particulates (Figure 4.1), the influent gas flow rate should be converted to the actual flow rate inside the electrostatic precipitator by taking the system pressure and temperature into consideration. The cross-sectional area, in this case, is $W \times H$.

The following equations can be used to convert between Q_v, Q_{mass}, and Q_{molar} (where MW is the molecular weight and MV is the molar volume at that temperature and pressure):

$$Q_{mass} = Q_{molar} \times \text{MW} = Q_v \times \text{Gas density} \tag{4.4}$$

$$Q_{molar} = Q_v \div \text{Molar Volume} = Q_{mass} \div \text{MW} \tag{4.5}$$

Example 4.1: Flow Rates and Velocity

An air stream containing benzene (C_6H_6) is treated by a pollution control system consisting of a hood, ductwork, activated carbon adsorbers, a fan, and an exhaust stack. The airflow rate is 200 standard m^3/min (7,060 standard ft^3/min). The ductwork is operated at $T = 50°C$ (122°F), with a static pressure of −8 in-H_2O, while the barometric pressure is 29.0 in-Hg. The ductwork dimensions are 0.5 m × 1 m. Find

a. Q_{mass} and Q_{molar} under standard conditions,
b. Q_v, Q_{mass}, and Q_{molar} under actual conditions, and
c. the air velocity inside the duct.

Solution:

a. Using Eq. 4.5,

$$Q_{molar} = Q_v \div \text{MV} = 200{,}000 \div 24.05 = 8{,}316 \frac{\text{g-mole}}{\text{min}} = 8.316 \frac{\text{kg-mole}}{\text{min}}$$

or

$$Q_{molar} = 7{,}060 \div 385 = 18.4\,\frac{\text{lb-mole}}{\text{min}}$$

Using Eq. 4.4,

$$Q_{mass} = Q_{molar} \times MW = \left(8.316\,\frac{\text{kg-mole}}{\text{min}}\right)(29) = 241\,\frac{\text{kg}}{\text{min}} = 530\,\frac{\text{lb}}{\text{min}}$$

b. $P = P_b + P_g = (29.0$ in-Hg)(407 in-H_2O/29.92 in-Hg) + (−8 in-H_2O) = 386.5 in-H_2O
   Using Eq. 4.2,

$$Q_{v,actual} = (200)\left(\frac{273+50}{273+20}\right)\left(\frac{407}{386.5}\right) = 232\,\text{m}^3/\text{min}$$

or

$$Q_{v,actual} = (7{,}060)\left(\frac{460+122}{460+68}\right)\left(\frac{407}{386.5}\right) = 8{,}195\,\text{ft}^3/\text{min}$$

Q_{molar} and Q_{mass} will be the same as those in part (a).

c. Using Eq. 4.3,

$$\text{Velocity} = Q_{v,actual}/A = \frac{232\ \text{m}^3/\text{min}}{(0.5\text{m} \times 0.1\text{m})} = 464\ \text{m/min} = 7.7\ \text{m/s}$$

Discussion:

1. The molar volume of the air @ T= 20°C and P=1 atm is assumed to be 24.05 L or 385 ft^3 here.
2. One atmosphere = 29.92 in-Hg = 407 in-H_2O.
3. Standard m^3/min is often abbreviated as Nm^3/min.
4. Typical air velocities in the ducts range from 6 to 33 m/s (or ~20 to 100 ft/s); see Section 4.5 for the ranges of transport velocities in the ductwork.

4.2 FLOW EQUATIONS

Flow equations are mathematical expressions that describe the behavior of fluids (i.e., liquids or gases) as they move through different system units (e.g., pipes or reactors). The common flow equations applicable to our scope of work include the continuity equation, Bernoulli's equation, and Darcy's law.

4.2.1 Continuity Equation

The *continuity equation* describes the conservation of mass in fluid flow, stating that the mass of fluid entering a system must equal the mass leaving the system (i.e., the mass flow rate stays the same). It can be expressed as:

$$\text{Mass flow rate} = \rho_1 \times A_1 \times u_1 = \rho_2 \times A_2 \times u_2 \tag{4.6}$$

where ρ is the fluid density, u is the fluid velocity, and A is the cross-sectional area.

The density stays the same for an incompressible fluid (e.g., water under ambient conditions) and Eq. 4.6 can be simplified to:

$$\text{Volumetric flow rate} = A_1 \times u_1 = A_2 \times u_2 \tag{4.7}$$

It should be noted that air is a compressible fluid. However, Eq. 4.7 applies when the air densities at two locations are same. The continuity equation is often used to relate the flow velocity and the cross-sectional area at different locations in a fluid flow system.

4.2.2 Bernoulli's Equation

The *Bernoulli equation* is based on energy conservation in fluid flow. It relates the pressure, velocity (u), and elevation (h) of a fluid flowing through a pipe or open channel, assuming steady, incompressible, and non-viscous (or "*inviscid*") flow between two points (where g is the gravitational constant):

$$P_1 + \frac{1}{2}\rho_1 u_1^2 + \rho g h_1 = P_2 + \frac{1}{2}\rho_2 u_2^2 + \rho g h_2 \tag{4.8}$$

A pitot tube is a device used to determine the fluid flow velocity (u). The basic *pitot tube* consists of a tube pointing directly into the oncoming fluid flow. It measures the total pressure (or the *stagnation pressure*, $P_{\text{stagnation}}$) at the tip of the pitot tube where the fluid is brought to rest. The velocity pressure (or the dynamic pressure, P_{dynamic}) is the difference between the stagnation pressure and the static pressure (P_{static}), the pressure in the fluid when the flow is not disturbed, as follows:

$$\text{Velocity Pressure} = P_{\text{dynamic}} = P_{\text{stagnation}} - P_{\text{static}} \tag{4.9}$$

To determine the fluid velocity (u) at the location where the pitot tube is placed, we can use the Bernoulli equation (i.e., Eq. 4.8) as follows:

$$\text{Velocity Pressure} = \frac{\rho u^2}{2} = P_{\text{stagnation}} - P_{\text{static}} \tag{4.10}$$

Then,

$$u = \sqrt{2\left(P_{\text{stagnation}} - P_{\text{static}}\right)/\rho} = \sqrt{2 \times \text{Velocity Pressure}/\rho} \tag{4.11}$$

Eq. 4.11 is general and valid when all the parameter values are in matching units.

There are three specific equations given in NCEES (2025) about "Pitot Tube Velocity Pressure":

i. "Standard air" (with dry air at 70°F and barometric pressure of 29.92 in-Hg):

$$u = 4{,}004.4 \times \sqrt{\text{Velocity Pressure } (h_v)} \tag{4.12}$$

ii. Other than "standard air":

$$\rho_{air} = 1.325 \times P_b/T \tag{4.13}$$

$$u = 1.096.7 \times \sqrt{h_v/\rho_{air}} \tag{4.14}$$

where ρ_{air} is the density of air (lb/ft³), u is the velocity (ft/min), P_b is the barometric (or absolute) static pressure (in-H_2O), T is the absolute temperature (in °R), and h_v is the velocity pressure (in-H_2O). Similar equations can be found in USEPA (1995).

Example 4.2: Pitot Tube Measurements to Determine Flow Velocity

A pitot tube was used to measure the airflow velocity in a duct (T = 20°C = 68°F = 528°R). Estimate the air velocity using the following measured pressure data:

- Total pressure (= Stagnation pressure) = 101,800 Pa (= 30.07 in-Hg)
- Static pressure = 101,300 Pa (= 29.92 in-Hg = 1 atm)

Solution:

a. From Example 2.2, ρ_{air} = 1.21 kg/m³ = 0.075 lb/ft³ (T = 20°C = 68°F = 528°R and P = 1atm)
b. Using Eq. 4.9,

$$\text{Velocity Pressure} = 101{,}800 - 101{,}300 = 500\text{ Pa} = 500\text{ N/m}^2 = 500\,\text{kg/s}^2 \cdot \text{m}$$

c. Using Eq. 4.11,

$$\text{Velocity} = \sqrt{\frac{2 \times 500\text{ kg/s}^2 \cdot \text{m}}{1.21\,\text{kg/m}^3}} = 28.7\text{ m/s}$$

rates, so the factor ($T_{dilution}/T_{exhaust}$) was used. The 774 m³/h flow rate is at $T = 20°C$.

2. The temperatures in the square bracket do not have to be absolute temperatures because the bracket is used to find the ratio of temperature differences.

4.3.3 Moisture, Carbon Dioxide, and Oxygen Corrections

Emissions of stack gases are usually expressed on a dry basis so that variations in the moisture content of the stack gas do not affect the assessment of the emissions. The conversion between the dry air flow rate (Q_{dry}) and wet air flow rate (Q_{wet}) is:

$$Q_{dry} = Q_{wet} \times \left(1 - y_{H_2O}\right) \tag{4.23}$$

The concentrations of pollutants in dry air (G_{dry}) and wet air (G_{wet}) are related by (NCEES, 2025):

$$G_{dry} = G_{wet} \times \left(\frac{1}{1 - y_{H_2O}}\right) \tag{4.24}$$

where y_{H_2O} is the mole or volume fraction of water in the air.

To correct an air stream ($Q = Q_{actual}$; $T = T_{actual}$ and $P = P_{actual}$) for dry and standard conditions ($Q_{dry,standard}$; $T = T_{standard}$ and $P = P_{standard}$), the following formula can be used (NCEES, 2025):

$$Q_{dry,\ standard} = Q_{actual} \times \left(\frac{P_{actual}}{P_{standard}}\right) \times \left(\frac{T_{standard}}{T_{actual}}\right) \times \left[1 - y_{H_2O}\right] \tag{4.25}$$

Example 4.6: Conversion Flow Rate to Dry and Standard Conditions

A moist air flow ($Q = 1{,}000$ m³/h, $T = 30°C$, $P = 101.2$ kPa, and RH = 60%) has a NO_2 concentration of 100 µg/m³. Find the following:

a. The dry air flow rate.
b. The NO_2 concentration in the dry air.
c. The dry air flow rate under the "standard" conditions ($T = 20°C$ and $P = 1$ atm).
d. The NO_2 concentration in the dry air under the "standard" conditions.

Solution:

a. Using a psychrometric chart (Figure 2.2), the humidity ratio of the moist air ($T = 30°C$ and RH = 60%) is ~0.013 kg H_2O/kg dry air. The molar/volume ratio of water can be found as

$$y_{H_2O} = (0.013/18)/[0.013/18 + 1/29] = 0.0205$$

Use Eq. 4.23 to find the dry air flow rate:

$$Q_{dry} = 1,000 \times (1 - 0.0205) = 979 \text{ m}^3/\text{h}$$

b. Use Eq. 4.24 to find the dry gas concentration:

$$G_{dry} = 100 \times \left(\frac{1}{1 - 0.025} \right) = 102.6 \ \mu\text{g/m}^3$$

c. Use equation Eq. 4.25 to find the dry air flow rate under the "standard" conditions:

$$Q_{dry,\ standard} = 1,000 \times \left(\frac{101.2}{101.325} \right) \times \left(\frac{293}{303} \right) \times [1 - 0.0205] = 946 \text{ m}^3/\text{h}$$

d. The dry gas concentration under the "standard" conditions can be found from the product of the dry gas concentration from part (b) and the ratio of the dry flow rate (from part (a)) and the dry air flow rate under the standard conditions (from part (c)). It is because the mass stays the same while the concentration is inversely proportional to the volumetric flow rate. Thus,

$$G_{dry,standard} = G_{dry} \times \left(\frac{Q_{dry}}{Q_{dry,standard}} \right) = 102.6 \times \left(\frac{979}{946} \right) = 106.2 \mu\text{g/m}^3$$

Discussion:

1. Q_{dry} should always be smaller than Q_{wet} (see Eq. 4.23 and part (a)), and G_{dry} should always be correspondingly larger than G_{wet} (see part (b)).
2. The two pressure items in Eq. 4.25 can be in any matching units, while the two temperature items need to be absolute temperatures and in matching units.

Emissions from combustion processes often vary in CO_2 and O_2 concentrations depending on the fuel type and combustion efficiency. Correcting concentrations in a gas stream for specific concentrations of CO_2 and O_2 is often done in emissions testing and environmental monitoring. The correction is to standardize measurements and account for variations in gas composition that may affect pollutant concentrations. Eq. 4.26 shows the general correction formula:

$$G_{dry,corrected} = G_{dry,\ measured} \times \left(\frac{CO_{2,\ ref}}{CO_{2,\ measured}} \right) \times \left(\frac{20.9 - O_{2,\ ref}}{20.9 - O_{2,\ measured}} \right) \quad (4.26)$$

velocity at 0.5 m and 1 m away from the hood entrance? If side baffles are added to the hood, what flow rates will be required?

Solution:

a. $A = \pi(0.25)^2 = 0.2\ m^2$

From Eq. 4.31,

$$Q_{hood} = (100)\left[10(0.5)^2 + 0.2\right] = 270\ m^3/min\ @\ x = 0.5\ m$$

$$Q_{hood} = (100)\left[10(1.0)^2 + 0.2\right] = 1{,}020\ m^3/min\ @\ x = 1.0\ m$$

b. With side baffles, the required flow rates will be 75% of the corresponding values in part (a), according to Eq. 4.32.

4.5.2 Ductwork

After the air captured by the hood enters the ductwork, a minimum transport velocity needs to be maintained to keep the particles from settling out of the gas stream and depositing onto the bottom of the ductwork; this is a practically important consideration when handling particle-laden gas streams. Table 4.1 shows commonly recommended transport velocities through ductwork (USEPA, 2012).

Proper duct size is key in addressing the minimum transport velocity. For a ductwork system with different cross-section sizes, the section with the largest cross-section will be most prone to particle settling due to the smaller transport velocity. However, the section with smaller cross-sections will incur more substantial pressure drops. The equation below can be used to determine the cross-sectional area of the ductwork (A_{duct}) with a given design transport velocity (v_{duct}):

$$A_{duct} = Q_{hood} \div v_{duct} \tag{4.33}$$

4.5.3 Fan System

Fans move the captured air through hoods, ducts, collection devices, and stacks. There are two main types of fans: axial and centrifugal (Figure 4.4). Most fans used in APC systems are centrifugal fans (i.e., blowers). A *centrifugal fan* has a fan wheel

TABLE 4.1
Transport Velocities through Ductwork

	Transport Velocity	
Type of Pollutant	**(ft/min)**	**(m/min)**
Gases	~1,000–2,000	~300–600
Light particle loading	~3,000–3,500	~900–1,070
Normal particle loading	~3,500–4,500	~1,070–1,370

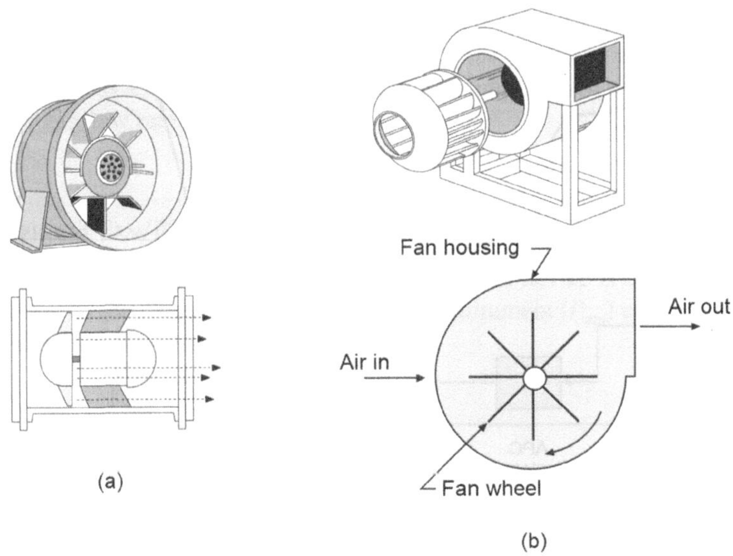

FIGURE 4.4 Types of fans: (a) axial and (b) centrifugal fans (Modified from USEPA, 2000).

composed of several fan blades mounted around a hub; it can generate high-pressure rises in the gas stream.

The rotational speed of centrifugal fans is one of the most critical operating variables. Most centrifugal fans can operate over a modest range of speeds. There are three fan laws we need to be familiar with. They are fundamental principles used to understand and predict the performance of fans/blowers in air handling systems (USEPA, 1995). They describe how changes in fan speed affect air flow rate, pressure, and power consumption.

The *first fan law* states that the volumetric gas flow rate is directly proportional to the fan wheel's rotational speed:

$$\frac{Q_2}{Q_1} = \frac{\text{RPM}_2}{\text{RPM}_1} \tag{4.34}$$

where RPM is the fan wheel's rotational speed in revolutions per minute. Thus, a 20% decrease in fan speed would result in a 20% decrease in the air flow rate through the system.

The *second fan law* states that the static pressure rise across the fan (ΔSP) is proportional to the square of the fan speed, as follows:

$$\frac{\Delta\text{SP}_2}{\Delta\text{SP}_1} = \left(\frac{\text{RPM}_2}{\text{RPM}_1}\right)^2 \tag{4.35}$$

The *static pressure rise* must be sufficient to draw the air into the hood to overcome the resistance from flowing through the hood, ductwork, APC systems, and stack at

5 Air Pollution Meteorology and Air Pollutant Concentration Models

5.1 THE TROPOSPHERE

As mentioned in Chapter 2, the atmosphere is classified into five layers based on its temperature. The troposphere is the part of the atmosphere where we live in. It is the lowest layer of the Earth's atmosphere. In the troposphere, its temperature and pressure decrease with altitude. According to the National Aeronautics and Space Administration (NASA), the temperature (T in °C) and pressure (P in kPa) at an altitude (h in meters) in the troposphere (for $h < 11{,}000\,\text{m}$) can be estimated by the following equations (https://www.grc.nasa.gov/www/K-12/airplane/atmosmet.html):

$$T = 15.04 - \left(6.49 \times 10^{-3}\right)h \tag{5.1}$$

$$P = 101.29 \times \left[\frac{T + 273.1}{288.08}\right]^{5.256} \tag{5.2}$$

Clouds and weather generally occur in the troposphere. While large-scale pressure systems control the prevailing meteorology of a region, local meteorological factors include horizontal airflow (i.e., wind speed and direction) and vertical air movement. These air movements affect the dispersion and transport of compounds of concern (COCs).

Example 5.1: Temperature and Pressure Profiles in the Troposphere

Use the two NASA equations to answer the questions below:

a. What is the temperature at the ground level assumed in Eq. 5.1?
b. What changes in temperature occur with every kilometer increase in elevation?
c. What would the temperature and pressure be at the peak of Mountain Everest?
d. What would the temperature and pressure be at the top of the troposphere?

DOI: 10.1201/9781003532705-5

Solution:

a. Use $h=0$ in Eq. 5.1; the ground-level temperature is 15.04°C.
b. As shown in Eq. 5.1, the relationship between T and h is linear, and the slope is −0.00649°C/m (or −6.49°C/km).
c. The elevation of Mount Everest is 8,848 m (29,029 ft), so

$$T = 15.04 - \left(6.49 \times 10^{-3}\right)(8{,}848) = -42.4°\text{C}$$

$$P = 101.29 \times \left[\frac{(-42.4) + 273.1}{288.08}\right]^{5.256} = 31.5\,\text{kPa} = 236\,\text{mm-Hg}$$

d. The model assumes the top of the troposphere is at 11,000 m:

$$T = 15.04 - \left(6.49 \times 10^{-3}\right)(11{,}000) = -56.4°\text{C}$$

$$P = 101.29 \times \left[\frac{(-56.4) + 273.1}{288.08}\right]^{5.256} = 22.7\,\text{kPa} = 170.4\,\text{mm-Hg}$$

5.2 HORIZONTAL MOVEMENT OF AIR

5.2.1 Horizontal Air Circulation

Everything on Earth absorbs, stores, and re-radiates the energy it receives from the Sun. Some parts of the Earth heat more readily than others, known as *differential heating*. Land surfaces heat and cool more rapidly than water bodies because water has a larger heat capacity. The equatorial regions receive more sunlight and heat compared to the polar regions. The tilt of the Earth's axis causes seasonal changes in solar radiation; solar heating would be the highest during midday.

Differential heating affects the air above the ground and drives many of the Earth's atmospheric and oceanic circulation patterns. The air above the warmer spot would be moist, lighter, and buoyant, forming a low-pressure area. On the other hand, the air in a high-pressure area is calm and heavy. Pressure differences create the wind. Meteorologists use four "*scales of motion*" to describe the air circulation in the atmosphere. *Microscale* is the smallest of these four, covering a vertical distance of ≤100 m and a horizontal distance of ≤2 km. Winds in local areas fall into the microscale. The *mesoscale* covers a vertical distance of ≤2 km and a horizontal distance of ≤200 km, and thunderstorms are mesoscale phenomena. The *synoptic scale* covers a vertical distance of ≤5 km and a horizontal distance of ≤1,000 km, and low- and high-pressure systems are synoptic-scale phenomena. The largest scale is the *macroscale* (planetary or global scale), which covers a large portion of the Earth.

Since the equator receives more solar irradiation than the poles, the air would be hotter, moister, and buoyant there. It will rise as the cold and dense air will move toward the equator. However, the Earth rotates from west to east, and its surface

b. For the unstable atmosphere ($p = 0.1$):

$$u_{30} = (3.0)\left(\frac{40}{10}\right)^{0.1} = 3.45\,\text{m/s}$$

5.3 VERTICAL MOVEMENT OF AIR

While horizontal movement of air is generally related to wind speed and direction, the forces about vertical movement of air are *atmospheric stability*, which is the atmosphere's tendency to resist or promote vertical air motion. Understanding atmospheric stability is essential for the transport of air pollutants. The mechanisms responsible for the vertical air movement are essentially temperature and pressure.

5.3.1 Lapse Rates

Keeping other variables constant, the air temperature decreases as atmospheric pressure decreases. *The lapse rate* is the rate of temperature decrease with altitude. An *air parcel* is a body of air with a constant number of molecules and acts as a whole. Under adiabatic conditions (i.e., no heat exchange with the surroundings), a rising warm air parcel would behave like a balloon. Due to the lower ambient pressure, the air within the parcel will expand until its density is equal to the density of the surrounding air. Due to its expansion, this air parcel's adiabatic cooling rate is 10°C/km (9.81°C/km, to be exact; or 5.4°F/1,000 ft). In this context, the air is considered dry if the air parcel's water remains gaseous. It is called the *dry adiabatic lapse rate* (*DALR*); it is a fixed rate independent of ambient air temperature and the starting position of the air parcel rising or descending. The DALR is central to the definition of atmospheric stability.

The actual temperature profile of the ambient air is called the *atmospheric lapse rate*, the *prevailing lapse rate*, or the *environmental lapse rate* (*ELR*). It is the result of complex interactions among meteorological factors. It is usually considered that the temperature decreases with height (like the adiabatic lapse rate), but the rate can be larger or smaller than the adiabatic lapse rate. In addition, the ambient temperature can sometimes increase with altitude. In this case, the atmospheric lapse rate will have a sign opposite to that of the adiabatic lapse rate. This layer is termed the *inversion layer*, and it is essential for air pollution because it limits vertical air motion. In other words, it can lead to the accumulation of pollutants near the ground, as it traps air and prevents dispersion. The atmospheric lapse rate is significant for vertical air movement because the surrounding air temperature determines the extent to which an air parcel rises or falls.

A rising parcel of dry air containing water vapor will continue to cool at the DALR until it reaches the dew point, where the water starts to condense. Condensation releases the latent heat of water within the parcel; consequently, the cooling rate of the air parcel decreases. It is called the *wet adiabatic lapse rate*. The wet adiabatic lapse rate depends on prevailing humidity and varies by location. However, in the middle latitudes, it is assumed to be about 6°C/km (USEPA, 1982), similar to the slope in NASA's equation (Eq. 5.1).

The height at which the air parcel's wet adiabatic lapse rate intersects with the atmospheric/environmental/prevailing lapse rate is the *mixing height,* which is the air parcel's maximum level of ascendance. In cases where no intersection occurs (e.g., when the atmospheric lapse rate is constantly greater than or equal to the dry adiabatic rate), the mixing height may extend to the tropopause. The atmosphere below the mixing height is the *mixing layer.* The larger the mixing layer, the greater the volume of air in which air pollutants can be diluted. The height of the mixing layer is not constant and depends on meteorological conditions. Above the mixing height, the atmosphere tends to be more stable, limiting vertical mixing.

5.3.2 Atmospheric Stability

The degree of atmospheric stability is determined by the temperature difference between an air parcel and its surroundings, which would cause the air parcel to move vertically (i.e., rise or descend). This movement is characterized by four primary conditions: unstable, neutral, stable, and inversion (Figure 5.3).

The solid straight line in Figure 5.3 represents the DALR. If the ELR of the surrounding atmosphere is greater than the DALR, the rising air parcel will continue to be warmer than its surroundings when it moves up. This is the *super-adiabatic rate* (the line below the dry adiabatic rate in Figure 5.2). As shown, the temperature difference between the air parcel and its surroundings increases with height and buoyancy. Consequently, it is an unstable condition. Unstable conditions most

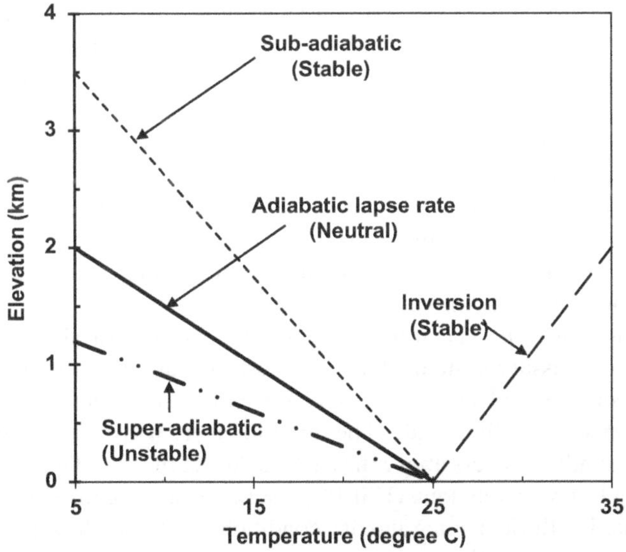

FIGURE 5.3 Basic conditions of atmospheric stability.

commonly develop on sunny days with low humidity and wind speed (USEPA, 1982). An unstable atmosphere would encourage mixing, often improving air quality and pollutants' dispersion.

The atmosphere is neutral when the ELR equals the DLAR (in Figure 5.2, these two lapse rates overlap). It means that air movement is neither encouraged nor hindered.

When the ELR < the DLAR, a rising warm air parcel would cool faster than its surroundings, reach the temperature of the surroundings at some point, and will not rise further. This is a *sub-adiabatic lapse rate* (the line above the adiabatic lapse rate in Figure 5.2), in which the air is stable and resists vertical motion of the air parcel. *Stable conditions* are most likely to occur on cloudy days with little wind and no intense surface heating or at night. A stable atmosphere will limit the mixing of air layers, which can lead to poor air quality.

As previously mentioned, the atmospheric temperature decreases as the altitude increases. An *inversion* occurs when the atmosphere's temperature increases with altitude (see Figure 5.2). The inversion layer acts as a barrier to the vertical movement of air, reducing air pollutants' dispersion, which results in elevated pollutant concentrations. Different atmospheric interactions cause inversions and can persist for various amounts of time.

5.3.3 Types of Inversion

By definition, a temperature inversion exists when warmer air overlies cooler air. Inversions can be due to (i) cooling of air at the surface, (ii) heating from above, (iii) a layer of warm air flowing over a layer of cold air, or (iv) a layer of cooling air flowing under a layer of warm air. There are four significant types of inversions – see below:

Radiation inversion. It is the most common *surface inversion* when the ground surface cools rapidly. As it cools, so does the layer of air close to the surface. If the temperature of this air layer becomes cooler than that of the air above, an inversion layer is formed. Radiation inversions usually occur late at night (i.e., *nocturnal radiation inversion*) and early morning under clear skies. During the day, the situation would be reversed. Diurnal cycles of daytime instability and nighttime inversions are standard. Therefore, the effects of radiation inversions are often short-lived. In locations where radiation inversions are common and relatively close to the ground surface, tall stacks emitting pollutants above the inversion layer can help reduce ground-level pollutant concentrations (USEPA, 1982).

Subsidence inversion. The *subsidence inversion* is almost always associated with high-pressure systems where air descends and flows outward. As the air descends (or subsides), it compresses and warms, causing the temperature to increase with altitude. The inversion layer will form and be elevated several hundred meters above the surface during the day. This type of inversion can last relatively long (USEPA, 1982). Subsidence inversions are often associated with clear skies and dry conditions as the stable atmosphere suppresses cloud formation and precipitation.

Frontal inversion. Frontal inversion is usually associated with cold or warm fronts. At the leading edge of either type of front, the warm air overrides the cold air, so slight vertical motion occurs in the chilly air closest to the ground. The strength of an inversion depends on the temperature difference between two air masses. This type of inversion is usually short-lived unless the front becomes stationary (USEPA, 1982). Frontal inversions can lead to cloud formation and precipitation as the warm air is lifted and cools.

Advective inversion. An *advective inversion* is associated with a horizontal flow of warm air. When warm air flows over a cold surface (e.g., snow cover or icy land), the air closest to the surface will be cooled to form a surface-based inversion. The advective inversion can be terrain-based, forcing warm air over a cooler air layer. It occurs when warm air from one side of the mountain overrides cooler air on the other. Advective inversions are vertically stable but may have strong winds underneath the inversion layer (USEPA, 1982). It also happens when warm air moves over cooler ocean/lake surfaces during the day, when land heats up faster than the sea/lake.

5.4 AIR POLLUTION PLUME

Regarding air pollution, COCs can enter the atmosphere differently (e.g., ground-level, stack, area source, or volume source). One method that has received much attention is the release of COCs from stacks (point sources). Although we generally say that dilution is not a solution for pollution control, the discharge of COCs into the atmosphere is commonly practiced to dilute the strength of a flue gas stream. In this practice, stacks release COCs at elevated locations above the ground surface level, allowing them to be sufficiently dispersed (diluted) before reaching the ground. The airspace that the COCs occupy is often called a *plume*. As the plume travels, it spreads and disperses both horizontally and vertically. For a continuous and constant discharge from a stack, the concentrations at the centers of cross-sections of the plume will decrease as the distance away from the stack increases. In addition, even though the total mass of the COCs of a cross-section stays the same, the plume's size will generally increase until it hits the ground as COC concentrations keep decreasing further from the center line due to dispersion. Eventually, the plume would hit the ground (to reach the receptors), but at lower COC concentrations than the flue gas.

Wind speed and direction, atmospheric stability, and topography are the key factors affecting the plume behavior.

5.4.1 Atmospheric Stability and Plume Behavior

Atmospheric stability and the resulting mixing height significantly impact the vertical movements of air pollutants in the atmosphere. It should be noted that, underneath the inversion layer, surface wind is the dominant factor in dispersing air pollutants. This subsection describes several plume types resulting from different stability conditions.

A series of loops or oscillations characterizes a looping plume (Figure 5.4) as it travels downstream from the source, resulting from turbulence caused by the overturning of air under unstable atmospheric conditions. It often occurs during sunny days with movement past an obstruction (e.g., structures and vegetation) or by wind shears (i.e., two adjacent air streams moving at different velocities). *Buoyant turbulence* is caused by heating or cooling air near the ground surface. For example, heating the ground surface during a sunny day will create an upward heat flux that heats the air layer adjacent to the ground. It creates an upward-rising thermal stream, creating positive buoyant turbulence (Turner, 1994). Both types of turbulence impact the dispersion of pollutants in the air.

5.4.2 Topographical Features

The physical characteristics of the Earth's surface are often referred to as topographical/terrain features. They can be grouped into four categories: (i) flat, (ii) mountain/valley, (iii) land/water, and (iv) urban (USEPA, 1982). They will exert different geometric and thermal effects on air movement.

Flat. Oceans and gently rolling features on land are good examples of this group. The geometric effect of flat terrain is limited to its differences in surface roughness, which will result in different wind speed profiles at various heights. The thermal effect of flat terrain is due to its capacity and rate of absorption and release of heat. For example, water does not absorb heat much, while concrete does. The concrete can then release large amounts of heat into the air (USEPA, 1982).

Mountain/valley. Atmospheric dispersion in mountains/valleys is more complex than that of flat terrain. It is also called *complex terrain*. The geometric and thermal effects of mountain/valley terrain are related to the features' size, shape, and orientation. Air will always find the least resistant path to travel. For example, if there is a temperature inversion layer on the top of the mountain, the air will move around the sides of the mountain. If the airflow is blocked, trapping or recirculating the air occurs. Regarding the thermal effect, different parts of a mountain/valley in a geological location are heated unevenly during the day because of the Sun's motion across the sky. In addition, different land features will absorb/release different amounts of heat; for example, tree-covered areas will receive less heat than bare ground (USEPA, 1982).

Land/water. Several large cities are situated close to bodies of water. With regard to geometric effect, a large water body can be considered flat terrain, while the land has many obstructions to airflow and its surface is considered less smooth than that of a large water body. The thermal properties of land and water are drastically different. Water heats and cools relatively slowly, while land and its objects heat and cool at various rates. Water temperatures are relatively stable and follow seasonal changes, while those of the land vary throughout the day. The warmer daytime temperatures over the land cause the air above it to become less dense and rise. The cooler air over the water body will be drawn inland, resulting in a *sea breeze*. At night, the air over the land cools rapidly, while the air over the water becomes relatively warmer, creating a *land breeze*. The wind speeds in a land breeze are light, while those in a sea breeze can be pretty fast (USEPA, 1982). Los Angeles is located in a coastal area; the daily sea breeze should help disperse automobile pollutants and other emissions. However, the area is also surrounded by high mountains, which hinder horizontal air movement. Instead, the emitted air pollutants can be trapped within the zone. It is a good example that topographical groups may have combined effects on air movement and pollutant dispersion.

Urban. Urban areas can exhibit all the characteristics of three topographical features discussed above. In addition, metropolitan areas add vast amounts of artificial pollutants to the atmosphere. With buildings and streets, the geometric effect of an urban area is much like complex terrain or even more complicated. Regarding the thermal effect, buildings absorb and hold lots of heat. The exchange of heat among the buildings is continuous. At night, some of the absorbed heat is transmitted to create a dome over the city called the *heat island effect*.

5.5 EFFECTIVE STACK HEIGHT

We usually notice smoke rising above the top of a smokestack, indicating smoke coming out of it (Figure 5.9). The vertical distance that the plume rises is called the *plume rise* (ΔH), which is calculated as the vertical distance from the emission point to the imaginary centerline of the plume.

The extent of the plume rise depends on the characteristics of the stack (e.g., the diameter of the stack) and the effluent gas stream (e.g., exit velocity, temperature), as well as the meteorological conditions (i.e., T, P, and wind speed) around the stack. *Holland's formula* is often used to estimate the plume rise:

$$\Delta H = \frac{v_s D}{u}\left[1.5 + \left(2.68 \times 10^{-5}\right)(P)(D)\left(\frac{T_s - T_a}{T_s}\right)\right] \quad (5.4)$$

where v_s is the exit velocity from the stack (m/s), u is the wind velocity at the point of release (m/s), P is the ambient pressure (Pa), D is the diameter of the stack tip, T_s is the stack gas temperature (K), and T_a is the ambient temperature (K). The calculated value from Eq. 5.4 is often adjusted for atmospheric stability by multiplying it by 1.1 or 1.2 for A and B stability and 0.8 or 0.9 for D, E, and F stability (Turner, 1994) [Note: see Section 5.6 for stability classes].

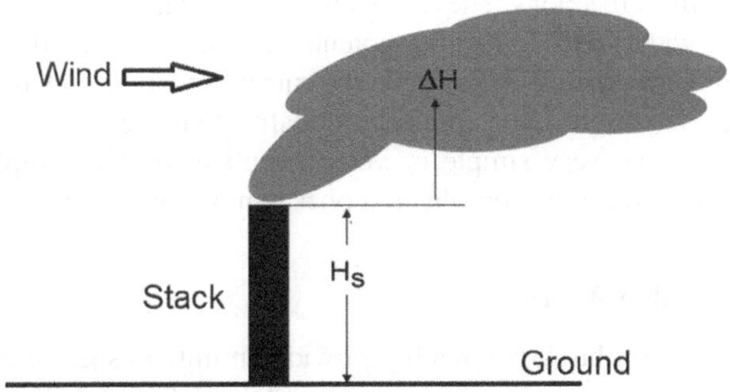

FIGURE 5.9 Plume rise (USEPA, 2003).

The plume centerline distance from the ground is called the *effective stack height*, which is the sum of the stack height (H_s) and the plume rise, as follows:

$$\text{Effective stack height } (H) = \text{Stack height } (H_s) + \text{Plume rise } (\Delta H) \quad (5.5)$$

Example 5.3: Effective Stack Height

Estimate the effective stack height of a 500 MW power plant (diameter of the stack = 4 m, stack height = 200 m, ambient temperature = 20°C, ambient pressure = 1 atm, exit gas velocity = 15 m/s, exit temperature = 140°C, and stack tip wind speed = 6 m/s).

Solution:

a. Use Eq. 5.4 to find the plume rise:

$$\Delta H = \frac{(15)(4)}{6}\left[1.5 + \left(2.68\times10^{-5}\right)\left(1.013\times10^{5}\right)(4)\left(\frac{413-293}{413}\right)\right] = 46.5 \text{ m}$$

b. Use Eq. 5.5 to find the effective stack height:

$$\text{Effective stack height } (H) = 200 + 46.5 = 246.5 \text{ m}$$

5.6 AIR POLLUTANT CONCENTRATION MODELS

Air concentration models are mathematical models used to estimate the distribution and concentration of pollutants in the atmosphere, and they simulate physical and chemical processes that affect the COCs in the plume. Using characteristics of the emission source (e.g., stack height, exit velocity, concentration, and temperature) and meteorological data (e.g., wind speed and direction, ambient temperature, and pressure), the models could estimate COC concentrations within the plume as well as the maximum concentrations where the plume hits the ground. When the primary air pollutants are not inert, the models can also include their decay and the formation of secondary pollutants through chemical reactions. These models are essential for air quality management because they are widely used to identify potential impacts of a new discharge, evaluate the contribution of an existing discharge to the current situation, and assist in designing effective strategies to improve the ambient air quality, to name a few.

The models can be very simple or very complicated. The simple ones can be solved analytically, while the complicated ones can be solved numerically.

5.6.1 A Simple Box Model

A *box model* is simple, but it can readily provide an initial estimate of concentration values based on the emission rate, wind speed, and wind direction. The box may represent a city, and the plan area over the city is defined by the length (L) and the width (W), while H is the vertical dimension of the airshed, which should be the mixing height. An *airshed* refers to a defined area where air quality is influenced by common meteorological conditions and the movement and dispersion of air pollutants are interconnected. The *mixing height*, also known as the *mixing layer height* or *boundary layer height*, refers to the height within the atmosphere where pollutants, heat, moisture, and other atmospheric constituents are mixed due to turbulence. Mixing height can vary significantly throughout the day. It is typically lower when the atmosphere is more stable (e.g., during the night and early morning) and becomes higher during the afternoon when the solar heating increases air turbulence. The mixing height is crucial in understanding how pollutants disperse and affect air quality (Figure 5.10).

It is the simplest model with the following simplified assumptions:

1. Wind comes perpendicular to one side of the box at a constant velocity (u), carrying a continuous background pollutant concentration (G_o).
2. The city continuously emits the pollutant at a constant flux rate of q.
3. The pollutant will not be destroyed or transformed inside the box.
4. The air in the box is entirely mixed.
5. No pollutant leaves or enters through the box's top side and sides parallel to the wind direction.

With these assumptions,

$$\text{Flow rate into and out of the box } (Q_{\text{air}}) = [(u)(W \times H)] \tag{5.6}$$

$$\text{Mass loading rate to the box} = (Q_{\text{air}})(G_o) = [(u)(W \times H)](G_o) \tag{5.7}$$

$$\text{Emission rate} = (\text{emission rate/area}, q)(\text{area}) = (q)(W \times L) \tag{5.8}$$

$$\text{Mass outflow rate} = \{[(u)(W \times H)](G_o)\} + [(q)(W \times L)] \tag{5.9}$$

Since the box is well mixed, the effluent pollutant concentration (G) will be the same as the concentration inside the box. It can be found by dividing the mass outflow rate (Eq. 5.9) by the airflow rate Q_{air} (Eq. 5.6), as follows:

$$G = G_o + \frac{qL}{uH} \tag{5.10}$$

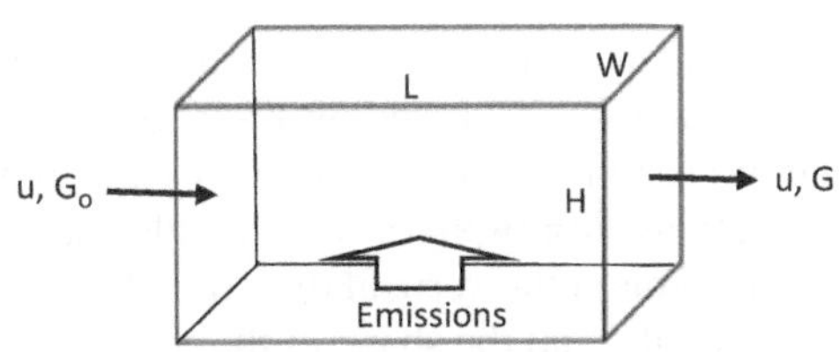

FIGURE 5.10 Schematic of a box model in an airshed.

Example 5.4: The Box Model

A city has an area of a rectangular shape (10 km × 20 km). The wind constantly blows from the west (perpendicular to the short side; i.e., the 10-km side) with a velocity of 4 m/s and carries a background SO_2 concentration of 10 μg/m³. The SO_2 emission rate from the city is 5.0×10^{-6} g/s·m².

a. What is the average SO_2 concentration over the city (assuming the mixing height = 500 m)?
b. If the concentration needs to be half the value in part (a), how should the SO_2 emission rate be reduced?

Solution:

a. Use Eq. 5.10 to find the effluent concentration:

$$C = C_o + \frac{qL}{uH} = 10 + \frac{\left(5\frac{\mu g}{s \cdot m^2}\right)(20{,}000\,m)}{\left(4\frac{m}{s}\right)(500\,m)} = 60\,\mu g/m^3$$

b. Use Eq. 5.10 again to find the reduced emission rate:

$$30 = 10 + \frac{(q)(20{,}000\,m)}{\left(4\frac{m}{s}\right)(500\,m)} \rightarrow q = 2\,\mu g/s \cdot m^2$$

Discussion: The emission rate needs to be dropped more than half (i.e., from 5 to 2 μg/s·m²) to reduce SO_2 concentration in the ambient air by half because there is SO_2 in the influent air.

5.6.2 Gaussian Plume Dispersion Model

The *Gaussian plume dispersion model* is the most widely used air concentration model to simulate pollutant dispersion from a point source. Pollutants emitted from the point source travel in the direction of the prevailing wind (i.e., the along-the-wind direction). Small-scale turbulent/eddy motions disperse the pollutants vertically and horizontally in the across-the-plume direction. These dispersion processes determine pollutant concentrations at any location within the plume and on the ground.

Meteorological conditions play an essential role in governing the dispersion. If the wind is strong, the pollutants will spread over a large area horizontally, yielding relatively low concentrations. Otherwise, the pollutants will remain near the release point with high concentrations. Atmospheric stability also plays a vital role in determining the pollutant dispersion in the vertical direction. Figure 5.11 shows the coordinate system used in the model and the shape of the plume.

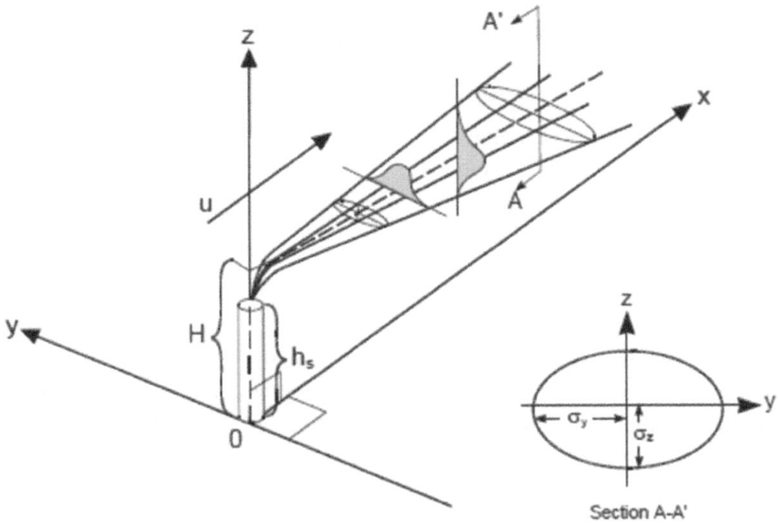

FIGURE 5.11 Coordinate system for Gaussian dispersion model (USEPA, 2014).

The following equation describes the three-dimensional concentration generated by a continuous source at a constant emission rate under steady-state meteorological conditions (USEPA, 2014):

$$G(x,y,z)=\frac{Q}{2\pi u\sigma_y\sigma_z}\exp\left[-\frac{1}{2}\left(\frac{y^2}{\sigma_y^2}\right)\right]\left\{\exp\left[-\frac{1}{2}\left(\frac{(z-H)^2}{\sigma_z^2}\right)\right]\right\} \tag{5.11}$$

Assuming the conservation of mass and reflection of all the pollutants' contact with the ground, an imaginary source needs to be added, resulting in the following equation:

$$G(x,y,z)=\frac{Q}{2\pi u\sigma_y\sigma_z}\exp\left[-\frac{1}{2}\left(\frac{y^2}{\sigma_y^2}\right)\right]\left\{\exp\left[-\frac{1}{2}\left(\frac{(z-H)^2}{\sigma_z^2}\right)\right]+\exp\left[-\frac{1}{2}\left(\frac{(z+H)^2}{\sigma_z^2}\right)\right]\right\} \tag{5.12}$$

where G is the steady-state concentration at a point (x, y, z) within the plume (g/m^3), x is the downwind distance along the plume center line (m), y is the horizontal distance along the plume center line (m), Q is the emission rate (g/s), u is the wind velocity at the release point (m/s), H is the effective stack height (m), and σ_y and σ_z are the horizontal and vertical dispersion coefficients (m).

Both *dispersion coefficients* (σ_y and σ_z), sometimes called *Pasquill–Gifford parameters*, depend on the downwind distance from the source and the atmospheric stability class. There are six Pasquill stability classes, ranging from A to F. Class A represents the most unstable condition, while class F represents the most stable condition:

- Class A = Strongly unstable
- Class B = Moderately unstable
- Class C = Slightly unstable
- Class D = Neutral
- Class E = Slightly stable
- Class F = Stable

Table 5.1 shows the relationship between the stability class and wind speed, the incoming solar radiation's strength (i.e., *insolation*), and the extent of cloud cover.

Figures 5.12 and 5.13 show the Pasquill–Gifford curves for σ_y and σ_z, respectively. The two dispersion coefficients can be read from the figure for a given downwind distance x and the criteria for stability class (Table 5.1). Three general trends can be readily observed from these two figures: (i) both dispersion coefficients increase with the downwind distance, (ii) their values increase as the air becomes more unstable, and (iii) σ_y values are larger than the corresponding σ_z values.

Values of σ_y and σ_z (in meters) can also be obtained using the curve-fitting equations below, whereas x must be expressed in kilometers (NCEES, 2025) (Table 5.2):

$$\sigma_y = ax^{0.894} \tag{5.13}$$

$$\sigma_z = cx^d + f \tag{5.14}$$

TABLE 5.1
Key to Stability Category

Wind Speed (m/s)	Day: Incoming Solar Radiation – Strong	Day: Incoming Solar Radiation – Moderate	Day: Incoming Solar Radiation – Slight	Night: Thinly Cast (≥4/8 Cloud)	Night: Clear (≤3/8 Cloud)
0–2	A	A–B	B	—	—
2–3	A–B	B	C	E	F
3–5	B	B–C	C	D	E
5–6	C	C–D	D	D	D
>6	C	D	D	D	D

Source: USEPA (1970).

Note: [1] Wind speed is measured at 10 m above the ground.

[2] "A–B" means to average the values obtained for each condition.

[3] Neutral class D should be assumed for overcast conditions, day or night.

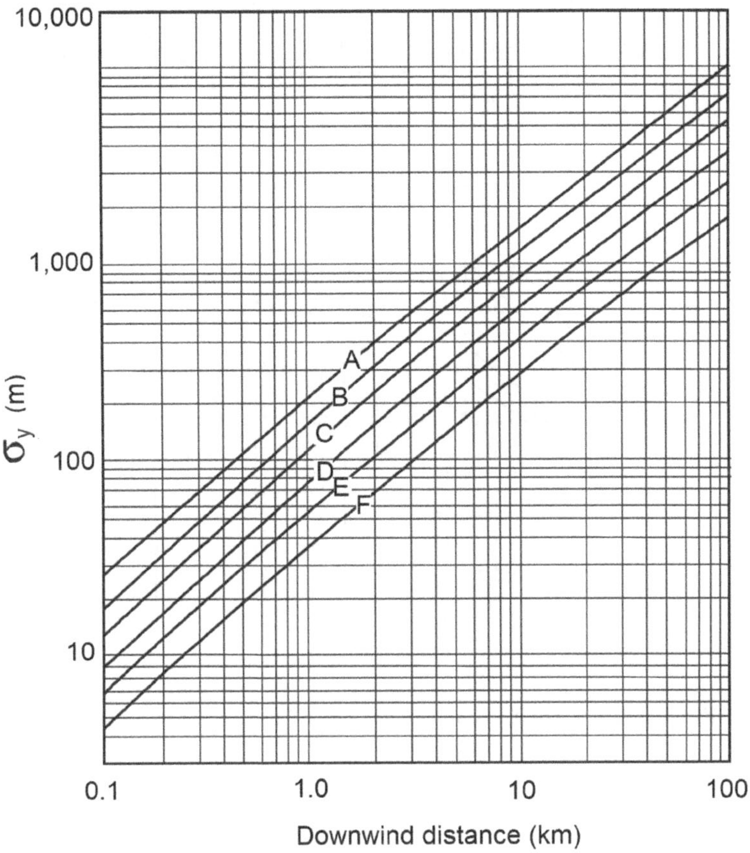

FIGURE 5.12 Pasquill–Gifford curves for the horizontal dispersion coefficient (USEPA, 1970; NCEES, 2025).

5.6.3 Special Cases for the Gaussian Dispersion Equation

There are several special cases of common interests that can be deduced from the general Gaussian dispersion equation (Eq. 5.12):

a. Concentrations at the plume centerline at $y=0$ and $z=H$:

$$G(x,0,H)=\frac{Q}{2\pi u\sigma_y\sigma_z}\left\{1+\exp\left[-\left(\frac{2H^2}{\sigma_z^2}\right)\right]\right\} \tag{5.15}$$

b. For receptors at the ground level ($z=0$):

$$G(x,y,0)=\frac{Q}{\pi u\sigma_y\sigma_z}\exp\left[-\frac{1}{2}\left(\frac{y^2}{\sigma_y^2}\right)\right]\left\{\exp\left[-\frac{1}{2}\left(\frac{H^2}{\sigma_z^2}\right)\right]\right\} \tag{5.16}$$

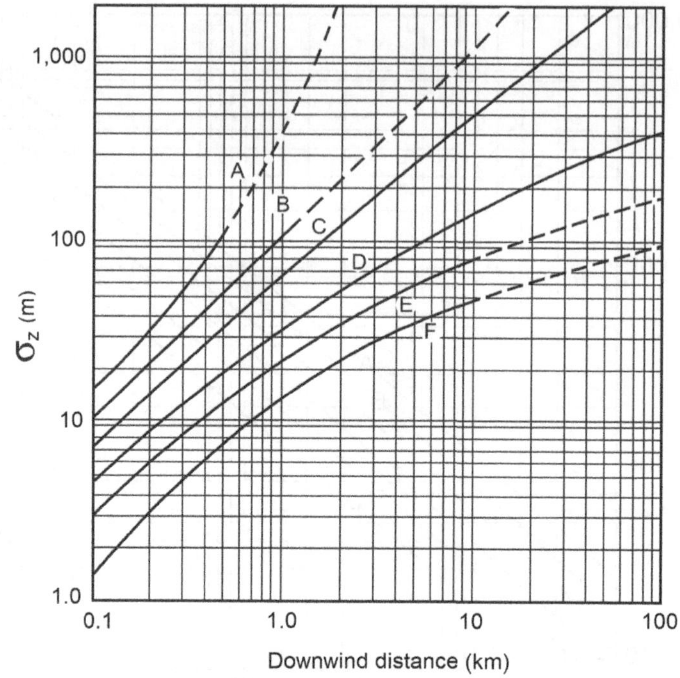

FIGURE 5.13 Pasquill–Gifford curves for the vertical dispersion coefficient (USEPA, 1970; NCEES, 2025).

TABLE 5.2
Curve-Fitting Constants for Pasquill–Gifford Curves

		$x<1$ km			$x\geq 1$ km		
Stability Class	a	c	d	f	c	d	f
A	213	440.8	1.941	9.27	459.7	2.094	−9.6
B	156	106.6	1.149	3.3	108.2	1.098	2.0
C	104	61.0	0.911	0	61.0	0.911	0
D	68	33.2	0.725	−1.7	44.5	0.516	−13.0
E	50.5	22.8	0.678	−1.3	55.4	0.305	−34.0
F	34	14.35	0.740	−0.35	62.6	0.180	−48.6

Source: NCEES (2025).

c. For receptors at the ground level and directly beneath the centerline ($z=0$ and $y=0$):

$$G(x,0,0)=\frac{Q}{\pi u\sigma_y\sigma_z}\left\{\exp\left[-\frac{1}{2}\left(\frac{H^2}{\sigma_z^2}\right)\right]\right\} \tag{5.17}$$

d. For a ground-level release ($H=0$), such as fires, explosions, fugitive emissions, or smoldering landfill sites:

$$G(x,y,z)=\frac{Q}{\pi u\sigma_y\sigma_z}\exp\left[-\frac{1}{2}\left(\frac{y^2}{\sigma_y^2}\right)\right]\left\{\exp\left[-\frac{1}{2}\left(\frac{z^2}{\sigma_z^2}\right)\right]\right\} \quad (5.18)$$

e. For a ground-level release and receptors at ground level ($z=0$, and $H=0$):

$$G(x,y,0)=\frac{Q}{\pi u\sigma_y\sigma_z}\exp\left[-\frac{1}{2}\left(\frac{y^2}{\sigma_y^2}\right)\right] \quad (5.19)$$

f. For a ground-level release and the maximum ground-level concentration ($y=0$, $z=0$, and $H=0$):

$$G(x,0,0)=\frac{Q}{\pi u\sigma_y\sigma_z} \quad (5.20)$$

Example 5.5: Gaussian Plume Dispersion Model for a Stack Source

It has been estimated that the SO_2 emission rate from a coal-fired plant is 500 g/s. The physical stack height is 30 m, and the plume rise is 6 m. On a shiny, sunny day (solar radiation is intense), the wind speed is 4 m/s. Determine:

a. the Pasquill stability class,
b. the horizontal dispersion coefficient at 3 km downwind from the stack,
c. the vertical dispersion coefficient at 3 km downwind from the stack,
d. the SO_2 concentration at the plume centerline and the effective stack height, 3 km from the stack,
e. the maximum SO_2 concentration on the ground, 3 km from the stack, and
f. the maximum SO_2 concentration on the ground is 3 km from the stack if the emission is released from the ground level, not from the tall stack.

Solution:

a. Stability class = B from Table 5.1 (Sunny day with strong insolation and wind speed of 4 m/s).
b. Stability class B and $x=3$ km; from Figure 5.12, $\sigma_y=420$ m.
c. Stability class B and $x=3$ km; from Figure 5.13, $\sigma_z=350$ m.
d. Effective stack height $=30+6=36$ m, use Eq. 5.13 to find the centerline SO_2 concentration at $x=3$ km:

$$G(3,0,36)=\frac{500}{2\pi(4)(420)(350)}\left\{1+\exp\left[-\left(\frac{2(36)^2}{(350)^2}\right)\right]\right\}=2.68\times10^{-4}\,\frac{\text{g}}{\text{m}^3}$$

e. Use Eq. 5.17 to find the maximum ground SO_2 concentration at $x = 3$ km:

$$G(3,0,0) = \frac{500}{\pi(4)(420)(350)}\left\{\exp\left[-\frac{1}{2}\left(\frac{36}{350}\right)^2\right]\right\} = 2.69 \times 10^{-4}\ \frac{g}{m^3}$$

f. Use Eq. 5.20 to find the maximum ground SO_2 concentration at $x = 3$ km from the ground-level release:

$$G(3,0,0) = \frac{500}{\pi(4)(420)(350)} = 2.71 \times 10^{-4}\ \frac{g}{m^3}$$

Discussion:

1. The effective stack height is relatively low, and the concentration differences between parts (d), (e), and (f) are insignificant.
2. The dispersion coefficients in stability class B and $x = 3$ km can also be found using Eqs. 5.13 and 5.14 and Table 5.2, as follows:

$$\sigma_y = ax^{0.894} = 156(3)^{0.894} = 416.5\,\text{m}$$

$$\sigma_z = cx^d + f = 108.2(3)^{1.098} + 2.0 = 363.5\,\text{m}$$

The values of dispersion coefficients read from the figures and from curve-fitting equations are comparable.

5.6.4 Location of the Maximum Ground-Level Concentration

Figure 5.13 plots the location of the maximum ground-level concentration downstream from the discharge (x_{max}) vs. $(Gu/Q)_{max}$ as a function of the effective stack height (H). $(Gu/Q)_{max}$ can then be used to find the maximum ground-level pollutant concentration.

Values of $(Gu/Q)_{max}$ can also be obtained using the equation below and constants in Table 5.3 (Ranchoux, 1976):

$$\left(\frac{Gu}{Q}\right)_{max} = \text{Exp}\left[a + b\ln H + c(\ln H)^2 + d(\ln H)^3\right] \tag{5.21}$$

Example 5.6: Maximum Ground Concentration from a Stack Source

For the same stack discharge and the same meteorological conditions as those of Example 5.5, determine the maximum ground-level concentration and the location of this concentration resulting from this discharge.

Solution:

a. Stability class B and $H = 36$ m; from Figure 5.14, $(Gu/Q)_{max} = 10^{-4}\,m^{-2}$.

b. $G = (10^{-4}\,\text{m}^{-2}) \times [500\ \text{g/s} \div 4\ \text{m/s}] = 0.0125\ \text{g/m}^3$.
c. Also from Figure 5.14, $x_{max} = 0.26\,\text{km}$.
d. Using Table 5.3 and Eq. 5.21,

$$\left(\frac{Gu}{Q}\right)_{max} = \text{Exp}\left\{-1.8060 - 2.1912\ln(36) + 0.0389\left(\ln(36)\right)^2\right\} = 1.05\times10^{-4}\,\text{m}^{-2}$$

Discussion: The $(Gu/Q)_{max}$ values obtained from using Figure 5.14 and from using Table 5.3 and Eq. 5.21 are essentially the same. However, Ranchoux (1976) did not provide an equation or a figure to find the location of the maximum ground-level concentration.

5.6.5 USEPA Models

The *Support Center for Regulatory Atmospheric Modeling* (SCRAM), https://www.epa.gov/scram, provides access to air quality models and other mathematical simulation techniques for assessing control strategies and source impacts. It covers three types of air quality models: (i) dispersion modeling, (ii) photochemical modeling, and (iii) receptor modeling.

Dispersion models are typically used in the permitting process to estimate the concentration of pollutants at specified ground-level receptors surrounding an emission source. *Photochemical models* simulate the changes in pollutant concentrations in the atmosphere using a set of mathematical equations characterizing the chemical and physical processes in the atmosphere over a large spatial scale. They are typically used in regulatory or policy assessments to simulate the impacts from all sources. *Receptor models* are observational techniques to identify the presence of air pollutants and quantify the contribution of emission sources to receptor concentrations. Chemical and physical characteristics of gases and particles measured at the sources and receptors are the input to this type of model.

TABLE 5.3
Curve-Fitting Constants for $(Gu/Q)_{max}$ Equation

	Constants			
Stability	**a**	**b**	**c**	**d**
A	−1.0563	−2.7153	0.1261	0
B	−1.8060	−2.1912	0.0389	0
C	−1.9748	−1.9980	0	0
D	−2.5302	−1.5610	−0.0934	0
E	−1.4496	−2.5910	0.2181	−0.0343
F	−1.0488	−3.2252	0.4977	−0.0765

Source: Ranchoux (1976).

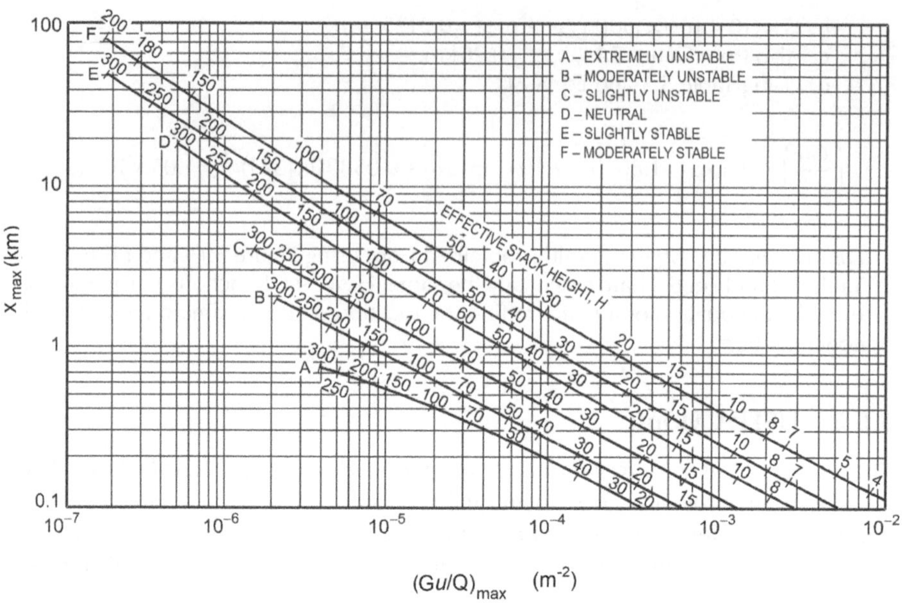

FIGURE 5.14 x_{max} vs. $(Gu/Q)_{max}$ (USEPA, 1970; NCEES, 2025).

Wind, atmospheric stability, plume rise, and topographical features interact in complex ways to cause the transport and dispersion of air pollutants in the atmosphere. Air quality dispersion modeling uses mathematical formulations to characterize the atmospheric processes that affect the dispersion of a pollutant emitted from a source. Using the source emission data (e.g., emission rate and stack height) and relevant meteorological inputs (e.g., temperature, wind direction and speed, cloud cover, mixing height), a dispersion model can predict the pollutant concentration at selected downwind receptor locations. The USEPA's *refined/recommended models* are air quality models listed in Appendix W of 40 CFR Part 51, and they are required to be used for State Implementation Plan (SIP) revisions for existing sources and National Source Review (NSR) and Prevention of Significant Deterioration (PSD) programs. *Screening tools* are models often applied to determine if refined modeling is needed. *Alternative models* are those not listed in Appendix W; however, they can be used in regulatory applications with a case-by-case justification to the reviewing authority.

The *Assessment System for Population Exposure Model* (ASPEN) and *Industrial Source Complex* (ISC) *Model* are frequently used by the USEPA in permitting processes for evaluating environmental health impacts. ASPEN is a screening tool that calculates ambient air concentration levels based on meteorology, chemistry, and emission rates of air toxics. The ISC model is a steady-state Gaussian plume model that can be used to estimate air pollutant concentrations from sources associated with industrial source complexes (USEPA, 2003).

5.7 SUMMARY

1. Meteorology plays an essential role in the dispersion and transport of air pollutants.
2. Dispersion of air pollutants depends on emission characteristics (e.g., emission rate, temperature, exit velocity, and effective stack height), meteorological conditions (e.g., temperature, pressure, wind direction and speed, and stability class), and topographical features.
3. Effective stack height is the sum of physical stack height and plume rise.
4. Air quality models use analytical or numerical techniques to simulate physical and chemical processes that affect the dispersion and reaction of air pollutants in the atmosphere.
5. The box model is the simplest air quality model to assess the effect of emissions on ambient air quality.
6. The Gaussian dispersion equation is commonly used to estimate a 3D concentration profile generated by a continuous source at a constant emission rate under stationary meteorological conditions.

EXERCISE QUESTIONS

1. A new city will be built downwind of an existing one. The planner has decided that the future city will have two rectangular zones of identical size (5 km × 2 km each). One will be residential, and the other will be industrial. The wind always blows from the west, carrying a background concentration of 2 mg/m^3. There are lots of lands available, and the two blocks can be put in any orientation, but they must have one adjacent side to each other (long to long, short to long, or short-short), and two sides of each block are perpendicular to the wind.
   - CO emission rate from the industrial zone = 400 g/s·km^2
   - CO emission rate from the residential zone = 100 g/s·km^2
   - Wind speed = 2 m/s
   - Mixing height = 250 m
   - You are hired as an environmental engineering consultant to use the box model to find the best and worst orientations of these two zones, giving the lowest CO concentration and the highest CO concentration in the residential area.
     a. Show these two orientations [Note: no calculations needed; show the orientations of these two blocks with the direction of the wind].
     b. What would be the highest CO concentration in the residential zone (i.e., the worst-case scenario)?
2. An industrial city is to be developed in a valley. The entire city area would be rectangular (5 km × 30 km). It is to have three zones of the same size (5 km × 10 km each), one for residential, one for light industrial, and one for heavy industrial. Use the area-source box model and simplified assumptions to answer the questions below:

- The wind blows continuously and parallel to the long side of the city block at 2 m/s and brings SO_2 from upstream at a concentration of 0.1 mg/m^3.
- The SO_2 emission rates are 0.2 and 2 g/km^2/d from the light and the heavy industrial zones, respectively. The mixing height is constant at 500 m.
  a. The sequences of these three zones would yield the best and the worst ambient SO_2 concentrations in the residential zone.
  b. The lowest potential ambient SO_2 concentration in the residential zone.
  c. The highest potential ambient SO_2 concentration in the residential zone.

3. An industrial boiler with a stack height of 80 m emits 4 g/m^3 SO_2 with a volumetric flow rate of 25 m^3/s on a sunny summer day (diameter of the stack = 2 m). The wind speed at the top of the stack is 4 m/s. The temperature of the exit gas is 120°C. The ambient temperature is 30°C, and the pressure is 750 mm-Hg.
   a. Determine the effective stack height.
   b. Determine the maximum ground-level concentration at one kilometer downstream of the stack.
   c. Determine the distance where the maximum ground-level concentration occurs.
   d. Determine the concentration at the location found in part (c).
4. The plant emits 300 g/s of particulates, and the effective stack height is 100 m. The wind speed is 3 m/s.
   a. What would be the locations of the maximum ground concentrations when the stability classes are C and D?
   b. What would the ratio of these two maximum ground-level concentrations be?
   c. What is the maximum ground concentration for air stability class C?
5. The SO_2 emission rate from a stack is 800 g/s (the effective stack height = 40 m). On a specific day, the Pasquill stability class of the atmosphere is D, and the wind speed = 3 m/s.
   a. What would be the maximum ground-level SO_2 concentration at 1 km downwind from the stack?
   b. What would be the maximum ground-level SO_2 concentration from this discharge?
   c. How far downwind will the maximum concentration in part (b) be located?

BIBLIOGRAPHY

NCEES. (2025). *PE/Environmental Reference Handbook (Version 1.6)*, National Council of Examiners for Engineering and Surveying (NCEES), Greenville, SC.

Ranchoux, R.J.P. (1976). "*Determination of Maximum Ground Level Concentration*," *J. Air Pollut. Control Assoc.* 26(11), 1088–9.

Turner, D.B. (1994). *Workbook of Atmospheric Dispersion Estimates – An Introduction to Dispersion Modeling*, 2nd edition, Lewis Publishers, Boca Raton, FL.

USEPA. (1970). *Workbook of Atmospheric Dispersion Estimates*, prepared by Turner, D. B. for Office of Air Programs, United States Environmental Protection Agency, Research Triangle Park, NC.

USEPA. (1981). *APTI 411: Air Pollution Meteorology - Student Workbook*, EPA 450/2-81-014, Air Pollution Training Institute, United States Environmental Protection Agency, Research Triangle Park, NC.

USEPA. (1982). *APTI SI 409: Basic Air Pollution Meteorology*, EPA 450/2-82-009, prepared by Northrop Services, Inc. for the United States Environmental Protection Agency, Research Triangle Park, NC.

USEPA. (1995). *APTI 325: Visible Emission Evaluation Procedures*, prepared by T.H. Rose (Eastern Technical Associates) for the United States Environmental Protection Agency.

USEPA. (2003). *APTI 452: Principles and Practices of Air Pollution Control - Student Manual*, 3rd edition, Air Pollution Training Institute, United States Environmental Protection Agency, Research Triangle Park, NC.

USEPA. (2014). *APTI 423: Air Pollution Dispersion Models*, Air Pollution Training Institute, United States Environmental Protection Agency, Research Triangle Park, NC.

6 Control of Particulate Emissions

6.1 INTRODUCTION

Particulate matter (PM) in air pollution is a mixture of tiny particles and liquid droplets suspended in the air. These particles can comprise various substances, including metals, organic compounds, soil, dust, and allergens. A noticeable effect of PM pollution is the reduction of visibility. Emissions from a stack or a tailpipe are visible mainly because of the PMs they carry. In addition, PMs may also have many toxic compounds with them. PM is a significant concern for public health and environmental protection due to its ability to penetrate deep into the respiratory system and cause adverse health effects.

6.1.1 Sources of Particulate Emissions

Particles are being emitted into the atmosphere from natural processes and human activities. Some particles are formed in the atmosphere. Solid particles can be generated from mechanical action (e.g., grinding), chemical reactions (e.g., metal oxides), and incomplete combustion (e.g., carbon black). Liquid particles can be generated from chemical reactions and the condensation of vapor.

Primary particles are those released directly from sources to the atmosphere. They include wind dust, ocean sprays, volcano ash, and carbon black from combustion sources. *Secondary particles* are formed in the atmosphere from chemical reactions involving primary gaseous pollutants. Examples of secondary particles are ammonium sulfate ($(NH_4)_2SO_4$) and ammonium nitrate (NH_4NO_3), formed from gaseous emissions from various combustion sources. Compared to primary particles, they are much smaller and represent a large portion of $PM_{2.5}$ in our ambient air.

Tables 6.1 and 6.2 show the PM_{10} and $PM_{2.5}$ emissions from the main source categories in the U.S. in 2024. The total emissions are 16.4 and 5.1 million tons for PM_{10} and $PM_{2.5}$, respectively. Contributions to PM_{10} emissions are 86.0%, 6.8%, 5.5%, and 1.8% from miscellaneous, industrial and other processes, stationary fuel combustion, and transportation, respectively. Contributions to $PM_{2.5}$ emissions are 67.4%, 16.5%, 12.9%, and 3.1% from miscellaneous, stationary fuel combustion, industrial and other processes, and transportation, respectively. Miscellaneous source category (e.g., wildfires, road dust, ocean sprays) accounts for most of the PM_{10} and $PM_{2.5}$ emissions (USEPA, 2025).

Figures 6.1 and 6.2 depict the trends of PM_{10} and $PM_{2.5}$ annual emissions in the U.S. since 1990. As shown, both emissions have been decreasing, except in recent years.

DOI: 10.1201/9781003532705-6

TABLE 6.1
PM_{10} Emission Rates and Source Categories in the U.S. in 2024

Source Category	(1,000 tons)	(%)
Stationary fuel combustion	**899**	**5.5**
Electric utility	94	0.6
Industrial	285	1.7
Others	520	3.2
Industrial and other processes	**1,118**	**6.8**
Chemical & allied product manufacturing	18	0.1
Metals processing	42	0.3
Petroleum & related industries	28	0.2
Other industrial processes	738	4.5
Solvent utilization	5	0.0
Storage & transport	34	0.2
Waste disposal & recycling	253	1.5
Transportation	**288**	**1.8**
Highway vehicles	189	1.2
Off-Highway	99	0.6
Miscellaneous	**14,114**	**86.0**
Wildfires	770	4.7
Others	13,344	81.3
Total	**16,419**	**100.0**

TABLE 6.2
$PM_{2.5}$ Emission Rates and Source Categories in the U.S. in 2024

Source Category	(1,000 tons)	(%)
Stationary fuel combustion	**838**	**16.5**
Electric utility	81	1.6
Industrial	243	4.8
Others	514	10.1
Industrial and other processes	**657**	**12.9**
Chemical & allied product manufacturing	13	0.3
Metals processing	30	0.6
Petroleum & related industries	23	0.5
Other industrial processes	347	6.8
Solvent utilization	4	0.1
Storage & transport	13	0.3
Waste disposal & recycling	227	4.5
Transportation	**160**	**3.1**
Highway vehicles	67	1.3
Off-Highway	93	1.8
Miscellaneous	**3,433**	**67.4**
Wildfires	504	9.9
Others	2.929	57.5
Total	**5,088**	**100.0**

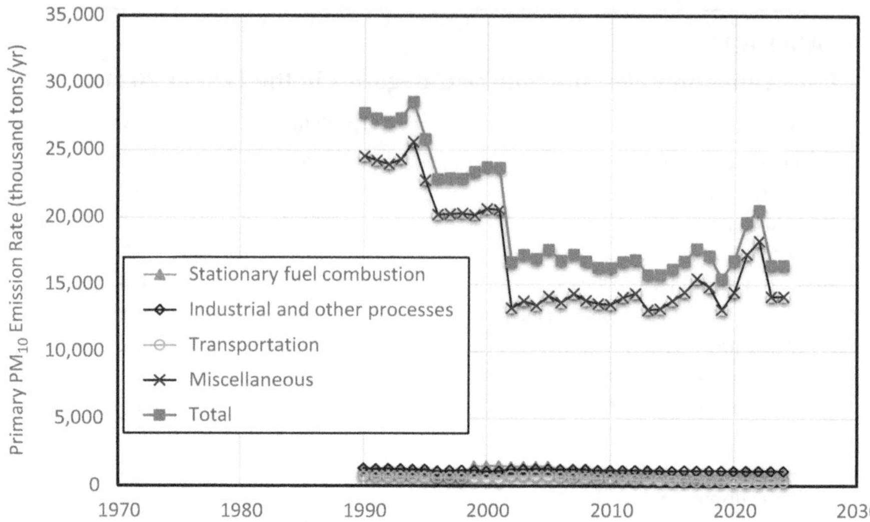

FIGURE 6.1 Annual trends of PM_{10} emissions from major source categories in the U.S. (1990–2024).

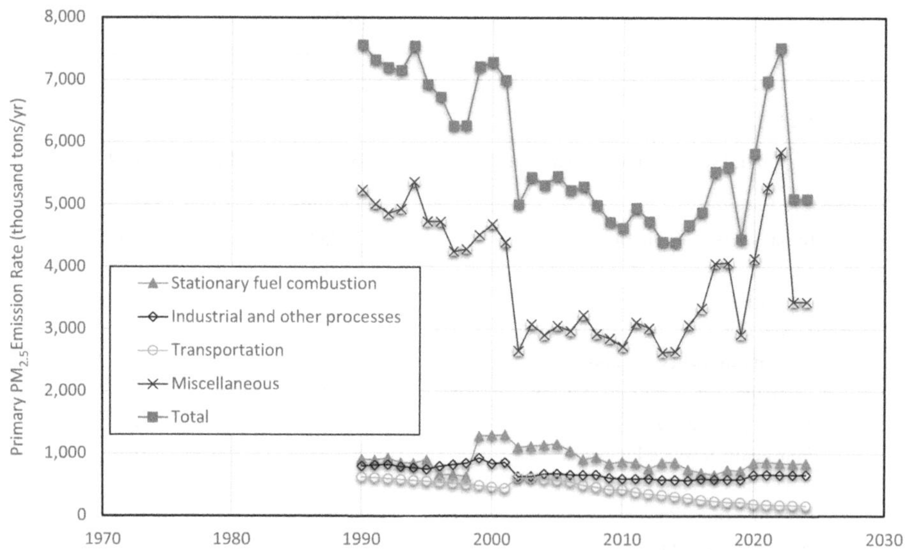

FIGURE 6.2 Annual trends of $PM_{2.5}$ emissions from major source categories in the U.S. (1990–2024).

6.1.2 Particulate Removal and Commonly Used Removal Devices

There is little we can do about reducing PM emissions from natural processes. Although we can employ devices to remove particulates in the ambient air, we do not do much, if any, to remove particulates already in the ambient air. Rainfall/snowfall is the primary natural mechanism that removes particulates in the ambient air. To reduce the PM concentrations in the ambient air, we need to minimize/remove the generation of particles from human activities and add control/removal devices to reduce their emissions. The discussion in this chapter focuses on the emission control of primary particles from stationary sources.

In general, there are five types of particle removal devices commonly used for source control of PM emissions, and they are:

i. Gravity settling chambers
ii. Cyclones
iii. Electrostatic precipitators (ESPs)
iv. Fabric filters
v. Wet scrubbers

Each type of device has its advantages and limitations. A description of the treatment process, removal mechanisms, and system components, as well as design considerations and calculations for each type of these devices, will be presented later in this chapter.

6.1.3 Characteristics of Particulates and Gas Stream Relevant to the Removal

To select a proper type of PM removal device, one must know the characteristics of particulates and the gas stream carrying them. Control regulations usually specify sizes (e.g., PM_{10} or $PM_{2.5}$) and mass concentrations of particles (e.g., mg/m^3 or µg/m^3) to be removed. Particle removal devices have their optimal (and limited) operational range relative to particle size. For example, one should not expect gravity settling chambers to remove $PM_{2.5}$, not even PM_{10}, effectively. The gravity settling chambers' gravitational force is the main driving force for particulate removal. An extremely long detention time would be needed to collect fine particles, requiring an impractical gravity settling chamber.

The main PM characteristics of concern in selecting the type of removal devices include particle size, particle size distribution (PSD), and other properties (e.g., shape, density, stickiness, oiliness, and acidity). For example, sticky particles may not be suitable for fabric filters because they may clog them.

A gas stream carrying particulates of concern must move through the PM removal device. The flow rate into the control device and its resultant velocity significantly affect the sizes of particles that can be removed at a desirable/design efficiency. The gas stream flow rate is positively related to the device's size and head loss (pressure drop) across the device. A higher rate would require a larger device and a larger blower, which would incur a higher energy cost.

The temperature of the influent gas stream is also an important parameter. For example, fabric materials used in some filters may not withstand elevated temperatures. High humidity in the influent gas stream may challenge some control devices (e.g., filters), but it is acceptable for wet scrubbers. In addition, some trace constituents may have significant adverse impacts. For example, the presence of hydrogen sulfide (H_2S) would lower the dew point of the gas stream, and acidic condensate may form when the gas stream is cooled in gravity chambers [Note: the temperature of a gas stream will drop upon expansion – the *Joule–Thomson effect*].

6.2 PARTICLE SIZE, SIZE DISTRIBUTION, AND REMOVAL MECHANISMS

6.2.1 Particle Size

Particles in the air can be of different shapes, such as perfect solid spheres, hollow spheres, irregular-shaped solids, flakes, fibers, condensation flocs, and aggregates (Figure 6.3).

For practical applications, we often assume that particles are spherical. An *equivalent particle diameter* ($D_{\text{p, equivalent}}$) is frequently used, considering that a spherical particle of this diameter would behave like the particle in question. For example, if the particle volume is the one of concern, its $D_{\text{p, equivalent}}$ can be found by using its volume (V_{particle}) as follows:

$$D_{p,\text{equivalent}} = \sqrt[3]{(6/\pi) \times V_{\text{particle}}} \tag{6.1}$$

Eq. (6.1) is based on the fact that the volume of a perfect sphere is equal to $(\pi/6)D_p^3$.

For air pollution control, the ranges of particle sizes we need to deal with are broad, typically in the range of hundredths to a few hundred micrometers. A micrometer (*μm* or *micron (μ)*) is one-millionth of a meter (i.e., 10^{-6}m) and is a standard unit for diameters of small particles. For example, PM_{10} represents particles with

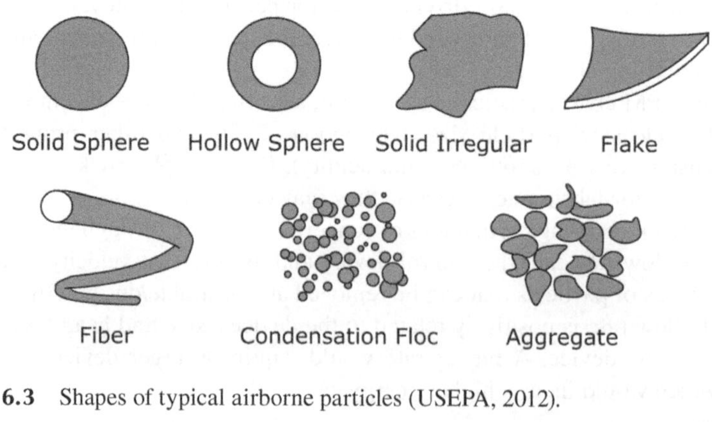

FIGURE 6.3 Shapes of typical airborne particles (USEPA, 2012).

aerodynamic diameters of ≤10 μ (aerodynamic diameter will be defined later) and is considered "respirable". Figure 6.4 shows the sizes of typical airborne particles compared to some common materials. Figure 6.5 illustrates the size ranges of particles generated by several particle formation mechanisms. Physical attraction and ash burnout are mechanisms for forming primary particles, while heterogeneous nucleation and homogeneous condensation are mechanisms for forming secondary particles. *Physical attraction* is when smaller particles are attracted to form larger particles without chemical bonding. *Ash burnout* is the oxidation and removal of carbon from ash during or after combustion. *Droplet evaporation* refers to the process where the liquid phase of a droplet (e.g., solvent or water) evaporates, leaving behind dissolved or suspended solutes that form solid particles. *Heterogeneous nucleation* is when a new particle begins forming on a surface or interface different from the original material (e.g., water condenses on dust to form cloud droplets). This contrasts with

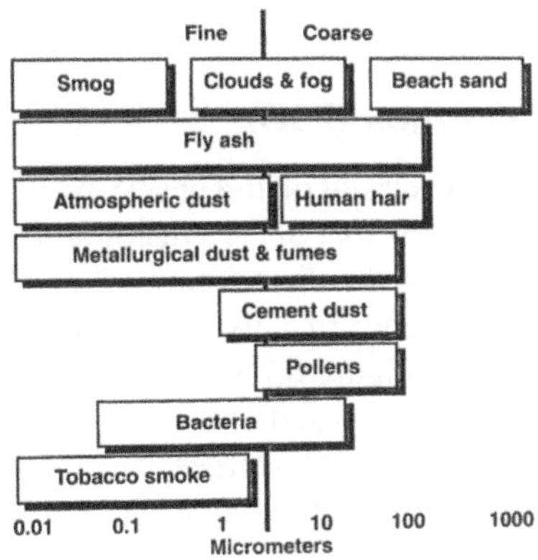

FIGURE 6.4 Sizes of airborne particles (USEPA, 2013).

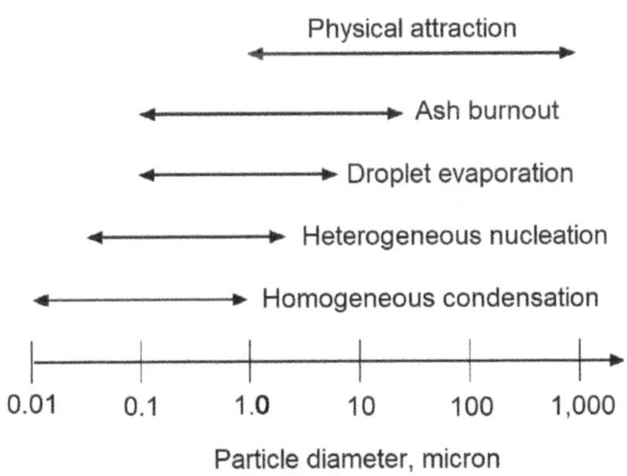

FIGURE 6.5 Size ranges of particles from several formation mechanisms (Modified from USEPA, 2012).

homogeneous nucleation, which occurs uniformly throughout the parent phase without any preferential surface.

Since a sphere's surface area and volume are πD_p^2 and $\pi D_p^3/6$, the surface and volume of a 10-μ particle are 100 and 1,000 times those of a 1-μ particle, respectively. The size of a particle will affect its behavior in both ambient air and removal devices. Particles with diameters between 1 and 10 μm are fundamental in air pollution control. Very fine particles (0.01–0.1 μm) can be generated in processes such as combustion. However, they often rapidly agglomerate into larger particles (>0.1 μm). Particles in the range of 0.1–1.0 μm can be a significant portion of particulate emissions from many sources and are relatively complex to collect (USEPA, 2012). Sometimes, the particles are classified into super-coarse, coarse, fine, and superfine categories for aerodynamic diameter >10, between >2.5 and ≤10, between >0.1 and ≤2.5, and ≤0.1μ, respectively.

Several particle size measurement devices include microscopes, optical counters, electrical aerosol analyzers, Bahco counters, and inertial impactors (Figure 6.6). Detailed descriptions of these instruments can be found in USEPA (2012).

6.2.2 Particle Size Distribution (PSD)

A gas stream usually carries particles of different sizes with different concentrations; one of the simplest ways to describe its PSD is a histogram (Figure 6.7). A *histogram* is a type of bar chart representing the distribution of data, commonly used in data analysis. The frequency of occurrence in a histogram can be the number of particles, surface area, volume, or mass. In Figure 6.7, the *mode* is the particle size that occurs most frequently. The *median* particle size divides the frequency distribution into two halves (50% of the particles with diameters larger than and the other 50% with diameters smaller than this diameter), corresponding to a cumulative fraction of 50%. The

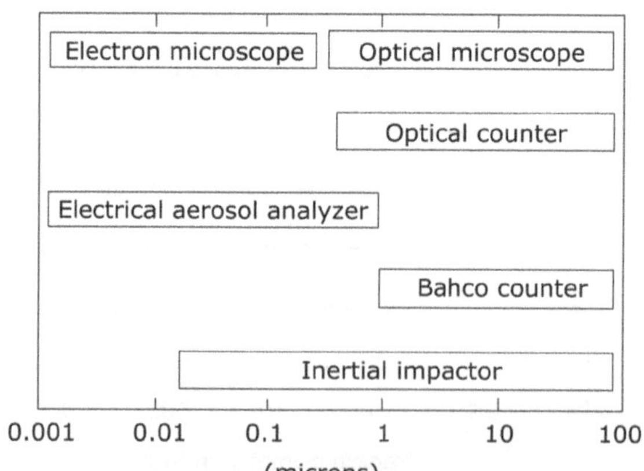

FIGURE 6.6 Ranges of particle size measuring devices (Modified from USEPA, 2012).

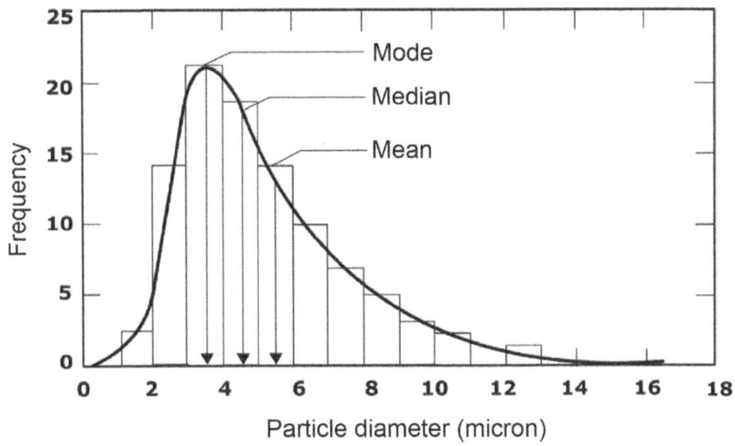

FIGURE 6.7 Particle size distribution (Modified from USEPA, 2012).

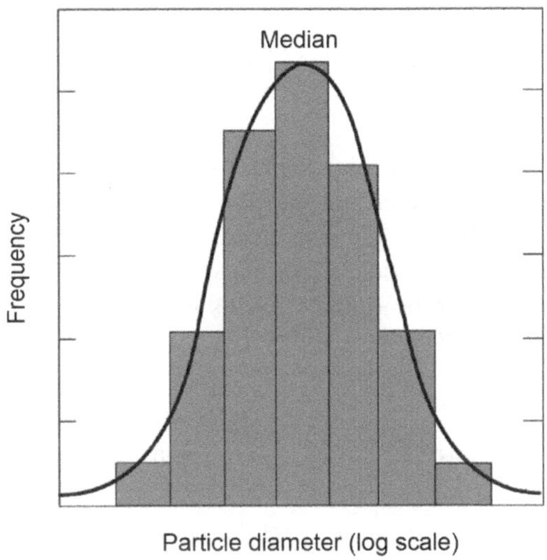

FIGURE 6.8 Log-normal particle size distribution (Modified from USEPA, 2012).

mean diameter is the arithmetic average of the diameters of all particles. The mean value (i.e., the arithmetic average) is more sensitive than the median to the quantities of the particles at the significantly lower and upper ends of the distribution. They are often considered "*outliers*", data points outside a dataset's overall distribution pattern.

A simplified saying is that if the curve shown in Figure 6.7 is a symmetrical bell curve, then the PSD has a *standard (or Gaussian) distribution*. The observed PSD of exhaust gas streams is *log-normally distributed* for many stationary and mobile sources. It means that when the frequency of occurrence is plotted against the logarithm of the particle diameter, a symmetrical bell curve will be generated (see Figure 6.8). The frequency on the vertical scale can be the percentage of mass divided by the difference in the logarithms of the particle sizes defining a specific interval, which is $[\log(D_{p,\max}) - \log(D_{p,\min})]$, or $\Delta\log D_p$. A plot of log-normally distributed PSD on a log-probability paper will be a straight line (Similarly, a plot of normally distributed PSD on a regular probability paper will also be a straight line). The data can be plotted based on the cumulative percentage of particles smaller or larger than a specific size. In this book, the case of "the percentage smaller" will be used.

Gaussian Particle Size Distribution

$$f(x) = \frac{1}{\sigma\sqrt{2\pi}} \exp\left[-\frac{1}{2}\left(\frac{D_p - D_{p,\text{mean}}}{\sigma}\right)^2\right] \tag{6.2}$$

Log-Normal Gaussian Particle Size Distribution

$$f(x) = \frac{1}{\sigma\sqrt{2\pi}} \exp\left[-\frac{1}{2}\left(\frac{\ln(D_p) - \ln(D_{p,\text{mean}})}{\sigma}\right)^2\right] \tag{6.3}$$

A straight line on a log-probability paper for a log-normally distributed particle mass can tell the mean diameter and the standard deviation of this set of particles. The mean diameter is the particle size corresponding to the 50% probability. For a log-normally distributed particle mass, its *standard deviation* (σ) and *geometric standard deviation* (σ_g) can be found from its geometric mean diameter (D_{50}) and the particle size at the 15.87% probability ($D_{15.87}$) or 84.13% probability ($D_{84.13}$), as follows:

$$\sigma = \ln(\sigma_g) = \ln\left(D_{84.13}/D_{50}\right) = \ln\left(D_{50}/D_{15.87}\right) \tag{6.4}$$

$$\sigma_g = D_{84.13}/D_{50} = D_{50}/D_{15.} \tag{6.5}$$

The standard deviation (σ) should be smaller than the geometric standard deviation (σ_g), and σ_g should be >1. For the case of $\sigma = 0$ (or $\sigma_g = 1$), all the particles in the entire particle set are the same size.

Example 6.1: A Log-Normally Distributed Particle Mass

A log-normally distributed particle mass has a geometric mean diameter of 15 μ and a geometric standard deviation of 5 μ. Determine the particle diameter so that 15.87% of the entire particle mass has diameters smaller than it.

Solution:

$$\ln(\sigma_g) = \ln(5) = \ln(15/D_{15.87})$$
$$D_{15.87} = \underline{3}$$

Discussion:

1. This particle set's standard deviation $(\sigma) = \ln(5) = 1.61\ \mu$.
2. Using a similar approach, one should find that 84.13% of the entire particle mass has diameters $\leq 75\ \mu$.

Example 6.2: Analysis of Particle Size Distribution

Determine if the sizes of particles in the table below are log-normally distributed. If so, find its geometric mean diameter and the geometric standard deviation.

Particle Size (μ)	Mass (mg)
<0.5	0.10
0.5–1.0	0.15
1.0–2.0	0.20
2.0–4.0	0.20
4.0–6.0	0.15
6.0–8.0	0.10
>8.0	0.10

Solution*:*

a. Calculate the fraction and the cumulative fraction of each size range as:

Particle Size (μ)	Mass (mg)	Fraction of Mass in this Range	Cumulative Fraction
<0.5	0.10	0.10	0.10
0.5–1.0	0.15	0.15	0.25
1.0–2.0	0.20	0.20	0.45
2.0–4.0	0.20	0.20	0.65
4.0–6.0	0.15	0.15	0.80
6.0–8.0	0.10	0.10	0.90
>8.0	0.10	0.10	1.00
Total	1.00		

b. Plot $D_{p,\,max}$ of each size range vs. its corresponding cumulative fraction smaller than it on a log-probability paper:

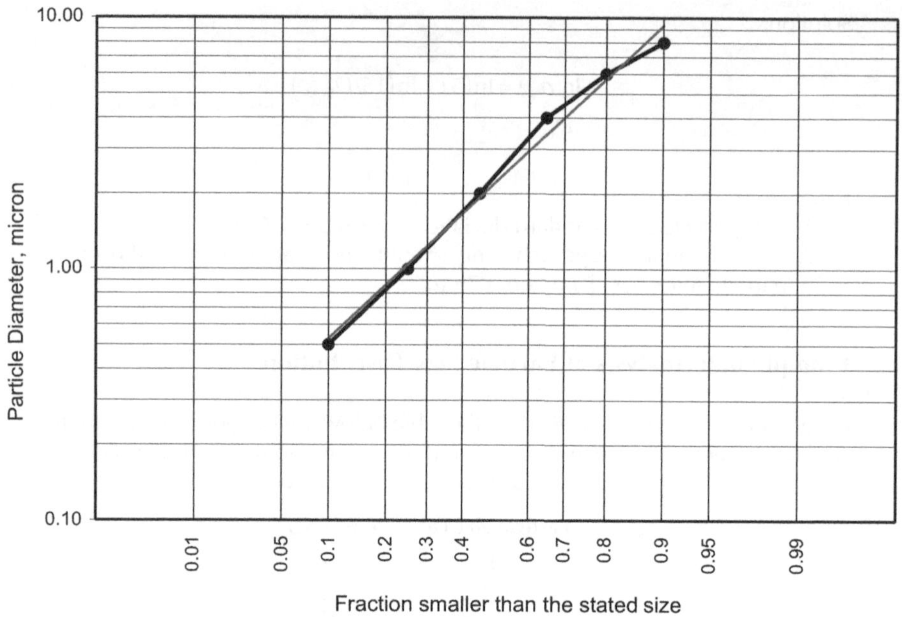

c. A straight line fits the data relatively well; thus, the sizes of this set of particles appear to be log-normally distributed.
d. From the plot, $D_{50}=2.3$ μ and $D_{84.13}=7.0$ μ, then $\sigma_g=(7.0/2.3)=\underline{3.05\ \mu}$.

Figure 6.9 illustrates the results of a study on PM concentrations in ambient air, and the measurements were made at a location with traffic. The x-axis is the particle diameter (in μ) on a logarithmic scale, while the y-axis is $\Delta V/\Delta \log D_p$, which is the volume/mass of the particles per unit volume of air with diameters in the size range from $\log D_p$ to $\log(D_p+\Delta D_p)$ in a relative scale. The dotted vertical line at $D_p=1$ μ (arbitrarily chosen) divides the particles into coarse and fine particles based on the particle size. The particles were also grouped into three modes: nuclei, accumulation, and coarse, based on their formation mechanisms. Particles of all three modes are geometrically distributed. The coarse-mode particles were mechanically generated (e.g., from wind erosion of crustal materials). For this mode's particles, the diameter of geometric mean by volume (DGV) is 4.9 μ, and the geometric standard deviation (σ_g) is 1.87 μ. Particles smaller than 1 μ were generated in combustion or formed from gas vapors through accumulation or condensation. The portion of the fine particles with ~0.1 μ$<D_p<$~1 μ was formed from accumulation (DGV=0.21 μ and $\sigma_g=1.8$ μ). The DGV and σ_g of the nuclei-mode particles are 0.018 μ and 1.6 μ, respectively. Nuclei-mode particles might grow from coagulation/flocculation into large sizes to be in the accumulation mode (USEPA, 2012).

The densities of a specific type of particle are often assumed to be the same. If this is the case, the volume fraction for a size range should be the same as the corresponding weight/mass fraction. We are more interested in removing the particulate

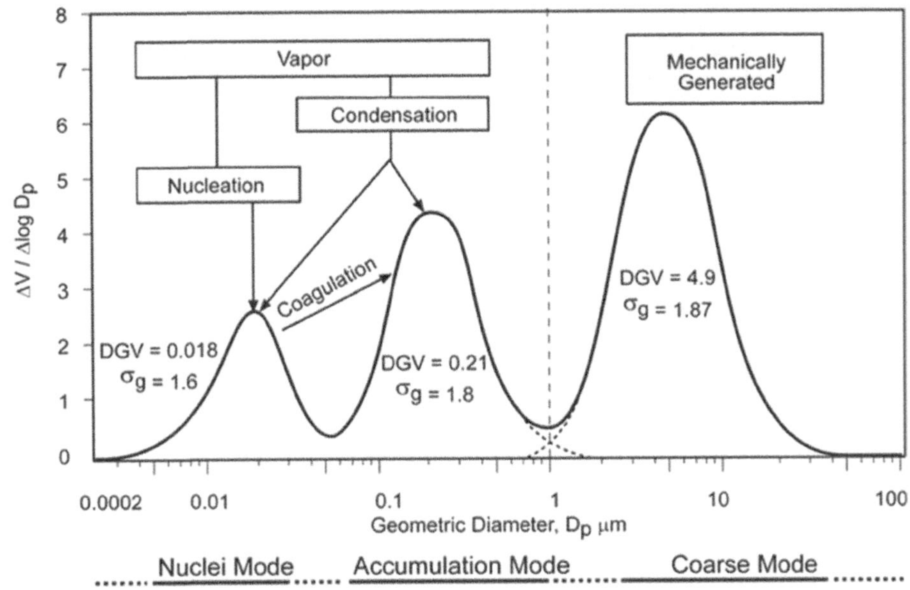

FIGURE 6.9 Size distribution and formation mode of PM in ambient air (Modified from USEPA, 2012).

mass than the number of particles for particulate control. However, we need to know the sizes of particles that need to be removed. Most, if not all, of the control devices are more effective in removing larger particles. Effectiveness in removing particles of specific size ranges is one of the primary considerations when selecting control devices for a pollution control project. Hereafter, discussions will focus more on eliminating mass than the number of particles.

6.2.3 Reynolds Number for Particulate Travel

Chapter 2 discussed the Reynolds number (*Re*) for a flowing fluid, sometimes termed the (*flow*) *Reynolds Number* (Re_f). To characterize conditions for particles flowing through a fluid or with a moving fluid, the (*particle*) *Reynolds Number* (Re_p) is defined as:

$$Re_p = \frac{D_p \times u_p \times}{\mu_f} \tag{6.6}$$

where D_p is the particle diameter, u_p is the particle velocity relative to the fluid, ρ_f is the density of the fluid, and μ_f is the dynamic viscosity of the fluid. It should be noted that the characteristic length here is the particle diameter, whereas the velocity is the particle migration velocity for the particle *Re*. In contrast, the characteristic length is the duct diameter (for a circular pipe), and the velocity is the fluid flow velocity for the flow Reynolds number (see Eq. 2.12).

The values of particle *Re* are usually much smaller than flow *Re*. Re_p values <1 indicate laminar conditions, which define the *Stokes flow region* (or the *laminar flow region*), in which viscous forces dominate, while Re_p values >1,000 indicate turbulent conditions and are termed the *tumultuous/turbulent region*. For $1 < Re_p < 1{,}000$, it is termed the *transition region*. Most particle movements in air pollution control devices occur in the Stokes/laminar flow or transition regions; particles <100 μ are typically in the Stokes/laminar flow range (USEPA, 2012).

Example 6.3: Particle Reynolds Number

A particle ($\rho_p = 2{,}000\,\text{kg/m}^3$) is moving at a speed of 0.6 m/min (0.01 m/s) relative to the gas stream ($T = 20°\text{C}$ (68°F), $P = 1$ atm). Determine its particle Reynolds number for each of the following cases:

a. $D_p = 10$ μ.
b. $D_p = 100$ μ.
c. $D_p = 10$ μ, but $\rho_p = 4{,}000\,\text{kg/m}^3$.
d. $D_p = 10$ μ, but $T = 200°\text{C}$ and $P = 1$ atm.

Solution:

a. Assuming the gas stream has the same density and viscosity as air, then

$$Re_p = \frac{\left(10 \times 10^{-6}\,\text{m}\right)\left(0.01\dfrac{\text{m}}{\text{s}}\right)\left(1.21\dfrac{\text{kg}}{\text{m}^3}\right)}{\left(1.81 \times 10^{-5}\dfrac{\text{kg}}{\text{m.s}}\right)} = 6.7 \times 10$$

b. Re_p is proportional to the particle diameter, so Re_p should be 10 times larger for $D_p = 100$ μ, 6.7×10^{-2}.
c. Re_p is independent of the density of the particle, so $Re_p = 6.7 \times 10^{-3}$.
d. Re_p is proportional to the density of the fluid but inversely proportional to the fluid's viscosity. Using the values of air density and viscosity from Examples 2.2 and 2.3,

$$\frac{Re_{p,2}}{Re_{p,1}} = \left(\frac{Re_{p,473\text{K}}}{6.7 \times 10^{-3}}\right) = \left(\frac{\rho_{f,2}}{\rho_{f,1}}\right)\left(\frac{\mu_{f,1}}{\mu_{f,2}}\right) = \left(\frac{0.75}{1.21}\right)\left(\frac{1.81 \times 10^{-5}}{2.61 \times 10^{-5}}\right)$$

$$Re_{p,473\text{K}} = \underline{2.88 \times 10^{-3}}$$

Discussion: The Re_p value decreases as the temperature increases because of the decrease in the air density and the increase in the air viscosity.

6.2.4 Particle Collection/Removal Mechanisms

Five types of particulate removal devices are described in this book: (i) gravity settling chambers, (ii) cyclones, (iii) fabric filters, (iv) ESPs, and (v) wet scrubbers. For each type, particles are collected/removed by one or more mechanisms. The removal mechanisms of one kind of device may be similar to or different from those of the others. In addition, one mechanism may play a more significant role than the others for a given type of device. This section discusses several important removal mechanisms individually. The particulate removal mechanisms involved in each type of device will be mentioned later. The primary particulate removal mechanisms include gravity settling, centrifugation, Brownian motion/diffusion, electrostatic attraction, and inertial impaction.

Gravity settling. For a discrete particle to settle down in a fluid, it would be subjected to three forces: gravitational force (F_G), buoyancy force (F_B), and drag force (F_D).

For a spherical particle, F_G can be expressed as:

$$F_G = m_p g = \left(\rho_p \times V_p\right) g = \left(\rho_p \times \frac{\pi D_p^3}{6}\right) g = \frac{\pi D_p^3 g}{6} \times \rho_p \tag{6.7}$$

where m_p, ρ_p, and V_p are the particle's mass, density, and volume, respectively, and g is the gravitational constant.

Against the particle's downward movement in the air, the buoyancy force would be

$$F_B = \frac{\pi D_p^3 g}{6} \times \rho_a \tag{6.8}$$

where ρ_a is the density of air.

The *drag force* (F_D) is the resistive force due to shearing stress as an object moves through a fluid. In general, it can be expressed as

$$F_D = C_D \times \rho_f \times \left(\frac{u_p^2}{2}\right) \times A \tag{6.9}$$

where C_D is the *drag coefficient*, ρ_f is the density of the fluid, and A is the characteristic frontal area (i.e., the projected area of an object perpendicular to the direction of flow) of the particle (= $\pi D^2/4$ for a sphere). Equations have been derived to relate Re_p with C_D. Most of them in the literature are the same for the laminar region and similar for the turbulent region, while they appear to be different for the transition region. As shown in USEPA (2012),

$$C_D = 24/Re_p \quad \text{for } Re_p < 1 \tag{6.10}$$

$$C_D = 18.5/Re_p^{0.6} \quad \text{for } 1 < Re_p < 1{,}000 \tag{6.11}$$

$$C_D = 0.44 \quad \text{for } Re_p > 1{,}000 \tag{6.12}$$

Under the laminar flow region ($Re_p < 1$), the drag force (F_D) can be expressed as (derived by Sir George Stokes in 1851):

$$F_D = 3\pi\mu D_p u_p \tag{6.13}$$

where u_p is the particle migration velocity.

When the system is in equilibrium, the sum of all three external forces (i.e., F_G, F_B, and F_D) equals zero. Thus,

$$\Sigma F = F_B + F_G + F_D = \frac{\pi D_p^3 g}{6} \times \rho_a - \frac{\pi D_p^3 g}{6} \times \rho_p + 3\pi\mu D_p u_p = 0 \tag{6.14}$$

The "minus" sign in Eq. 6.14 accounts for the gravitational force being in the opposite direction of the buoyancy and drag forces.

In equilibrium, the particle will stand still or settle down at a constant velocity, which is often termed the *terminal velocity* (u_t):

$$u_t = \frac{g \times (\rho_p - \rho_a) \times D_p^2}{18\mu} \approx \frac{g \times \rho_p \times D_p^2}{18\mu} \tag{6.15}$$

The equation is often called the *Stokes Law* or the *Stokes equation* [Note: $\rho_p >> \rho_a$].

If a particle has $D_p > 3$ µ, the gas/air would appear "continuous" around the particle. This particle will not be affected by collisions with individual gas molecules because this type of collision occurs frequently on all sides of this particle. On the other hand, if a particle has $D_p < 3$ µ, the fluid will appear to be "discontinuous", and this often occurs for small particles in the laminar flow region ($Re_p < 1$). In this case, the collisions with gas molecules would cause the particle to move in a direction related to the resultant force acting on the particle. The particle will "*slip*" between the gas molecules. For gravitational settling, the particle will move faster than what Eq. 6.15 would predict. To correct for this, the *Cunningham slip correction factor* (C_c) is often used to reduce the drag coefficient in the laminar flow region, as follows:

$$C_D = \left[24/Re_p\right]/C_c \tag{6.16}$$

The correction factor will also reduce the drag force in Eq. 6.13, as follows:

$$F_D = \frac{3\pi\mu D_p u_p}{C_c} \tag{6.17}$$

Consequently, the terminal velocity in Eq. 6.15 can be modified to:

$$u_t = C_c\left[\frac{g\times(\rho_p-\rho_a)\times D_p^{\,2}}{18\mu}\right]\approx C_c\left[\frac{g\times\rho_p\times D_p^2}{18\mu}\right] \tag{6.18}$$

It should be noted that the Stokes Law (Eq. 6.15) will underestimate the terminal velocities of sub-micron particles (as shown in Eq. 6.18 with $C_c > 1$), but overestimate them for particles having diameters of tens of microns or larger.

An empirical equation can be used to estimate the values of the Cunningham slip correction factor (USEPA, 2012):

$$C_c = 1+\frac{(6.21\times10^{-4})\times T}{D_p} \tag{6.19}$$

where T is the absolute temperature (in K) and D_p is the particle diameter (in μ).

As shown in Eq. 6.19, the values of C_c are always larger than unity. They increase with T but decrease with D_p. At $T=293$ K (20°C), the values of C_c are 2.8, 1.18, and 1.06 for $D_p=0.1$, 1, and 3 μ, respectively, using Eq. 6.19. The Cunningham slip correction factor becomes increasingly significant when the particle diameter becomes <3 μ.

Example 6.4: Terminal Settling Velocity of a Particle

A spherical particle ($D_p=2$ μ, $\rho_p=2{,}000$ kg/m³) is settling in still air ($T=20$°C (68°F), $P=1$ atm) by gravity.

a. Estimate the value of the Cunningham slip correction factor.
b. Estimate this particle's terminal settling velocity (u_t).
c. What would be the u_t value if $D_p=100$ μ?
d. What would be the u_t value if $D_p=2$ μ and 4,000 kg/m³?

Solution:

a. From Eq. 6.19,

$$C_c = 1+\frac{(6.21\times10^{-4})\times T}{D_p} = 1+\frac{(6.21\times10^{-4})(293)}{2} = 1.09$$

b. From Eq. 6.15,

$$u_t = (1.09)\left[\frac{(9.81)(2{,}000-1.21)(2\times10^{-6})^2}{(18)(1.81\times10^{-5})}\right] = 2.62\times10^{-4}\ \text{m/s}$$

c. For a 100-μ particle, its u_t value would be approximately 0.66 m/s, which is 2,500 times that of a 2-μ particle, because u_t is proportional to D_p^2.
d. The u_t value will be approximately doubled (i.e., 5.24×10^{-4} m/s), because it is proportional to $(\rho_p - \rho_a)$ or proportional to ρ_p since $\rho_a << \rho_p$.

Discussion:

1. Eq. 6.19 is an empirical equation; the parameters' values must be in specified units.
2. Settling velocities of particles of micron sizes due to gravitation are small, as shown in part (a), but they become much more significant for a 100-μ particle.

To evaluate the motion of a particle in air or a control device, the *aerodynamic diameter* (D_a) is often used as:

$$D_a = D_p \sqrt{\rho_p C_c} \tag{6.20}$$

where D_p is the particle diameter in μ and ρ_p is the particle density in g/cm³ (or in 1,000 kg/m³) for typical uses. With these, the aerodynamic diameter will have a unit of μ_a, which stands for "*micron, aerodynamic*". Particles of the same aerodynamic diameter will behave similarly in several types of particulate removal devices. The aerodynamic diameter of a non-spherical particle is defined as the diameter of a spherical particle with a density of 1 g/cm³ (1,000 kg/m³) and the same settling velocity as the non-spherical particle. It should be noted that the diameters used for definitions of $PM_{2.5}$ and PM_{10} are aerodynamic, not physical ones.

Example 6.5: Aerodynamic Diameter

Determine the aerodynamic diameter of a particle (a) $D_p = 20$ μ and $\rho_P = 1{,}000$ kg/m³ and (b) $D_p = 10$ μ and $\rho_P = 4{,}000$ kg/m³, assuming $C_c = 1$.

Solution:

a. From Eq. 6.20,

$$D_a = D_p \sqrt{\rho_p C_c} = (20)\sqrt{(1)(1)} = 20\ \mu_a$$

b. Also, from Eq. 6.20,

$$D_a = (10)\sqrt{(1)(4)} = 20\ \mu_a$$

Discussion:

1. Density values used in Eq. 6.20 should be in g/cm³ (or in 1,000 kg/m³).
2. These two particles have the same aerodynamic diameter, even though one physical diameter is twice as large.

Centrifugation. A particle within a gas stream traveling in a circular motion is experiencing a centrifugal force (F_c), which can be expressed as:

$$F_c = m\left(\frac{v^2}{r}\right) \tag{6.21}$$

where v is the tangential velocity of the gas stream and r is the radial position of the particle. With $r = 0.5$ m and $v = 10$ m/s, the acceleration (v^2/r) would be 200 m²/s, which is 20.4 (= 200/9.81) times the gravitational acceleration.

For particles in the laminar flow region and assuming a particle is moving at the same speed as the gas stream (v_{gas}) in a circular motion, the tangential velocity of this particle (u_p) can be determined by:

$$u_p = C_c\left[\frac{\rho_p D_p^{\ 2}}{18\mu}\right]\left(\frac{v_{gas}^{\ 2}}{r}\right) \tag{6.22}$$

The expressions of particle velocity in Eqs. 6.18 (for gravity settling) and 6.22 (for centrifugation) are essentially the same, except that the acceleration terms are different (i.e., g vs. v_{gas}^2/r).

Example 6.6: Tangential Velocity of a Particle in a Cyclone

Referring to the particle in Example 6.4, if it enters a cyclone (diameter = 1 m) with a gas stream at a velocity of 10 m/s. Determine the tangential velocity of this particle.
Solution:

Using Eq. 6.22 and radius (r) = $D/2$ = 0.5 m,

$$u_p = (1.09)\left[\frac{(2{,}000)(2\times10^{-6})^2}{(18)(1.81\times10^{-5})}\right]\left(\frac{10^2}{0.5}\right) = 5.35\times10^{-3}\ \text{m/s}$$

Discussion: The particle's tangential velocity caused by the centrifugal force is 5.35×10^{-3} m/s, which is 20.4 times the terminal velocity due to the gravitational force (2.62×10^{-4} m/s).

Electrostatic attraction. Electrostatic attraction is the primary mechanism responsible for particulate removal by ESPs. An ESP can be envisioned as a large number of parallel plates with discharge electrodes in the center plane between two plates. The plates are electrically grounded so that electrical fields are established between a discharge electrode and each pair of plates. The gas stream moves between the plates, and the particles in the gas stream acquire charges from the electrons discharged from the electrodes or the gas ions formed from collisions with the electrons. The particles carrying the charges will then move toward the plates. Once they hit the plates, they are considered collected.

For discussion, let us envision a space between these two plates. The *secondary voltage* in an ESP refers to the high voltage applied across the discharge electrode and the collection plates, usually 20–100 kilovolts (kV). This voltage is crucial for generating the *corona discharge*, which ionizes the gas and charges the particles. A robust electrical field (typically with strength exceeding a critical value in air) is established between the discharge electrode and the plates, which are electrically grounded. With the corona discharge, released electrons are accelerated by the strong electric field and move toward the grounded collection plates. On their way, some of these electrons are captured by gas molecules, and these gas molecules become negatively or positively charged; they are often called *gas ions*. On their way to the collection plates, these gas ions may be intercepted/captured by larger particles because these particles locally disrupt the electrical field. The negatively charged particles will then start moving toward the collection plates. They may quickly reach the *saturation charge* (*or* the *equilibrium charge*), which refers to the maximum charge that can be imparted to particles in the gas stream. At this time, the charge becomes strong enough to deflect additional gas ions. The magnitude of the saturation charge is a function of the particle surface area, electrical field strength, residence time, and gas composition. Large particles would accumulate higher electrical charges because of their larger surface areas, and consequently, they will be more strongly affected by the applied electrical field. This is *field charging* (or *contact charging*), and it is the dominant charging mechanism for particles with $D_p > 2\ \mu$ (USEPA, 2003, 2012).

Field charging becomes less critical as particle size decreases because it needs more mass/size to disrupt the electrical field. For particles with $D_p < 0.4\ \mu$, the diffusional charging becomes more critical. *Diffusional charging* results from collisions caused by the random Brownian motion of gas ions and particles, not due to the electrical field (USEPA, 2003, 2012).

The strength of an electrical field (E) can be defined as the applied voltage (V) over the distance (x), as follows:

$$E = \frac{\partial V}{\partial x} \tag{6.23}$$

A typical unit for E is V/m (or kV/cm).

The saturation charge (q_{sat}) of a spherical particle could be expressed as:

$$q_{\text{sat}} = 3\pi\left[\frac{\kappa}{\kappa + 2}\right]\varepsilon_0 D_p^2 E_0 \tag{6.24}$$

where κ is the relative permittivity of the particle, ε_0 is the permittivity of air, and E_0 is the strength of the electric field.

In electromagnetism, *permittivity* measures a material's ability to transmit an electrical field in a particular medium. Opposite to the literal meaning of "permittivity", a charge will yield more electric flux in a medium with a lower permittivity. The lowest permittivity is that of a vacuum (free space), which is 8.854×10^{-12} Coulomb/V·m (or Farad/meter; F/m) [Note: the Farad is the SI unit of capacitance.

A capacitance of 1 Farad means the capacity can store one Coulomb of charge when a potential difference of one Volt is applied; i.e., 1 F = 1 C/V]. *Relative permittivity* (or *dielectric constant*) is usually used for a medium other than the vacuum, which is the ratio of the permittivity of the medium (ε) to that of the vacuum (ε_0). It measures a material's ability to store electrical energy in an electric field relative to the storage ability in a vacuum. The dielectric constant of air is 1.0006, which is very close to that of the vacuum, which is 1.0 [Note: the dielectric constant of water @ 20°C (68°F) = 80.4; This high value is due to water's polar nature and its ability to form hydrogen bonds, which enhance its capacity to store electrical energy]. Typical dielectric constant values for solid particles are from 2 to 12; for example, PTFE (Teflon) = ~2.1 (because of its excellent insulating properties), quartz = ~4, alumina (Al_2O_3) = ~9.5, and silica = ~12.

Coulomb is the SI unit for electric charge, and one *Coulomb* (*C*) equals the quantity of electricity conveyed by a current of one ampere in 1 second (= 6.24×10^{18} electrons). Consequently, the charge of one electron is 1.602×10^{-19}C.

Example 6.7: Saturation Charge of a Particle in an Electrical Field

The particle in Example 6.4 has reached its saturation charge in an ESP where the field strength is 320 kV/m. Assuming the dielectric constant of this 2-μ particle is 5, how many electronic charges are on the surface of this particle at saturation?

Solution:

Using Eq. 6.24,

$$q_{\text{sat}} = 3\pi\left[\frac{5}{5+2}\right]\left(8.854\times10^{-12}\frac{\text{C}}{\text{V}\cdot\text{m}}\right)\left(2\times10^{-6}\text{m}\right)^2\left(320{,}000\frac{\text{V}}{\text{m}}\right)$$

$$= \left(7.63\times10^{-17}\text{C}\right) = \left(7.63\times10^{-17}\text{C}\right)\times\left(\frac{6.24\times10^{18}}{\text{C}}\right) = 476\,\text{electrons}$$

Discussion: Since the saturation charge is proportional to D_p^2, the number of electrons on a 1-μ particle will be 119 (i.e., one-quarter of 476).

The electric force on a particle (F_E) is the product of its surface charge and the local electric field strength (E_p), as follows:

$$F_E = q\times E_p = \left\{3\pi\left[\frac{\kappa}{\kappa+2}\right]\varepsilon_0 D_p^2 E_0\right\}\times E_p \tag{6.25}$$

It is common practice to use an average E to replace both the field strength at the time when the particle is being charged (E_0) and at the location where the particle is currently located (E_p); Eq. 6.25 can then be simplified to:

$$F_E = q \times E_p = 3\pi \left[\frac{\kappa}{\kappa+2}\right] \varepsilon_0 D_p^2 E^2 \tag{6.26}$$

If the Stokes drag force (Eq. 6.13) is the only resistance to the particle movement induced by the electrostatic force, then the terminal velocity (u_t) will occur when the electrostatic force and the drag force are equal. Then,

$$F_E = E_D = 3\pi \left[\frac{\kappa}{\kappa+2}\right] \varepsilon_0 D_p^2 E^2 = \frac{3\pi \mu D_p u}{C_c} \tag{6.27}$$

The terminal velocity can be readily derived by rearranging Eq. 6.27:

$$u_t = \frac{\left[\frac{\kappa}{\kappa+2}\right] \varepsilon_0 D_p E^2}{\mu} \times C_c = w \tag{6.28}$$

Instead of using the name of terminal velocity, *drift velocity* is commonly used in the ESP literature, and the symbol w is widely used for the drift velocity. It refers to the average velocity at which charged particles move toward the collection plates under the influence of the electric field.

Example 6.8: Saturation Charge of a Particle in an Electrical Field

Referring to the particle in Example 6.7, find its drift velocity.

Solution:

Using Eq. 6.28,

$$w = \left\{\left[\frac{5}{5+2}\right]\left(8.854 \times 10^{-12} \frac{\text{C}}{\text{V} \cdot \text{m}}\right)\left(2 \times 10^{-6}\text{m}\right)\left(320{,}000 \frac{\text{V}}{\text{m}}\right)^2\right\} \div$$

$$\left(1.81 \times 10^{-5} \frac{\text{N} \cdot \text{s}}{\text{m}^2}\right) \times 1.09 = 7.8 \times 10^{-2} \text{m/s}$$

Discussion:

1. Note: 1 C·V = 1 N·m.
2. The drift velocity of the 2-μ particle in this ESP is 7.8×10^{-2} m/s, which is ~15 times its tangential velocity in a cyclone (5.35×10^{-3} m/s) and ~300 times the terminal velocity due to gravity (2.62×10^{-4} m/s) – see Examples 6.4 and 6.6.

Inertial impaction. As a gas stream moves toward a stationary object (e.g., a *collection target* in a particulate removal device), it will tend to deflect its flow path around the object (Figure 6.10). For particles in the gas stream,

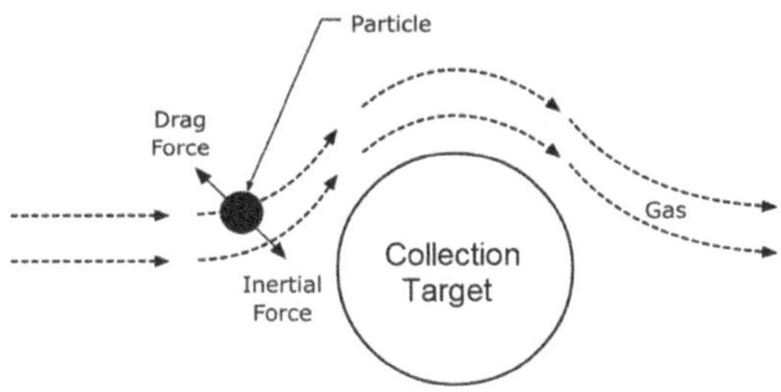

FIGURE 6.10 Internal impaction (Modified from USEPA, 2012).

smaller particles having less inertia are likely to move around the object with the gas stream and continue in the path of the gas stream. *Inertia* is a property of matter that remains in its existing state of rest or at a constant speed in the direction of motion (i.e., its resistance to changes in its state of motion). Particles with sufficient inertia will be displaced across the gas streamlines, move toward the collection target, and get collected/ captured.

For particles having Re_p <1, the effectiveness of impaction can be related to the *inertial impact number* (Ψ_I), which is defined as:

$$\psi_I = \left[\frac{\left(C_c D_p^{\,2} \rho_p\right) u_p}{18\mu}\right] \div D_c \tag{6.29}$$

where u_p is the particle velocity relative to the collection target and D_c is the diameter of the collection target. It is a dimensionless number used to describe particles' behavior in a fluid when interacting with obstacles (i.e., the collection targets). It should be noted that the terms inside the parentheses in Eq. 6.29 are the square of the aerodynamic diameter (see Eq. 6.20). The term inside the square bracket in Eq. 6.29 is often termed *Stokes stopping distance,* which represents the distance a particle can travel before it is stopped by viscous friction when the gravitational force is ignored.

The inertial impact number is also called the *separation number* or *Stokes number* in the literature. As the value of the inertial impact number increases, particles tend to move onto the collection target and get collected. On the other hand, the particles will stay on the gas streamlines and move around the target if this parameter approaches zero. One may think that a larger target will yield a more effective impact. However, as shown in Eq. 6.29, the inertial impact number (Ψ_I), which is related to the effectiveness of impaction, is inversely proportional to the diameter of the collection target. Smaller collection targets will be more effective regarding inertial impaction.

Example 6.9: Stokes Stopping Distance

Referring to Example 6.4, if 2-μ particles are ejected at a velocity of 5 m/s, how far will they travel before the viscous friction stops them? What would be the Stokes' stopping distance for a 4 μ particle (ρ_p=2,000 kg/m³) ejected at the same velocity of 5 m/s?

Solution:

a. From Eq. 6.29,

$$\text{Stokes stopping disctance} = \left[\frac{\left(C_c D_p^{\;2} \rho_p\right) v_p}{18\mu}\right] = \frac{(1.09)\left(2\times10^{-6}\right)^2 (2{,}000)(5)}{18\left(1.81\times10^{-5}\right)}$$

$$= 1.34\times10^{-4} = 134\,\mu$$

b. The Stokes stopping distance is proportional to $D_p^{\;2}$. Therefore, the Stokes stopping distance of the 4-μ particle should be four times that of the 2-μ particle, which would be 536 μ.

Discussion: The Stokes stopping distances for both cases are relatively short, implying that the air is very viscous for particles to move within it.

Brownian motion. Tiny particles in a gas stream will move randomly, resulting from their collisions with fast-moving molecules in the gas stream (Figure 6.11); it is often termed *Brownian motion*. The *diffusion coefficient* (*Diff*) represents the rate at which particles move due to Brownian motion, as follows:

$$\text{Diff} = \left[\frac{k_B T C_c}{3\pi\mu D_p}\right] \tag{6.30}$$

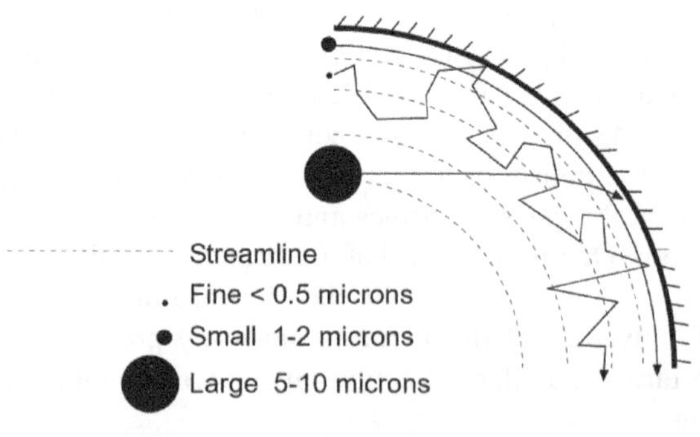

FIGURE 6.11 Particle motion vs. gas streamlines (USEPA, 2012).

where k_B is the Boltzmann constant (1.38×10^{-23} kg·m²/s²·K =1.38×10^{-23} J/K) and T is the absolute temperature (in K). As shown, particle diffusivity increases with temperature and decreases with its size or as the viscosity of the gas stream increases.

The effect of Brownian motion on particle collection efficiency is related to the *diffusional collection parameter* (Ψ_D), which is defined as (USEPA, 2012):

$$\psi_D = \text{Diff} \div \left(D_c u_p\right) = \left[\frac{k_B T C_c}{3\pi\mu D_p}\right] \div \left(D_c u_p\right) = \left[\frac{k_B T C_c}{3\pi\mu D_p D_C u_p}\right] \tag{6.31}$$

As Ψ_D increases, the tendency for particles to collect by Brownian motion increases. Smaller particle size (D_p), collection target (D_c), and relative particle velocity (u_p) values will lead to larger Ψ_D values (i.e., better collection by diffusion). However, Brownian motion is only essential for particles with D_p <0.3 μ, and it is only responsible for slight increases in the overall collection efficiency of particulate removal devices in this size range. In other words, particle diffusion is not a significant particle removal mechanism for most of the control devices (USEPA, 2012).

Example 6.10: Particle Diffusivity

Referring to Example 6.4, what is the particle diffusivity of the 2-μ particle?
Solution:
From Eq. 6.30,

$$\text{Diff} = \left[\frac{kTC_c}{3\pi\mu D_p}\right] = \frac{\left(1.38\times10^{-23}\right)(293)(1.09)}{3\pi\left(1.81\times10^{-5}\right)\left(2\times10^{-6}\right)} = 1.29\times10^{-11}\ \text{m}^2/\text{s}$$

Discussion: The diffusivity of this 2-μ particle, 1.29×10^{-11} m²/s, is much smaller than the diffusivities of gases in air (~10^{-5} m²/s typically) or those of solutes in water (~10^{-9} m²/s typically).

6.3 GRAVITY SETTLING CHAMBER

Gravity settling chambers (or *gravity settlers*) are essentially expansion chambers in which the gas flow velocity is reduced through expansion into an ample space to allow time for particles to settle out by gravity (Figure 6.12). They have the advantages of simple construction, low capital, operational and maintenance (O&M) costs, and easy removal of the collected particles. However, the removal efficiency depends mainly on the particle's aerodynamic diameter. The larger the aerodynamic diameter, the better the removal would be. It is typically used to remove particles with D_p> 50 μ and is commonly used as pretreatment for other particulate removal devices.

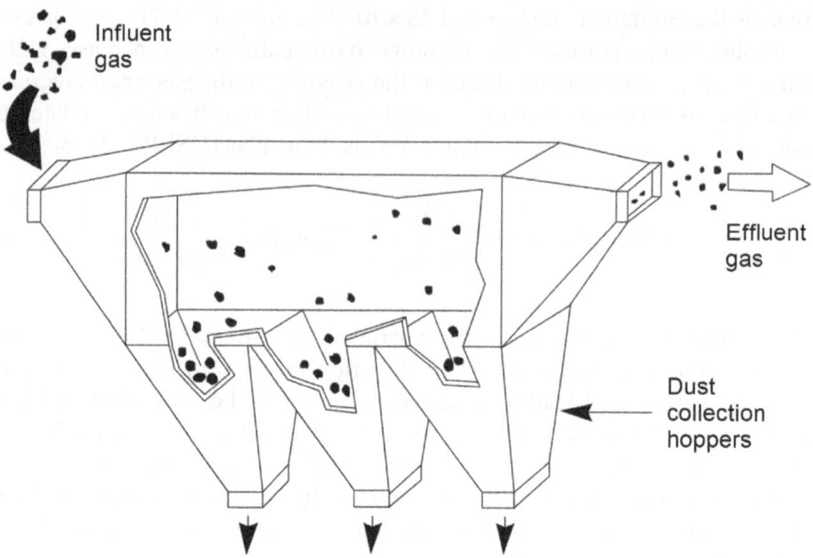

FIGURE 6.12 Schematics of a horizontal-flow gravity settler (Modified from USEPA, 2012).

6.3.1 Principles of Gravity Settling

A gravity settling chamber is typically a long horizontal rectangular chamber (Length × Width × Height). The cross-sectional area of the chamber ($W \times H$) is much larger than that of the inlet duct, so the gas stream will expand and move slowly through the chamber to allow particles to find time to move downward to the bottom of the chamber by gravity (Figure 6.13).

Assuming the gas velocity (v_{gas}) is uniform through the chamber, it can be readily determined as:

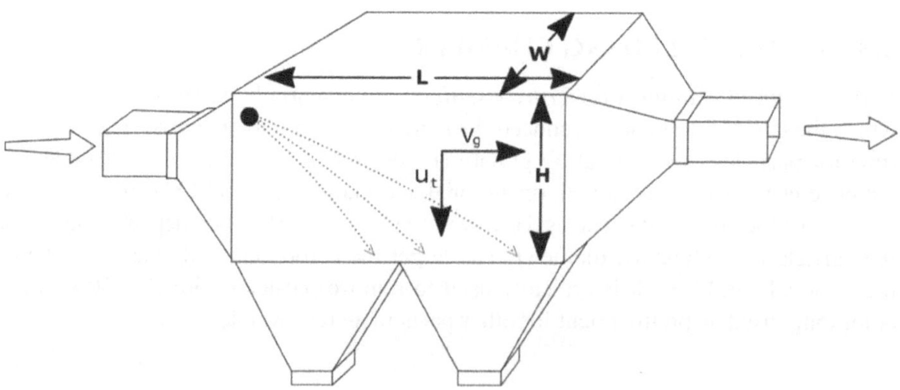

FIGURE 6.13 Particle settling in a gravity settler (Modified from USEPA, 2012).

$$v_{\text{gas}} = Q_{\text{gas}}/(W \times H) \tag{6.32}$$

where Q_{gas} is the flow rate of the gas stream. A particle entering the chamber will have a horizontal velocity (u_h), the same as v_{gas}, and a vertical settling velocity is its terminal velocity (u_t). The residence time of the gas stream in the chamber (τ) can be readily calculated as:

$$\tau = \frac{L \times W \times H}{Q_{\text{gas}}} = \frac{L}{Q_{\text{gas}}/(W \times H)} = \frac{L}{v_{\text{gas}}} \tag{6.33}$$

The time needed for a particle to settle to the bottom of the tank (t_s) can be found by dividing the height of the chamber by the terminal velocity as:

$$t_s = \frac{H}{u_t} \tag{6.34}$$

A particle will be considered removed if it can settle to the tank bottom before the gas exits the chamber. For a *block flow pattern*, we assume that the particles are distributed uniformly across the cross-section at the inlet (i.e., some particles will enter at the very top of the chamber, while others will enter closer to the bottom). We also assume that there is no mixing and interactions among the particles in both the transverse and the longitudinal directions of the gas flow. For this gas flow pattern, the removal efficiency (η_{block}) for particles of a specific terminal velocity is essentially the ratio of the gas residence time (τ) and the traveling time needed for these particles to settle to the bottom of the tank (t_s):

$$\eta_{\text{block}} = \frac{\tau}{t_s} = \frac{(L/v_{\text{gas}})}{(H/u_t)} \tag{6.35}$$

Using Eq. 6.18 for the terminal velocity (u_t), Eq. 6.31 becomes

$$\eta_{\text{block}} = \left(\frac{L}{H \times v_g}\right)(u_t) = \left(\frac{L}{H \times v_g}\right)\left(\frac{C_c \times g \times \rho_p \times D_p^2}{18\mu}\right) \tag{6.36}$$

If the calculated value for a particle is larger than unity, it means that its removal is 100% (i.e., the particle will settle to the bottom of the tank before the gas stream leaves the chamber).

For an extreme case opposite to the block flow model, we assume that the gas stream is well mixed in the transverse direction, but not in the longitudinal direction. For this case, blending in the transverse direction would reduce the removal efficiency, and the removal efficiency based on the *mixed flow model* (η_{mixed}) can be found from that of the block flow model (η_{block}) as follows:

$$\eta_{\text{mixed}} = 1 - \exp\left[-\eta_{\text{block}}\right] = 1 - \exp\left[-\left(\frac{L}{H \times v_g}\right)\left(\frac{C_c \times g \times \rho_p \times D_p^2}{18\mu}\right)\right] \quad (6.37)$$

In real-life situations, gas flows will be somewhat turbulent, and mixing in the transverse direction will reduce the removal efficiency. On the other hand, the blending in the longitudinal direction will allow the particles to move both upstream and downstream, which will have insignificant impacts on the overall removal efficiency. Consequently, the removal efficiency should be between the values predicted by the block and mixed flow models (i.e., Eqs. 6.36 and 6.37).

6.3.2 Gravity Settling Systems and Their Components

As shown in Eq. 6.34, the removal efficiency is inversely proportional to the height of the chamber (*H*). This makes sense because the particles would not need to settle a long distance to reach the bottom of the chamber for a smaller *H*. Therefore, multi-tray settling chambers are available, with several horizontal trays within the rectangular chamber, so the settling distances are much shorter (Figure 6.14).

To minimize mixing in the transverse direction of the gas stream flow, inlet and outlet baffles are often added to guide the flow to be more parallel in the axial/longitudinal direction.

As shown in Figures 6.12 and 6.13, in addition to the rectangular portion for gas expansion, there are collection hoppers at the bottom of the settling chamber, and they are another key component of the removal device. Hoppers are often designed

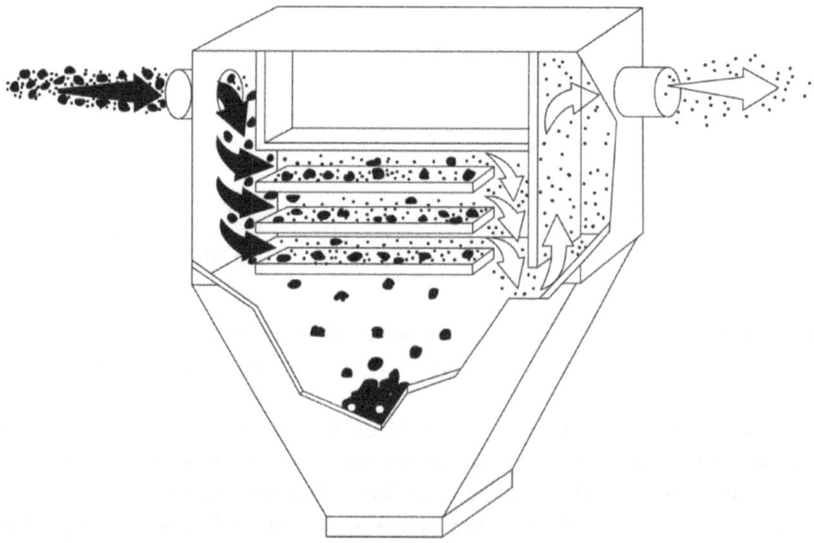

FIGURE 6.14 Schematics of a multi-tray gravity settler (USEPA, 2012).

with sealing valves, and the collected particles need to be emptied frequently to avoid dust re-entrainment. The potential dust re-entrainment constrains the spacing between the trays for multi-tray gravity settlers (USEPA, 2012).

6.3.3 Factors Affecting Particle Removal by Gravity Settling

The critical parameter for designing a settling chamber is the terminal velocity derived from the balance of the gravitational, buoyancy, and drag forces. The gas residence/detention time (i.e., the ratio of the chamber volume and the gas flow rate) should be long enough to allow sufficient time for target particles to settle. With that, the design variables include the length, width, and height of the rectangular portion of the chamber.

For a given flow rate, the chamber's height and width (i.e., the cross-sectional area) will determine the velocity at which the gas moves through the chamber, v_g, as in Eq. 6.32. This velocity is often called the *throughput velocity*. The throughput velocity should be smaller than the *pickup velocity,* which will re-entrain settled particles into the moving gas stream. The pickup velocity depends mainly on the size and density of the particles, ranging from 5.8 to 25.0 ft/s (1.8 to 7.6 m/s) for some typical applications. Without specific data, the throughput velocity should be <10 ft/s (3 m/s).

Although the height of the chamber should be as shallow as possible so that the target particles do not need to travel a long distance downward for removal, a sufficient height is required to avoid/minimize re-entrainment of settled particles and reduce the frequency of dust removal from the hopper(s).

It should be noted that the temperature of the gas stream would drop as it enters the chamber. The decrease in temperature is due to gas expansion, which is called the *Joule–Thomson effect* in thermodynamics.

6.3.4 Design Calculations for Gravity Settlers

When selecting any pollution control device, its removal efficiency is often one of the significant considerations. Sometimes, we may need to put two or more devices (they may be the same or different) in series to increase the overall removal efficiency. At other times, we may need to operate control devices in parallel to accommodate maintenance needs or fluctuations in the influent flow rates. *Removal efficiency* (η) is the fraction/percentage of the influent pollutants removed by the device. At the same time, *penetration* (p) is the fraction/percentage of the influent pollutants that are not removed (i.e., escape from the control device). The relationship between them is very straightforward:

$$\text{Penetration } (p) = 1 - \text{Removal efficiency } (\eta) \tag{6.38}$$

If we employ removal devices in series, using penetration will simplify the calculation of the overall removal efficiency. For example, if we have two control devices in series with removal efficiencies of η_1 and η_2, respectively, the removal efficiency can be found by:

$$\eta_{\text{overall}} = 1-(1-\eta_1)(1-\eta_2) = 1-(p_1)(p_2) \tag{6.39}$$

The example below illustrates the design calculations for a horizontal gravity settling chamber.

Example 6.11: Gravity Settling Chamber

A gravity settling chamber ($L \times W \times H = 10$ m$\times$4 m$\times$3m) is used to treat a gas stream ($Q = 20\,\text{m}^3/\text{s}$; $T = 20°\text{C}$; $P = 1$ atm), which contains particles of three distinct sizes ($D_p = 2$, 50, and 100 μ). The densities of all particles are the same, 2,000 kg/m^3. The mass fractions of the 2-, 50-, and 100-μ particles are 5%, 25%, and 70%, respectively. Determine:

a. the throughput velocity of the gas stream,
b. the terminal velocities of the particles,
c. the removal efficiency for each particle size, and
d. the overall removal efficiency of all particles.

Solution:

a. From Eq. 6.32,

$$\text{Throughput velcoity} = v_{\text{gas}} = 20/(4 \times 3) = 1.67 \text{ m/s}$$

b. $(u_t)_{2\mu} = 2.62 \times 10^{-4}$ m/s (from Example 6.4)
For the 50- and 100-μ particles, their Cunningham slip correction factors are essentially equal to one. Since the terminal velocity is proportional to D_p^2, we can find their terminal velocities by using that of the 2-μ particles.

$$(u_t)_{50\mu} = \left[(u_t)_{2\mu}\right] \times (1/1.09)(50/2)^2 = [(2.62 \times 10^{\blacklozenge 4})/1.09](25)^2 = 0.151 \text{ m/s}$$

$$(u_t)_{100\mu} = \left[(u_t)_{50\mu}\right] \times (100/50)^2 = (0.151)(4) = 0.603 \text{ m/s}$$

c. From the 2-μ particles and Eq. 6.36,

$$\eta_{\text{block}} = \left(\frac{L}{H \times v_g}\right)(u_t) = \left(\frac{10}{3 \times 1.67}\right)\left(2.62 \times 10^{-4}\right) = 0.0005$$

From Eq. 6.37,

$$\eta_{\text{mixed}} = 1 - \exp\left[-\eta_{\text{block}}\right] = 1 - \exp(-0.0005) = 0.0005$$

The removal efficiencies for 50- and 100-μ particles can be determined using the approach above by inserting the corresponding u_t value. The results are tabulated below for comparison:

Particle Size (μ)	Terminal Velocity (m/s)	η (Block Flow)	η (Mixed Flow)
2	0.000263	0. 0005	0.0005
50	0.151	0.30	0.26
100	0.603	1.00	0.70

As expected, the removals using the mixed flow model are smaller than those of the block flow model. The actual removals should be between the values predicted by these two models. This device can effectively remove particles having a D_p >100 μ.

d. Using η_{mixed} as the worst-case scenario to estimate the total mass removal: Basis: total input particle mass = 100 kg

Particle Size (μ)	Inlet Mass (kg)	η	p	Outlet Mass (kg)
2	5	0.0005	0.9995	4.9975
50	25	0.26	0.74	18.5
100	70	0.70	0.30	21.0
Total	**100**			**44.5**

$$\text{Total removal} = 1 - p = 1 - (44.5/100) = 55.5\%$$

Discussion:

1. Since the question is to find the extent of removal, not the exact mass in the outlet, the "basis" for calculations can be arbitrarily chosen to facilitate the calculations. Although a total mass of 100 kg is unreasonably high, it will not affect the results (i.e., percentage removals).
2. As shown, this gravity settling chamber can remove many large particles and most of the inlet mass, but the removal of 2 μ particles is insignificant. Other particle removal devices, such as fabric filters, ESPs, or wet scrubbers, should be employed if 2-μ particles need to be removed significantly.

6.4 CYCLONES

Cyclones have a cylindrical shape that causes the gas stream to move in high-speed turns so that only gas molecules and tiny particles with small inertia can move with it. The large particles' large inertia will prevent them from making turns, so they will hit the cyclone's inner wall and get collected.

Figure 6.15 illustrates the *fractional efficiency curves* of cyclones. A typical cyclone's cut size (defined later) is about 5 μ. If finer particle removal is needed, cyclones can be used as pre-collectors ahead of other devices capable of removing finer particles, such as fabric filters and electronic precipitators.

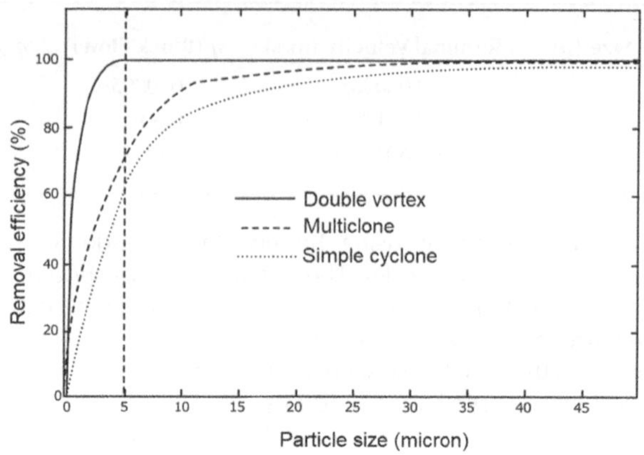

FIGURE 6.15 Fractional efficiency curves of cyclones (Modified from USEPA, 2012).

6.4.1 Principles of Cyclones

As discussed in the previous section for the gravity settling chambers, the removal efficiency for a particle depends on the ratio of the gas residence time (τ) and the time needed for the particle to reach the bottom of the chamber (t_s). The concept is essentially the same for cyclones, except that the particle does not need to travel to the bottom of the cyclone to get removed. Instead, it only needs to hit the cyclone's inner wall to get removed when the gas is circulating downward. The gas stream then spins upward to the outlet, carrying the uncollected particles. Assuming the block flow model, the maximum distance a particle needs to travel to the wall is the width of the rectangular inlet (W_{inlet}) vs. the height (H) of the settling chamber for gravitational settling. In the meantime, the length of the gas travel in which the particles could hit the wall is $N(\pi D_{\text{cyclone}})$, compared to the length of the gravity settling chamber (L). N is the number of turns the gas stream makes before it spins upward, and D_{cyclone} is the diameter of the cyclone. The removal efficiency can then be derived, similar to Eq. 6.36, as:

$$\eta_{\text{block}} = \frac{\tau}{t_{\text{wall}}} = \frac{\left(N\pi D_{\text{cyclone}}/v_{\text{gas}}\right)}{\left(W_{\text{inlet}}/u_t\right)} = \left[\frac{N\pi D_{\text{cyclone}}}{\left(W_{\text{inlet}}\right)\left(v_{\text{gas}}\right)}\right] \times u_t \tag{6.40}$$

where t_{wall} is the time needed for the particle to travel to the wall.

The tangential velocity (u_p) in Eq. 6.22 is the terminal velocity of the particles in a cyclone. The radius of the cyclone (r) can be replaced with ($D_{\text{cyclone}}/2$) as follows:

$$u_t = u_p = C_c\left[\frac{\rho_p D_p^{\ 2}}{18\mu}\right]\left(\frac{v_{\text{gas}}^{\ 2}}{r}\right) = C_c\left[\frac{\rho_p D_p^{\ 2}}{9\mu}\right]\left(\frac{v_{\text{gas}}^{\ 2}}{D_{\text{cyclone}}}\right) \tag{6.41}$$

The gas residence time can be found by dividing the length of the gas flow inside the cyclone before ascending by its flow velocity, as follows:

$$\tau = \frac{N\pi D_{\text{cyclone}}}{v_{\text{gas}}} \tag{6.42}$$

where N is the number of gas turns in the cyclone before ascending.

From Eqs. 6.40 and 6.41,

$$\eta_{\text{block}} = \left[\frac{N\pi D_{\text{cyclone}}}{(W_{\text{inlet}})(v_{\text{gas}})}\right] \times \left\{C_c\left[\frac{\rho_p D_p^{\ 2}}{9\mu}\right]\left(\frac{v_{\text{gas}}^{\ 2}}{D_{\text{cyclone}}}\right)\right\} = C_c\left[\frac{\pi N \rho_p D_p^{\ 2} v_{\text{gas}}}{9\mu W_{\text{inlet}}}\right] \tag{6.43}$$

Although an equation similar to Eq. 6.37 can estimate removal efficiencies based on the mixed flow model, the block flow model has been more widely used because its predictions are reasonably good (de Nevers, 2000).

As shown in Eq. 6.43, the removal of a specific particle by a cyclone is proportional to the density of the particle (ρ_p), its diameter to the power of two ($D_p^{\ 2}$), and inversely proportional to the viscosity of the gas stream (μ). The removal is also proportional to the number of turns (N) and inversely proportional to the width of the rectangular inlet (W_{inlet}); these two variables (i.e., N and W_{inlet}) depend on the physical configuration of the cyclone. The removal is also proportional to the velocity of the gas stream (v_{gas}). For a given gas flow rate (Q_{gas}), v_{gas} will be larger for a rectangular inlet having a smaller cross-section ($H_{\text{inlet}} \times W_{\text{inlet}}$). If the aspect ratio of the rectangular inlet is fixed, v_{gas} would be inversely proportional to $(W_{\text{inlet}})^2$, and the removal will then be inversely proportional to $(W_{\text{inlet}})^3$, as follows:

$$\eta_{\text{block}} \sim \left[\frac{v_{\text{gas}}}{W_{\text{inlet}}}\right] \sim \left[\left(\frac{Q_{\text{gas}}}{L \times W_{\text{inlet}}}\right) / W_{\text{inlet}}\right] \sim \left[\frac{Q_{\text{gas}}}{L \times W_{\text{inlet}}^2}\right] \sim \left[\frac{Q_{\text{gas}}}{W_{\text{inlet}}^3}\right] \tag{6.44}$$

Although a larger v_{gas} or a smaller W_{inlet} will yield a higher removal efficiency, it will incur significantly larger head loss and energy costs.

6.4.2 Cyclones and Their Components

Cyclones for industrial applications can be classified into two groups: (i) large-diameter cyclones and (ii) small-diameter multi-cyclones.

Figure 6.16 shows a schematic of large-diameter cyclones. They have a cylindrical portion on the top and a conical portion at the bottom. Cyclone diameter (D) is the key design parameter determining cyclone capacity. Designs of cyclones are relatively standardized, and the cyclone ratios of dimensions are typically based on the cyclone diameter. The ratios are different for "*high-efficiency*", "*conventional*", and "*high-throughput*" cyclones (see Table 6.3). Dirty gas can come from the cylindrical portion's top or bottom. The gas enters the rectangular inlet; usually, the height of the inlet is twice its width (i.e., $H/W=2$). The inlet is typically arranged tangentially to the cylindrical portion of the cyclone. The typical inlet height for conventional cyclones with top inlets (Figure 6.17 and Table 6.3) is half the cyclone diameter

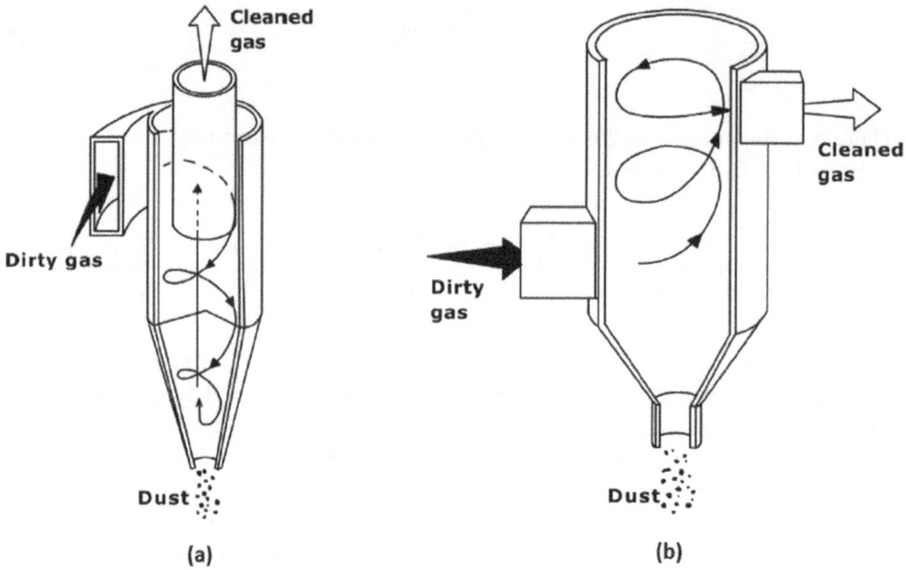

FIGURE 6.16 Large-diameter cyclones: (a) top inlet and (b) bottom inlet (Modified from USEPA, 2012).

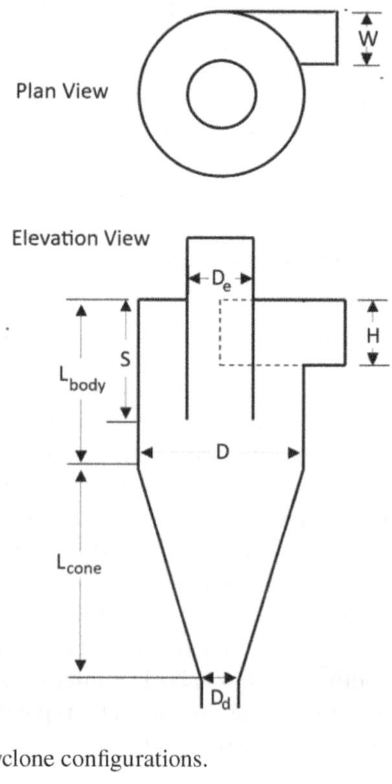

FIGURE 6.17 Typical cyclone configurations.

TABLE 6.3
Cyclone Ratios of Dimensions to Body Diameter (*D*)

Dimension	High Efficiency	Conventional	High Throughput
Gas inlet height (H)	0.44	0.50	0.80
Gas inlet width (W)	0.21	0.25	0.35
Vortex finder diameter (D_{exit})	0.40	0.50	0.75
Vortex finder length (S)	0.50	0.60	0.85
Dust outlet diameter ($D_{discharge}$)	0.40	0.40	0.40
Body length (L_{body})	1.40	1.75	1.70
Cone length (L_{cone})	2.50	2.00	2.00

Source: NCEES (2005).

(i.e., $H/D = 0.5$). After a few turns, the gas stream gets cleaner and exits from the top (through the vortex finder), while the collected particles are removed from the conical bottom of the cyclone (through the dust outlet). The vortex finder extends slightly into the cylindrical section to avoid pulling particles into it. The conventional cyclones' vortex finder length (S) and diameter (D_{exit}) are 0.60 D and 0.50 D, respectively. The dust outlet diameter ($D_{discharge}$) has a size of 0.4 D. The body length (L_{body}) and cone length (L_{cone}) of the conventional cyclones are 1.75 D and 2.00 D, respectively. The diameters of the large cyclones range from approximately one foot (0.3 m) to >12 ft (3.6 m), and typical operating head loss is from 2 to 4 in-H_2O (0.5 to 1.0 kPa). In general, high-efficiency cyclones have a narrow inlet and a long body and are suitable for removing fine particles, but they incur a larger pressure drop. On the contrary, high-throughput cyclones have a wider body, lower pressure drop, and lower particle removal efficiency. Conventional cyclones strike a balance between pressure drop and removal efficiency.

A multi-cyclone collector is a group of small-diameter cyclones, typically 6 to 12 inches (0.15 to 0.3 m) in diameter (Figure 6.18). It has better particulate removal

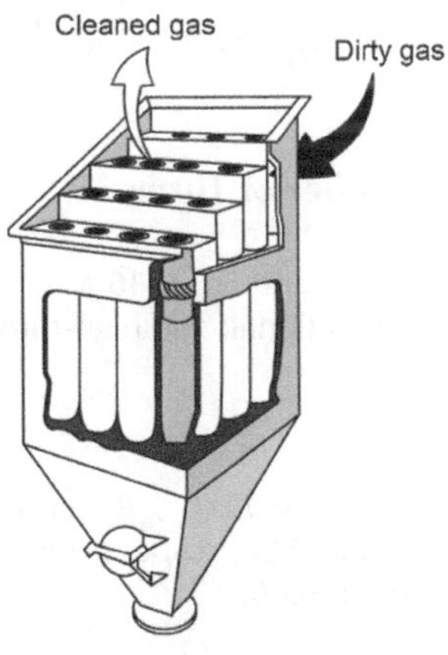

FIGURE 6.18 Schematics of a multi-cyclone collector (Modified from USEPA, 2012).

capability than large-diameter cyclones (see Figure 6.16) but is more expensive and would incur a larger head loss, >4-in H_2O (1.0 kPa) (USEPA, 2003).

6.4.3 Factors Affecting the Performance of Cyclones

Important factors affecting the performance of a cyclone collector include the size and density of the particles, the gas velocity through the cyclone, the residence time of the gas, and the configurations of the cyclone (i.e., dimensions of the body and those of the inlet).

6.4.4 Design Calculations for Cyclones

Cut diameter (or *cut size*), D_{cut}, has been widely used to describe the size ranges of particles that a particulate removal device can collect. It is defined as the diameter of the particles collected with 50% efficiency. The particles with $D_p > D_{\text{cut}}$ will be collected by more than 50% and vice versa.

By letting $\eta_{\text{block}} = 50\%$ and $D_p = D_{\text{cut}}$, Eq. 6.43 can be rearranged to:

$$D_{\text{cut}} = \left(\frac{9W_{\text{inlet}}\mu}{2\pi C_c N v_{\text{gas}} \rho_p} \right)^{0.5} \cong \left(\frac{9W_{\text{inlet}}\mu}{2\pi N v_{\text{gas}} \rho_p} \right)^{0.5} \tag{6.45}$$

Since the typical D_{cut} values for cyclones are a few microns and the Cunningham slip correction factors (C_c) for particles in this size range are essentially equal to 1, C_c is often omitted in Eq. 6.45. As shown in Eq. 6.45, the cut size is proportional to the square root of the size of the inlet and inversely proportional to the square root of the number of turns, gas velocity, and density of the particles. The smaller cut size means better particle removal.

The *effective number of turns* (N_e) in a cyclone separator refers to the number of spiral rotations particles make as they travel through the cyclone. It is commonly used as N in Eq. 6.45, and it can be approximated by the following equation (NCEES, 2025):

$$N_e = \frac{1}{H}\left[L_{\text{body}} + \frac{L_{\text{cone}}}{2} \right] \tag{6.46}$$

Example 6.12: Effective Number of Turns

Use the aspect ratios in Table 6.3 and Eq. 6.46 to estimate the effective number of turns for high-efficiency, conventional, and high-throughput cyclone separators, respectively.

Solution:

a. From Table 6.3, the ratios of H/D, L_{body}/D, and L_{cone}/D for conventional cyclones are 0.50, 1.75, and 2.00, respectively.
b. Inserting the values into Eq. 6.46,

$$N_e = \frac{1}{H}\left[L_{body} + \frac{L_{cone}}{2}\right] = \frac{1}{0.50}\left[1.75 + \frac{2.00}{2}\right] = 5.5$$

c. Similarly, the effective numbers of high-efficiency and high-throughput cyclones can readily be calculated as 6.02 and 3.375, respectively.

Discussion:

1. The values of N_e for efficiency, conventional, and high-throughput separators are 6.02, 5.5, and 3.375, respectively.
2. The particle removal of the high-efficiency separators would be the best because of the largest effective turns. However, the head loss would consequently be the largest.

Example 6.13: Residence Time and Cut Diameter of a Cyclone

A cyclone has a diameter of 1 m and a rectangular inlet (H=0.4 m and W=0.2 m). The gas flow rate is 2.4 m³/s. Determine:

a. the gas flow velocity,
b. the gas residence time in the cyclone for particulate removal (N=5), and
c. the cut diameter if the density of the particles is 2,000 kg/m³.

Solution:

a. $v_{gas} = (Q_{gas})/(H \times W_{inlet}) = (2.4)/[(0.4)(0.2)] = 30$ m/s
b. From Eq. 6.42, $\tau = (N\pi D_{cyclone})/v_{gas} = [5\pi(1)]/(30) = 0.52$ s
c. From Eq. 6.45,

$$D_{cut} = \left(\frac{9W_{inlet}\mu}{2\pi N v_{gas}\rho_p}\right)^{0.5} = \sqrt{\frac{9(0.2)(1.81\times10^{-5})}{2\pi(5)(30)(2,000)}} = 4.2\times10^{-6}\,\text{m} = 4.2\mu$$

Discussion:

1. The gas residence time for this operation is only 0.52 s, much shorter than that of a gravity settling chamber.
2. The cut size is 4.2 μ, which falls into the range of typical cyclones.

Example 6.14: Cut Diameters of Two Cyclones

Referring to Example 6.13, if we feed the same air stream (same Q, particle loading, PSD) to a cyclone of the same design, except that all the dimensions are half as big, what will be the new cut diameter (using the block flow model)? What are the implications?

Solution:

a. Assuming N stays the same, from Eq. 6.45,

$$\frac{D_{\text{cut, new}}}{D_{\text{cut, old}}} = \sqrt{\left(\frac{W_{\text{inlet, new}}}{W_{\text{inlet, old}}}\right)\left(\frac{V_{\text{gas, old}}}{V_{\text{gas, new}}}\right)} = \sqrt{\left(\frac{1}{2}\right)\left(\frac{1}{4}\right)} = 0.35$$

$$D_{\text{cut,new}} = (4.2)(0.35) = \underline{1.5\,\mu}$$

b. D_{cut} will become smaller (from 4.2 to 1.5 μ). However, because the dimensions are reduced by half, the gas velocity will be four times larger. The corresponding head loss would be much higher (discussed later in this section).

Since a typical gas stream should contain particles of different sizes, an empirical equation (Eq. 6.47) has been developed to determine the removal efficiency (η) for each size, using the ratio of D_p and D_{cut} (i.e., D_p/D_{cut}). The equation has been found to fit the actual performance of cyclones well.

$$\eta = \frac{1}{1+\left(D_{\text{cut}}/D_p\right)^2} = \frac{\left(D_p/D_{\text{cut}}\right)^2}{1+\left(D_p/D_{\text{cut}}\right)^2} \tag{6.47}$$

Using Eq. 6.47, a collection efficiency can be generated, as shown in Figure 6.19. The removal efficiency equals 50% when the D_p/D_{cut} ratio = 1.

Example 6.15: Removal Efficiencies of a Cyclone

A particle stream has 50% by wt. of particles having D_p = 1 μ and the other 50% (by wt.) with D_p = 25 μ. If we pass this stream through a cyclone with D_{cut} = 5 μ, how much of the particles (by wt.) will be removed by this cyclone?

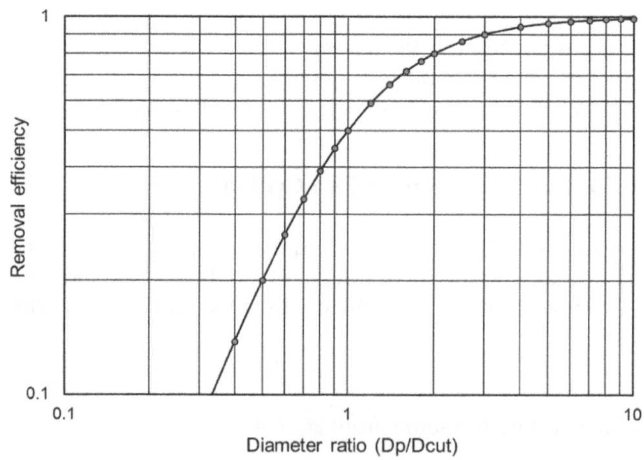

FIGURE 6.19 Removal efficiency vs. diameter ratio (D_p/D_{cut}).

a. Using the approach of USEPA (2012) and assuming $K = 0.048$ (the high end of the range),

$$\Delta P = (0.048)\left(\frac{(0.075)(50)^2}{2}\right) = 4.5 \text{ in-H}_2\text{O}$$

b. Using the approach of de Nevers (2000),

$$\Delta P = (8)\left(\frac{1}{2}\right)\left(0.075\frac{\text{lb}_\text{m}}{\text{ft}^3}\right)\left(50\frac{\text{ft}}{\text{s}}\right)^2\left(\frac{\text{lb}_\text{f}}{32.2\text{lb}_\text{m}\cdot\frac{\text{ft}}{\text{s}^2}}\right) = 23.3\frac{\text{lb}_\text{f}}{\text{ft}^2}$$

$$= \left(23.3\frac{\text{lb}}{\text{ft}^2}\right)\left(\frac{\text{ft}^2}{144 \text{ in}^2}\right)\left(\frac{407 \text{ in-H}_2\text{O}}{14.7 \text{ psi}}\right) = 4.48 \text{ in-H}_2\text{O}$$

Discussion: A K value of 8, as suggested by de Nevers (2000), is essentially equivalent to 0.048 of USEPA (2012) with specific units for the three parameters.

Example 6.17: Pressure Drop of a Cyclone (SI Units)

Referring to the cyclone in Example 6.16, estimate the pressure drop using SI units.

Solution:

Assuming $K = 8$, insert the given values into Eq. 6.43:

$$\Delta P = (8)\left(\frac{1}{2}\right)\left(1.21\frac{\text{kg}}{\text{m}^3}\right)\left(15.24\frac{\text{m}}{\text{s}}\right)^2\left(\frac{\text{N}}{\text{kg}\cdot\frac{\text{m}}{\text{s}^2}}\right) = 1{,}126 \text{ N/m}^2 \ (= 4.52 \text{ in-H}_2\text{O})$$

Example 6.18: Electricity Cost of a Cyclone (SI and the U.S. Customary Units)

Referring to the cyclone in Examples 6.16 and 6.17, if it operates continuously and the blower efficiency is 85%, estimate the annual electricity cost. The flow rate is $10 \text{ m}^3/\text{s}$ ($353 \text{ ft}^3/\text{s}$), and the electricity cost is \$0.20 per kilowatt hour (\$0.20/kWh).

Solution:

a. Using the SI units,

$$\text{Power} = \frac{\left(10\frac{\text{m}^3}{\text{s}}\right)\left(1{,}126\frac{\text{N}}{\text{m}^2}\right)}{0.85} = 13{,}247\frac{\text{N}\cdot\text{m}}{\text{s}} = 13{,}247 \text{ W}$$

Solution:

a. Using Eq. 6.47 to calculate the removal efficiencies:

$$\eta_{1\mu} = \frac{(1/5)^2}{1+(1/5)^2} = 0.038;\ \eta_{25\mu} = \frac{(25/5)^2}{1+(25/5)^2} = 0.96$$

b. $(\eta_{removal})_{total} = (0.038) \times (50\%) + (0.96) \times (50\%) = 49.9\%$

Discussion:

1. With $D_{cut} = 5\mu$, the removal for 25-μ particles approaches 100%, wh that for 1-μ particles is <4%.
2. The total removal comes mainly from the removal of the large particle

Pressure drop is an essential operational parameter for cyclones. As shown in I the cut size (D_{cut}) is inversely proportional to the square root of the gas veloci This means that larger v_{gas} will reduce D_{cut} and improve the removal effici cyclones. However, a larger v_{gas} will incur a more significant pressure drop (loss), ΔP, which is the difference between the inlet pressure (P_{in}) and the out sure (P_{out}). A more significant ΔP demands a more powerful air movemen and incurs higher energy costs to move the gas stream through the cyclone. T sure drop can be estimated based on the quantity of ($\rho_{gas} \times v_{gas}^2/2$), which is as *velocity head*, as follows:

$$\Delta P = (P_{in} - P_{out}) = K\left(\frac{\rho_{gas} v_{gas}^2}{2}\right)$$

In fluid mechanics, this approach is commonly used to estimate the head loss t a fluid transportation system. Different K values exist for different componen as elbows, valves, and flow meters. Since each cyclone has its specific design have its own particular K, which can be derived from experimental data. Sin an empirical value, the reported K values are often tied to the units of ΔP, ρ v_{gas} used in the equation. According to USEPA (2012), typical K values rang 0.026 to 0.048 if ΔP is in inches of water, ρ_{gas} in lb/ft³, and v_{gas} in ft/s. de (2000) stated that most cyclones have pressure drops of about eight velocity (i.e., $K = 8$).

Example 6.16: Pressure Drop through a Cyclone

Estimate the head loss through a cyclone if the gas velocity is 50 ft/s (15.24 r and $T = 68°F$ (20°C).

Solution:

From Example 2.2, $\rho_{air} = 0.075$ lb/ft³ (=1.21 kg/m³) at $T = 68°F$ (20°C)

b. Using the U.S. customary units,

$$\text{Power} = \frac{\left(353\frac{\text{ft}^3}{\text{s}}\right)\left(23.3\frac{\text{lb}_f}{\text{ft}^2}\right)}{0.85} = \left(9{,}676\frac{\text{ft}\cdot\text{lb}_f}{\text{s}}\right)\left(\frac{745.7\ \text{W}}{550\ \text{ft}\cdot\text{lb}_f/\text{s}}\right) = 13{,}120\ \text{W}$$

c. $$\text{Annual Enegy Cost} = (13{,}247\ \text{W})(365\ \text{d})\left(24\frac{\text{hr}}{\text{d}}\right)\left(\frac{\text{kW}}{1{,}000\ \text{W}}\right)(\$0.20\ \text{kWh})$$

$$= \$23{,}210$$

6.5 ELECTROSTATIC PRECIPITATORS

Removal of particles by ESPs is based on applying a surface force (i.e., electrostatic) instead of a body force (e.g., gravitational, centrifugal) to the particles. The process applies to solid and liquid particles.

6.5.1 Principles of Electrostatic Precipitators

Similar to the terminal velocity for particles in gravity settling chambers and the tangential velocity of particles in cyclones, the drift velocity (w) is how fast a particle will move toward a collection plate of an ESP. To determine the removal efficiency, let's return to a simplified system – two identical parallel plates (with length L and height H) with a separation distance of S. An electrode is placed in the center plane between these two plates. The gas stream will flow through the channel formed by these two parallel plates.

Let's work the channel's right (or the left) half. The gas flow rate (Q_{gas}) through this half of the channel will be the product of the cross-sectional area and the gas velocity (v_{gas}), as follows:

$$Q_{gas} = \left(\frac{S}{2}\times H\right)\times v_{gas} \tag{6.49}$$

The collecting area (A) will be the area of one single plate, which is:

$$A = L\times H \tag{6.50}$$

Similar to the approach we made for the gravitational settling chambers with the block flow pattern, the removal efficiency (η_{block}) for particles of a specific drift velocity (w) is essentially the ratio of the gas residence time (τ) and the traveling time needed for these particles to reach the collection plate (t):

$$\eta_{block} = \frac{\tau}{t} = \frac{(L/v_{gas})}{\left(\frac{S}{2}\right)/w} = \frac{w\times L}{\left(\frac{S}{2}\right)\times v_{gas}} = \frac{w\times(L\times H)}{\left(\frac{S}{2}\times H\right)\times v_{gas}} = \frac{wA}{Q} \tag{6.51}$$

Similar to gravitational settling chambers, the removal efficiency of the mixed flow mode (η_{mixed}) can be derived as:

$$\eta_{\text{mixed}} = 1 - \exp\left[-\eta_{\text{block}}\right] = 1 - \exp\left[-\frac{wA}{Q}\right] \tag{6.52}$$

Eq. 6.52, instead of Eq. 6.51, is widely used in practice for conceptual design and analysis of ESPs. It has a name, the *Deutsch–Anderson equation*. The ratio of the collection surface and the gas flow rate (i.e., *A*/*Q*) is often called *specific collection area* (SCA). It should be noted that any consistent set of units can be used for *w* and (*A*/*Q*) as long as "*wA*/*Q*" is dimensionless.

As mentioned in Section 6.2.4, contact and diffusional are two charging mechanisms for particles in an ESP. The combined effect of these two mechanisms creates an efficiency–particle size curve, as shown in Figure 6.20. The efficiencies are high for particles with $D_p > 1$ μ because the effectiveness of contact charging increases with particle size. On the other hand, removal efficiencies for particles with $D_p < 1$ μ increase as the particle size decreases because of the rise in diffusional charging for smaller particles. The removal efficiencies for particles in the size range between 0.1 and 1 μ are lower because of the size-dependent limitations of both charging mechanisms (USEPA, 2012).

6.5.2 ESP Systems and Their Components

ESPs can be used to collect solid as well as liquid particles. *Dry ESPs* operate at high temperatures, and particles are dry; collected particles on the plates are removed by rapping (i.e., mechanical vibration). *Wet ESPs* are used for sticky or moist particles,

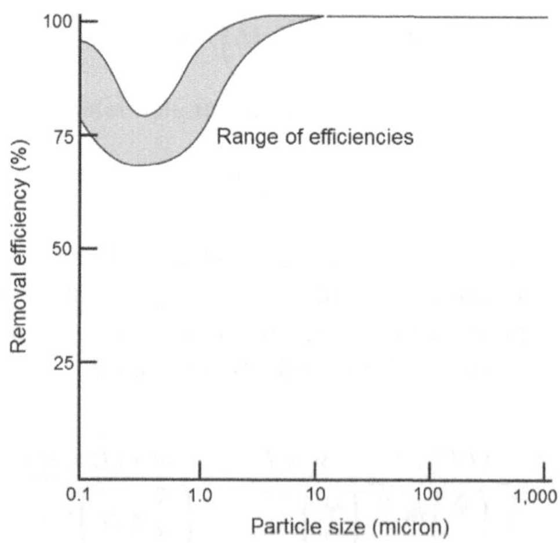

FIGURE 6.20 Fractional efficiency curve of ESPs (Modified from USEPA, 2012).

and the collection plates are continuously or intermittently washed with water. To meet the different needs of industrial applications, there are three categories of ESPs: (i) dry, negative corona, (ii) wet, negative corona, and (iii) wet, positive corona (USEPA, 2012).

Dry, negative corona-type ESPs are most commonly used. The collected solids are removed from the collection plates as a "dry" material. They are termed "*negative corona*" because the particles are forced to move from a negatively charged zone to the electrically grounded collection plates. As shown in Figure 6.21, the influent gas is guided through the perforated screen to go through passages formed by many parallel collection plates at a velocity between 3 and 6 ft/s (1 and 2 m/s). The typical spacing between two plates is 12 to 18 inches (0.3 to 0.45 m). High-voltage discharge electrodes are centered between each pair of plates. The electrodes can have small diameters, as shown in Figure 6.22 [Note: some newer ESPs use rigid support frames for discharge electrodes]. The high voltages applied to the electrodes create a negative corona that would charge most particles negatively. The discharge electrodes are divided into fields operated in series, and the ESPs can also be divided into separate chambers parallel to a solid wall. The system shown in Figure 6.21 appears to have two chambers, and each chamber has three fields, as indicated by the number of hoppers shown at the bottom. The dust layer accumulated on the collection plates is periodically removed by mechanical hammers called *rappers*. They fall into hoppers by gravity as particle agglomerates and/or as dust layer sheets (USEPA, 2012).

Wet, negative corona-type ESPs are used where mists or fogs need to be removed or when solid particles to be removed have undesirable electrical or physical properties, so that a washing system, instead of rappers, is used to remove particles accumulated on the collection plates (USEPA, 2012).

Wet, positive corona-type ESPs remove organic droplets or mists from small industrial applications. The positive, instead of negative, voltages are applied to the

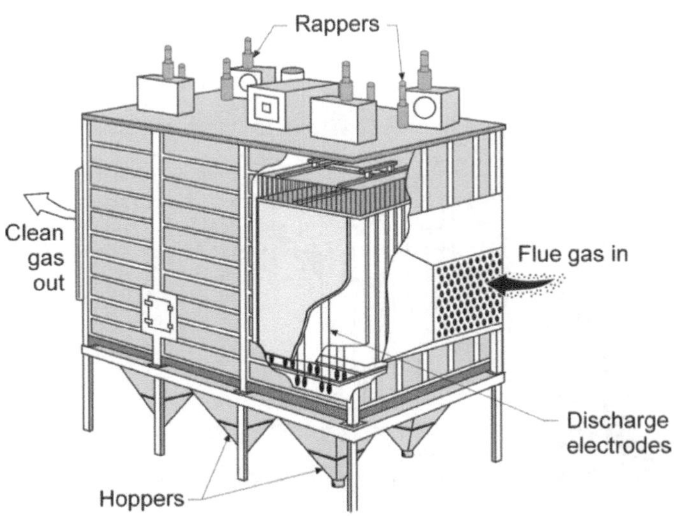

FIGURE 6.21 Schematic of a dry, negative corona type of ESP (USEPA, 2012).

discharge electrodes, which are housed in a pre-ionizer section ahead of the electrically grounded collection plates downstream. Since the ESPs only collect liquid particles, no rappers or washing liquid distribution systems are needed (Figure 6.23). The plates are cleaned periodically (USEPA, 2012).

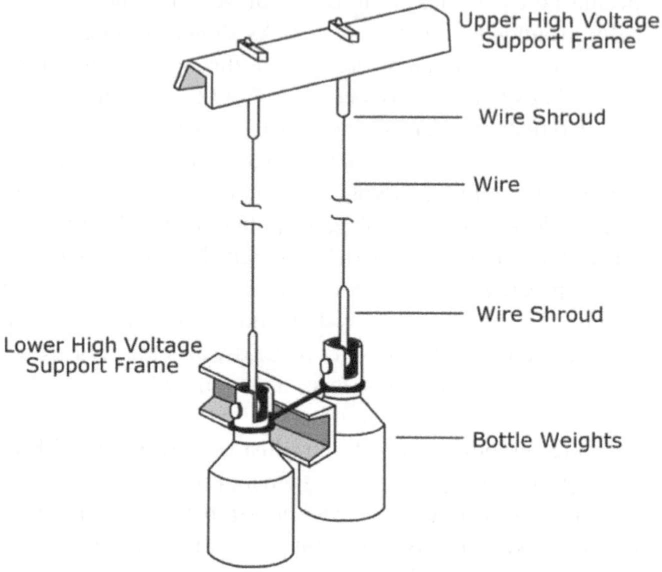

FIGURE 6.22 Wire-type discharge electrodes (USEPA, 2012).

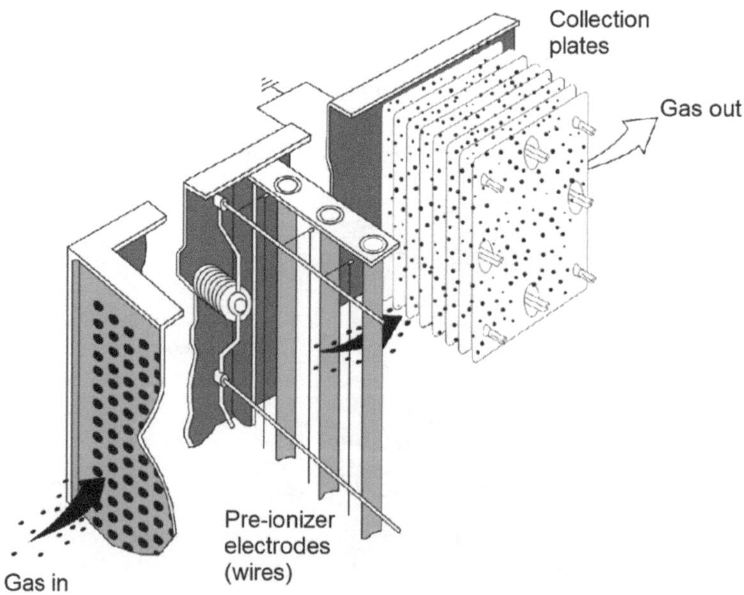

FIGURE 6.23 Wet, positive corona-type ESP (Modified from USEPA, 2012).

6.5.3 Factors Affecting the Performance of ESPs

From the Deutsch–Anderson equation (Eq. 6.52), the drift velocity (w) and the SCA (the A/Q ratio) are the only apparent variables affecting the performance of an ESP. The larger the SCA, the higher the removal efficiency would be. Therefore, for an ESP with a fixed total surface area of collection plates, lowering the gas flow rate into the treatment zone by reducing its temperature should raise the removal efficiency.

The larger the drift velocity, the higher the corresponding removal efficiency because the target particles will move to the collection plates faster. As shown in Eq. 6.28, the drift velocity is proportional to the particle diameter (D_p) and the square of the electric field strength (E). A stronger field will make the drift velocity larger. The electricity field strength is nothing but the voltage gradient with distance. It seems to be the simple ratio of the voltage difference between the discharge electrodes and the grounded collection plates and the distance between them. However, in actual operations, the cakes accumulated on the collection plates will affect the strength of the electricity field where the particles are being charged.

The voltage on the surface of an electrically grounded plate is zero; and the electrostatic voltage difference across the dust cake holds the cake on the vertical surface of the collection plate. The electrostatic voltage of the outermost surface of the dust layer, where new particles and gas ions are arriving, can be >10,000 V. The resistivity of the dust layer measures the ability of electrical charges to move through a dust layer. A commonly used resistivity unit is Ohm-cm. When the resistivity is lower, the electrons are conducted more readily; the charge difference across the dust layer will be smaller. The resistivity depends on the composition of the cake materials and the operating conditions (e.g., temperature and moisture content).

ESPs typically work best when the resistivity of the dust cake is in the moderate range (5×10^6 to 5×10^{10} Ohm-cm). For the case of low-resistivity dust cakes, they are weakly held. During rapping, many of the particles or their small agglomerates may be released back into the gas stream instead of moving downward to the collection hoppers. It will result in a temporary emission spike, called a *puff*.

If the resistivity of the cake is too high, the voltage at the outermost surface of the cake would become much larger. Consequently, the voltage difference between the discharge electrode and the cake's outermost surface becomes smaller, as does the electrostatic field strength. The migration speed of the particles toward the collection plate will become slower because of the smaller electrostatic field strength. The *back corona* is the second adverse effect of high-resistivity cakes, adversely impacting removal efficiency. This is because corona discharges in the gas trapped inside the cake will occur if the voltage drop across the dust layer becomes sufficiently large. Some gas molecules will become positively charged after a collision with other ions. The positively charged gas ions will move backward toward the discharge electrode. On their way, they will neutralize some negative charges on the particles in the dust layer and some negative ions on the particles approaching the cake. The third adverse impact is that the frequency of electrical sparking within the high-resistivity cake layer may become too frequent to be desirable. In addition to all these, it is also difficult to dislodge the dust cakes because the stronger electrostatic fields hold them more strongly.

The resistivity of high-resistivity dust can decrease as the gas temperature drops because of the increased adsorption of electrically conductive vapors. Some common chemicals, such as sulfur trioxide and ammonia, can also be injected into the gas stream inlet to improve the particles' *hygroscopic* property (i.e., a tendency to absorb moisture from the air). On the other hand, not much can be done about the dust cake's resistivity when it is below the moderate/preferred range.

6.5.4 Design Calculations for ESPs

For the conceptual design of an ESP, the drift velocity (w) and SCA need to be specified to estimate the removal efficiency. The drift velocity depends on the characteristics of particles (e.g., dielectric constant and particle size) as well as those of the system (e.g., the strength of the electrostatic field). USEPA (2012) states that typical drift velocity values range from 0.13 to 0.67 ft/s (4.0 to 20.4 cm/s). The SCA value tells the total surface area needed for a gas flow rate. The SCA values range from 300 to 1,400 ft^2 per 1,000 scfm (0.3 to 1.4 min/ft) or 1.0 to 4.6 min/m (USEPA, 2012). Although the ranges of w and A/Q are provided here, the proper values to achieve a specific particle removal would depend on many site-, unit-, and particle-specific factors such as PSD and resistivity, aspect ratio of the plates, and electric field strength, to name a few.

Example 6.19: Performance of ESPs

Referring to the 2-μ particles in Example 6.4, if an ESP is used to achieve a 95% removal efficiency,

a. what would the required surface area of the collection plates be?
b. what would the removal efficiency be if the flow rate is doubled?
c. what would be the removal efficiency for 1-μ particles?
[Note: $T=20°C$; $Q=20\ m^3/s$ (706 ft^3/s); $w=7.15$ cm/s]

Solution:

a. Using Eq. 6.52,

$$\eta = 99\% = 1 - \exp\left[-\frac{wA}{Q}\right] = 1 - \exp\left[-\left(7.15\frac{\text{cm}}{\text{s}}\right)\left(\frac{A}{Q}\right)\right]$$

$$(A/Q) = 0.644 \text{ s/cm} = 64.4 \text{ s/m} = 1.07 \text{ min/m}$$

$$A = (64.4 \text{ s/m})(20 \text{ m}^3/\text{s}) = \underline{1{,}288 \text{ m}^2}$$

b. If the flow rate is doubled, then $(A/Q)=(0.644/2)=0.322$ s/cm. Assuming that w is independent of Q (not a bad assumption), then

$$\eta = 1 - \exp[-(7.15)(0.322)] = 90\%$$

c. Since w is proportional to D_p, the drift velocity of the 1-μ particle will be half that of the 2-μ particle. With the same (A/Q) as part (a), the removal efficiency will be the same as that of (b), 90%, since they have the same $[w(A/Q)]$ value.

Discussion:

1. In this example, drift velocity (7.15 cm/s) and SCA (1.07 min/m) are within the typical ranges. This example illustrates that ESPs effectively remove fine particles (D_p=2 μ in this example).
2. The units of w and (A/Q) should match.
3. Efficiency decreases from 99% to 90% if the flow rate doubles.
4. The removal efficiency of 1-μ particle is lower (90%), but still relatively high when compared to typical cyclones.

To take a more conservative approach, some would like to use the *modified Deutsch–Anderson equation*:

$$\eta = 1 - \exp\left[-\left(\frac{wA}{Q}\right)^k\right] \tag{6.53}$$

where k is an empirical constant, ranging from 0.4 to 0.6.

Example 6.20: The Modified Deutsch–Anderson Equation

Referring to Example 6.19, use the modified Deutsch–Anderson equation with k=0.5 to estimate the removal efficiencies for the 2- and 1-μ particles by an ESP with A/Q=0.644 s/cm.

Solution:

a. The drift velocity of the 2-μ particles = 7.15 cm/s.
   Using Eq. 6.53,

$$\eta = 1 - \exp\left[-(7.15 \times 0.644)^{0.5}\right] = 88.3\%$$

b. The drift velocity of the 1-μ particles = (7.15/2) = 3.575 cm/s.

$$\eta = 1 - \exp\left[-(3.575 \times 0.644)^{0.5}\right] = 78.1\%$$

Discussion: As expected, the estimated removal efficiencies are lower when using the modified *D–A* equation.

6.6 FABRIC FILTERS

All three types of particulate removal devices mentioned earlier (gravity settling chambers, cyclones, and ESPs) share something in common. In these devices, a force is imposed on a particle in the gas stream to move it out of the gas streamline to hit a wall and get retained by the wall. The forces are the gravitational force for gravity settling chambers, the centrifugal force for cyclones, and the electrostatic force for ESPs. The last two types of particulate removal devices in this chapter (i.e., fabric filters and wet scrubbers) work differently. In these devices, the gas streams are divided by the collection targets (i.e., fabric media for filters and water droplets for wet scrubbers), and the collection targets retain the particles as the gas stream flows through the openings of the filter media or around the water droplets.

6.6.1 Principles of Filtration

A fabric filter system consists of a housing unit that hosts lots of fabrics to provide a large surface area for the gas stream to go through. Most of the particles in the incoming gas stream are retained by the fabric, and the dust cake is gradually formed by the particles collected on the fabric. As the gas stream moves through, both the dust accumulated in the cake and the filter fabric become the collection targets. The particle captured by these collection targets can be attributable to several mechanisms, including inertial impaction, Brownian diffusion, and electrostatic attraction (moderate electrostatic charges, negative or positive, can build on the surfaces of the dust cake, fabric, and particles). Sieving of particles by the pores of the cake is another removal mechanism.

Inertial impaction is the dominant removal mechanism for particles with $D_p > 1\mu$, while Brownian diffusion is moderately effective for collecting sub-micron particles. Different from most of the other particle removal devices, there are multiple opportunities for particles to encounter the collection targets. The fractional efficiency curve of fabric filters (Figure 6.24) illustrates that they are effective even in the difficult-to-control range of 0.2 to 0.5 μ. It should also be noted that fabric filters have a much wider applicable range of 0.1 to 1,000 μ than most particulate control devices (USEPA, 2012).

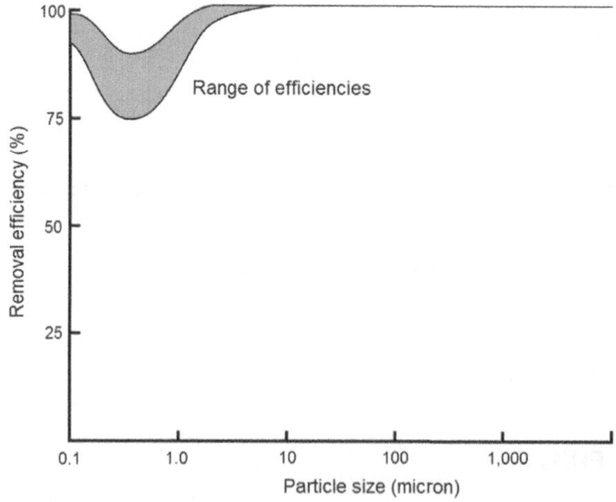

FIGURE 6.24 Fractional efficiency curve of fabric filters (Modified from USEPA, 2012).

6.6.2 Fabric Filtration Systems and Their Components

Two main functions of fabric material in a filtration system are to provide air passage and to support the filter cake. They need to tolerate temperature, chemical attack, and physical abrasion. They may also need to be flexed in some types of filter systems during the cleaning cycle. The two most common types of fabrics used are (i) woven and (ii) felted. A *woven fabric* is a type of textile produced by interlacing two sets of yarns/threads at right angles to each other. The yarns in the "warp" direction provide the strength, while those in the "fill" (or the weft) direction determine the characteristics of the fabric. The sizes of the pores can be >50 μ (Figure 6.25).

A *felted fabric* (or *felt*) is a type of non-woven textile and comprises randomly oriented fibers supported by a very open weave. The felted fabrics are usually thicker than the woven ones because of a layer of felt on both sides of the weave support (Figure 6.26). The fibers on the dirty air side and the dust cake provide numerous targets for incoming particles. In addition to woven and felted fabrics, membrane fabric, sintered metal fibers, and ceramic cartridges are also used (see USEPA (2012) for details).

As the cake builds up on the fabric's surface by removing more dust from the gas stream, the head loss (pressure drop) across the filter will go up accordingly. The filter material must be cleaned periodically to avoid excessive pressure drops that challenge the blower to deliver a specific gas stream flow rate and incur more energy costs. Based on the cleaning methods, the commonly used filters are grouped into (i) shaker, (ii) reverse-air, and (iii) pulse-jet fabric filters.

Figure 6.27 shows a schematic of a *shaker fabric filter*. The open bottoms of the filter bags are attached to a tube sheet, which provides a seal to separate the upper portion of the bags from the dust hopper. The closed tops of the bags are connected to a shaker mechanism. The particle-laden gas stream enters from the hopper, where some larger particles may settle out. The remaining particles will move upward with the gas stream, and many will deposit on the inner surface of the bags as the gas stream moves through the bag. The dirty gas inlet to the fabric filter is closed during

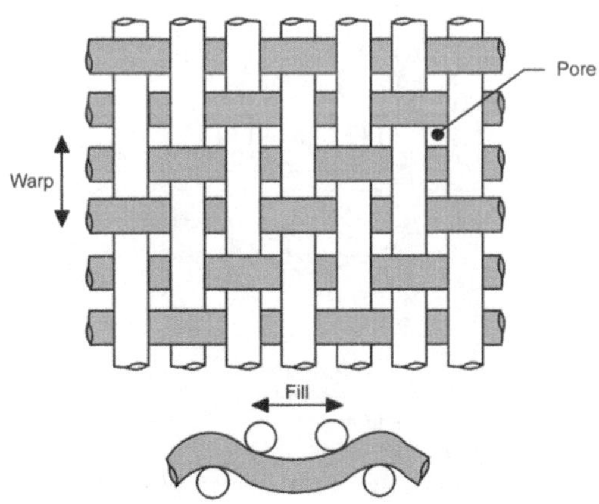

FIGURE 6.25 Woven fabric (USEPA, 2012).

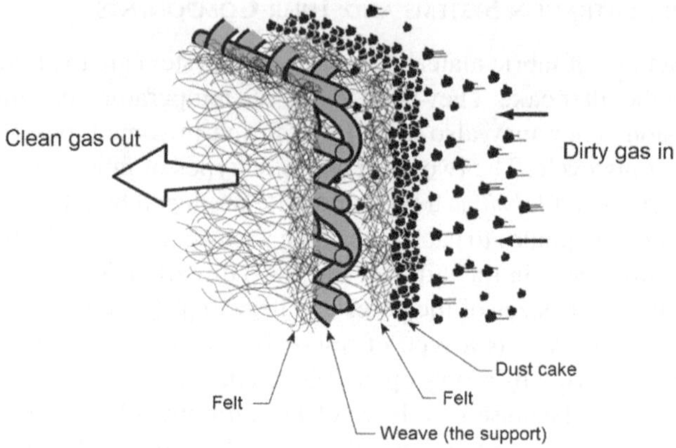

FIGURE 6.26 Felted fabric (Modified from USEPA, 2012).

the cleaning cycle. The bags are then mechanically shaken to dislodge the dust cakes into the collection hopper (USEPA, 2012).

Construction of *reverse-air fabric filters* is very similar to that of the shaker fabric filters, except that the close tops are attached to a support structure, instead of a shaker mechanism (see Figure 6.27). For the service cycle, the operations of these two types of filters are essentially the same. The cleaning cycle starts by stopping

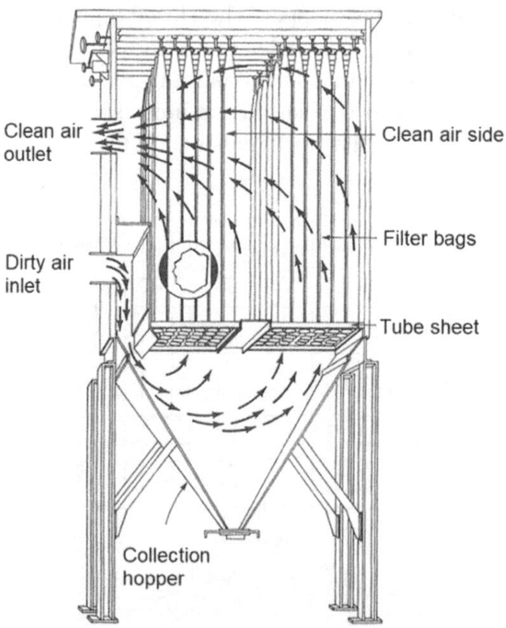

FIGURE 6.27 Shaker fabric filter (Modified from USEPA, 2012).

the dirty gas stream's influent and closing the filter outlet. These filter systems usually have several compartments operated in parallel. Part of the clean gas stream effluent(s) from the other compartment(s) in service is directed to pass from the outside to the inside of the bags to dislodge the cakes and then move through the dirty air inlet. Since the flow path of the gas in the cleaning cycle is opposite to that in the service cycle, that is where the name "reverse air" came from (USEPA, 2012).

Figure 6.28 shows a *pulse-jet fabric filter* in which a tube sheet is located close to the top of the filter unit to hold the tops of the bags. The gas stream enters the unit from the side of the casing or the hopper, flows into the bags, and leaves the dust cake outside the bags before moving upward into the discharge outlet. The bags are also supported by metal cages inside to prevent collapsing. The cleaning cycle can be done concurrently within the service cycle on a row-by-row basis. A high-pressure pulse of compressed air is introduced at the top of each bag, generating a pressure wave traveling downward inside the bag. The pressure wave will also induce some filtered gas to flow downward. The combined action will force the bags to expand outward to crack the dust cake on the outside of the bags, causing the dust to fall into the collection hopper. Offline cleaning is more common for filter systems with multiple compartments (USEPA, 2012).

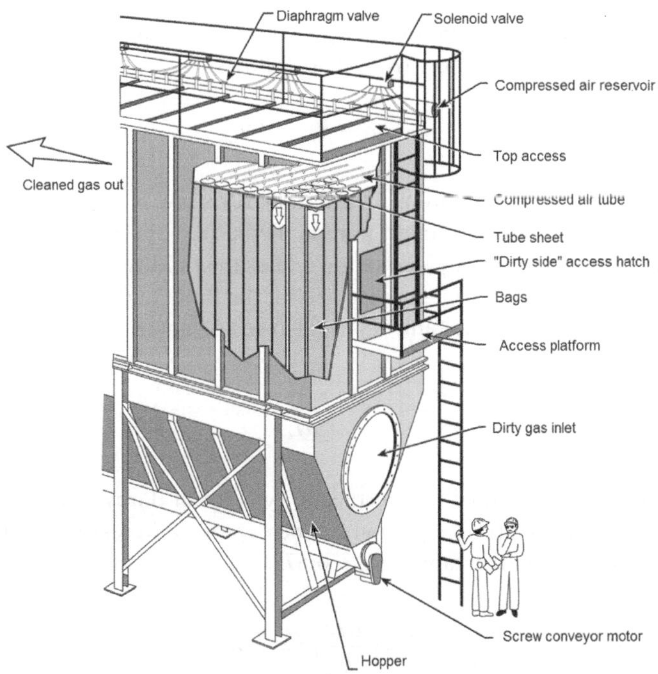

FIGURE 6.28 Pulse-jet fabric filter (Modified from USEPA, 2012).

6.6.3 Factors Affecting the Performance of Fabric Filters

Dust cake is the most critical component of fabric filtration, governing the performance and operation of a given system. A new filter bag's initial particle removal efficiency is not as good before a sufficiently thick dust layer is built on the fabric. The poorer performance may also appear immediately after the bags are returned to service after cleaning. Excessive intensity, frequency, and duration of cleaning can increase particle emissions, at least for a short duration. On the other hand, a dust cake that is too thick may result in a larger pressure drop, which means a higher energy cost to push a specific gas flow rate through the filter.

The total press drop of air flow through bag houses (ΔP_{total}) is the sum of the pressure drop across the filter (ΔP_{filter}) and the pressure losses in system components, e.g., ductwork, dampers, and hoppers before and after the baghouse (ΔP_{system}):

$$\Delta P_{total} = \Delta P_{filter} + \Delta P_{system} \tag{6.54}$$

ΔP_{filter} is the sum of the pressure drop across the clean filter fabric (ΔP_{fabric}) and that of the dust cake (ΔP_{cake}), as follows:

$$\Delta P_{filter} = \Delta P_{fabric} + \Delta P_{cake} \tag{6.55}$$

The pressure drop across the fabric (ΔP_{fabric}) is proportional to the gas velocity (v_{gas}) and the fabric's resistance to flow (R_{fabric}), as follows:

$$\Delta P_{fabric} = R_{fabric} \times v_{gas} \tag{6.56}$$

A dust cake is nothing but a porous medium. Darcy's law is commonly used in fluid mechanics to describe flow through porous media (e.g., groundwater aquifers, soil columns, and filter beds). Darcy's equation relates the fluid velocity (v) to the intrinsic permeability of the medium (k), the viscosity of the fluid, and the pressure drop (ΔP) across the thickness of the medium (L) as (also see Chapter 4):

$$v = \frac{k\Delta P}{\mu L} \tag{6.57}$$

The common unit for the intrinsic permeability is m^2 or darcy [Note: 1 darcy = $9.87 \times 10^{-13}\,\text{m}^2 \cong 10^{-12}\,\text{m}^2$]. Darcy's equation can be expressed in resistivity (R), instead of intrinsic permittivity, by rearranging Eq. 6.57:

$$\Delta P = \left[\frac{\mu L}{k}\right] v = R \times v \tag{6.58}$$

Consequently, the pressure drop across the fabric would become:

$$\Delta P_{\text{fabric}} = R_{\text{fabric}} \times v_{\text{gas}} = \left[\frac{\mu_g L_{\text{fabric}}}{k_{\text{fabric}}}\right] \times v_{\text{gas}} = \frac{\mu_g L_{\text{fabric}} v_{\text{gas}}}{k_{\text{fabric}}} \tag{6.59}$$

where L_{fabric} is the thickness of the fabric and k_{fabric} is the intrinsic permittivity of the fabric. Similarly, the pressure drop across the dust cake:

$$\Delta P_{\text{cake}} = R_{\text{cake}} \times v_{\text{gas}} = \frac{\mu_g L_{\text{cake}} v_{\text{gas}}}{k_{\text{cake}}} \tag{6.60}$$

where R_{cake} represents the resistance of the cake to flow. As shown, the pressure drop is proportional to the gas velocity and the thickness of the cake (and the filter fabric).

The thickness of the dust cake will increase with the time of the filter in service (Δt) and the particulate mass loading, which is proportional to v_{gas} and the particulate mass concentration in the gas stream (C_{PM}), thus:

$$L_{\text{cake}} = (\text{constant})(C_{\text{PM}})(v_{\text{gas}})(\Delta t) \tag{6.61}$$

The constant in Eq. 6.61 accounts for unit conversions and the conversion of the retained mass into the volume of the dust cake. If the bulk density of the cake (ρ_b) is known and stays constant, Eq. 6.61 can be expressed as:

$$L_{\text{cake}} = (C_{\text{PM}})(v_{\text{gas}})(\Delta t) \div \rho_b \tag{6.62}$$

The total pressure drop across the system is:

$$\Delta P_{\text{system}} = R_{\text{fabric}} v_{\text{gas}} + R_{\text{cake}} v_{\text{gas}} \tag{6.63}$$

Since R_{cake} is proportional to L_{cake}, which increases with particulate concentration, the service time, and gas stream velocity (v_{gas}), the pressure drop across the cake is proportional to the square of v_{gas}.

The gas stream velocity (v_{gas}) is the ratio of the gas flow rate (Q_{gas}) and the surface area of the bags (A_{bag}) as:

$$v_{\text{gas}} = \frac{Q_{\text{gas}}}{A_{\text{bag}}} \tag{6.64}$$

This (Q/A) ratio in Eq. 6.64 is often called the *air-to-cloth ratio, superficial velocity,* or *face velocity*. It sizes the bags needed for a given gas flow rate. Using a larger air-to-cloth ratio to design a fabric filter would make the required surface area smaller. However, as the air-to-cloth ratio increases, the *interstitial gas velocities* (the actual gas velocities) through the cake and the fabric increase. These higher velocities may push some smaller particles through these two barriers (i.e., the cake

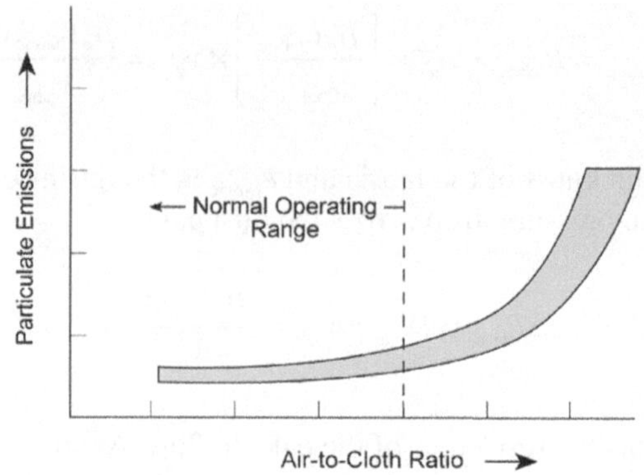

FIGURE 6.29 Effect of air-to-cloth ratio (Modified from USEPA, 2012).

and the filter fabric) and emit from the system. In addition, the higher air-to-cloth ratios may adversely crush the existing cake structure so that uncovered openings on the fabric may allow particles to go through.

For a filter system, there is a threshold air-to-cloth ratio. Below this threshold value, the particulate emission rates are relatively constant. Above this threshold, the emissions can increase sharply (Figure 6.29). The shaker fabric filters often use woven fabrics and operate with an air-to-cloth ratio of 2 to 4 ft/min (0.6 to 1.2 m/min). Reverse-air collectors usually use woven fabrics (membrane and felted bags are used in some applications) and operate with an air-to-cloth ratio of 1.5 to 3.5 ft/min (0.45 to 3.5 m/min). Pulse-jet collectors use felted fabrics and operate with an air-to-cloth ratio of 3 to 10 ft/min (0.9 to 3.1 m/min) (USEPA, 2012).

Example 6.21: Thickness of a Dust Cake

Suppose the face velocity (i.e., the air-to-cloth ratio) to a fabric filter is 1.0 m/min. The dust concentration is 10 g/m^3, and the filtration time before cleaning is 300 seconds (5 minutes). Assuming the density of dust cake = 700 kg/m^3, estimate the thickness of the cake at the end of the service cycle.

Solution:

Using Eq. 6.62,

$$L_{\text{cake}} = (C_{\text{PM}})(v_{\text{gas}})(\Delta t) \div \rho_b = (10 \times 10^{-3}\ \text{kg/m}^3)(1.0\,\text{m/min})(5\ \text{minutes}) \div 700\ \text{kg/m}^3$$
$$= 7.1 \times 10^{-5}\text{m}$$

Discussion:

1. Matching units need to be used.
2. NCEES (2025) includes an equation similar to Eq. 6.62 for the calculation of filter cake thickness. However, the unit for dust loading is in gr/ft^3 and that for bulk density is in gr/ft^3.
3. 1 grain = 64.79891 mg ≈ 64.8 mg (or 0.0648 g); 1 g ≈ 15.432 grains.
4. 1 grain/ft^3 ≈ 2.29 × 10^{-3} kg/m^3; 1 kg/m^3 = 437 grain/ft^3.

Example 6.22: Pressure Drop across a Dust Cake

Suppose the face velocity (i.e., the air-to-cloth ratio) of a fabric filter is 1 m/min. In that case, the thickness of the filter cake is 2 mm, and the intrinsic permittivity of the cake is one darcy. Estimate the pressure drop across the cake ($T=20°C$).

Solution:
Using Eq. 6.60,

$$\Delta P_{cake} = \frac{\mu_g L_{cake} v_{gas}}{k_{cake}} = \frac{(1.81\times10^{-5}\text{kg/m}\cdot\text{s})(0.002\,\text{m})\left(\frac{1}{60}\text{m/s}\right)}{(9.87\times10^{-13}\text{m}^2)} = 611\frac{\text{N}}{\text{m}^2} = 2.45\,\text{in-H}_2\text{O}$$

6.6.4 Design Calculations for Fabric Filters

As mentioned in the previous section, the air-to-cloth ratio can be used to estimate the required surface area of a fabric filter system; typical ranges were also given. More suggested air-to-cloth ratios can be found in the literature. These general values should only be used as a starting point. A safety factor should always be used, especially when the dust loading is higher or the particle sizes are small. In addition, the calculated surface area is for the bags in service; the surface area of the bags, if they are offline for cleaning or regular maintenance, is not part of the calculated value (USEPA, 2012). Note: NCEES (2025) includes (i) a table of air-to-cloth ratios for different types of dusts (e.g., alumina, carbon black, fly ash) for two different groups of bag houses (shaker and reverse air (woven) vs. pulse jet (felt); (ii) a table for maximum filtering velocities for various dusts in shaker or reverse-air baghouses, and (iii) temperature and chemical resistance of some common industrial fabrics.

Example 6.23: Surface Area of Fabric Filters

A shaker fabric filter has six compartments. During normal operation, five compartments are in service while the other one is offline for cake removal or maintenance. Each compartment contains 100 bags that are 0.4 m in diameter and 6 m long. The air-to-cloth ratio is 1.2 m/min.

a. Calculate the available surface area of each bag.
b. What is the air flow rate treated by this baghouse?

Solution:
a. $A=\pi Dh=\pi(0.4)(6)=7.54\,\text{m}^2$
b. Number of bags in service = (5)(100) = 500

$$\text{Surface area of bags in service} = (500)(7.54) = 3{,}770\ \text{m}^2$$

From Eq. 6.64,

$$\frac{Q}{A} = 1.2\ \text{m/min} = \frac{Q_{gas}}{3{,}770\ \text{m}^2}$$

$$Q_{gas} = (1.2)(3{,}770) = 4{,}520\ \text{m}^3/\text{min}$$

6.7 WET SCRUBBERS

Wet scrubbers use a liquid stream, typically water, to collect particulates and/or gaseous pollutants from a gas stream. Wet scrubbers are often the control device of choice if the gas stream to be treated contains explosive gases or vapors. Chemicals can also be added to the scrubbing fluid to adjust certain characteristics of the gas stream (e.g., corrosiveness). It can also be a good alternative if the simultaneous removal of particulates and gaseous pollutants is desirable.

6.7.1 Principles of Wet Scrubbing

Although wet scrubbers can collect both particles and gases at the same time, they usually do not have high efficiency for both simultaneously. This is mainly because they are designed to employ different removal mechanisms for particulates and gaseous pollutants. For effective removal of gaseous pollutants, wet scrubbers need to provide large air–water interfacial areas and relatively long residence times to facilitate the absorption of gaseous pollutants into liquid droplets (more details in the next chapter).

The main removal mechanism involved in collecting PM by wet scrubbers is impaction, in which water droplets serve as the collection targets. Due to the limited residence time in most wet scrubbers, Brownian motion is typically not a significant removal mechanism. Some wet scrubbers use enhancements to improve their removal of fine particles by electrostatic charging and promotion of condensation (USEPA, 2012).

There are many types of wet scrubbers, and many of them have limited efficiencies for particles in the difficult-to-control range of 0.1 to 1.0 μ. The removal efficiency in this size range depends primarily on the intensity of contact between the gas (and the particles) and the water in the scrubber. Some types of scrubbers (e.g., Venturi scrubbers) use high energies to develop large relative velocities between particles and collection targets (i.e., water droplets) to develop excellent inertial impaction efficiencies for particles in this size range. Figure 6.30 is a typical fractional efficiency curve that illustrates the range of performance for various types of wet scrubbers.

6.7.2 Wet Scrubbing Systems and Their Components

There are several types of scrubber designs to bring contact between liquid droplets and particulates in the gas stream. Since inertial impaction is the primary removal mechanism in scrubbers for particulate removal, the capability of a scrubber can be

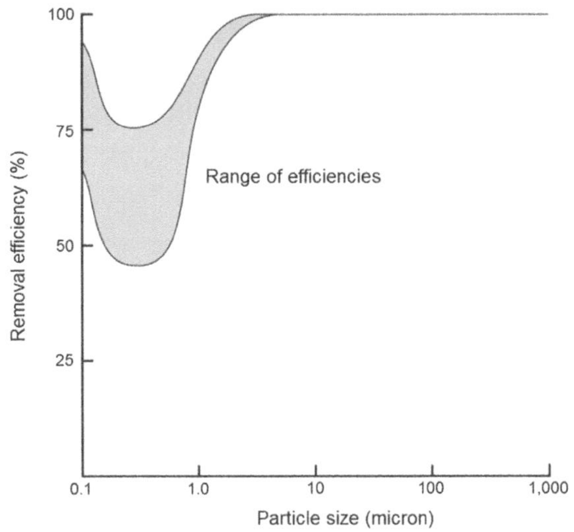

FIGURE 6.30 Fractional efficiency curve of wet scrubbers (Modified from USEPA, 2012).

approximated from the pressure drop of the gas stream across the scrubber. A higher pressure drop implies more aggressive contact between the liquid and gas streams, resulting in better removal of fine particles.

Scrubbers with static pressure drops of < ~5 in-H_2O (1.25 kPa) are referred to as *low-energy wet scrubbers*. They are, in general, capable of removing particles with diameters >5 to 10 μ. *Medium-energy wet scrubbers* are capable of removing particles with sizes in microns and have pressure drops of 5 to 25 in-H_2O (1.25–6.25 kPa). Sub-micron particles can be effectively removed by high-energy wet scrubbers, but the pressure drop can be greater than 100 in-H_2O (25 kPa), depending on the particle size to be removed (USEPA, 2012).

Spray tower scrubbers consist of a chamber, cylindrical or rectangular, with an array of spray nozzles mounted on multiple heads that are often spaced about one meter apart (Figure 6.31). Water droplets typically have a mean size of several hundred microns. It is an example of low-energy wet scrubbers, which is simple and inexpensive. Because of the relatively low velocity between the particles and the water droplets, they are only effective for particles larger than 5 μ. They can be alternatives when control of both PM and gaseous pollutants is needed (USEPA, 2012).

Packed-bed wet scrubbers are medium-energy wet scrubbers. They consist of a chamber containing a layer of packing. The packing is held in place by a wire-mesh retainer and supported by a plate. Scrubbing liquid is evenly distributed above the packing and moves downward through the packing bed to form a liquid film on the packing. The gas stream is forced to move upward through the packing to get in contact with the liquid (Figure 6.32). Packing materials come in different sizes and materials; some common types are shown in Figure 6.33. Although ceramic and metal packings are available, plastic packings are used in most of the air pollution applications (USEPA, 2012).

If the upward air velocity is too large through the packing bed, the water may not be able to travel downward. This limitation will result in a reduction in removal

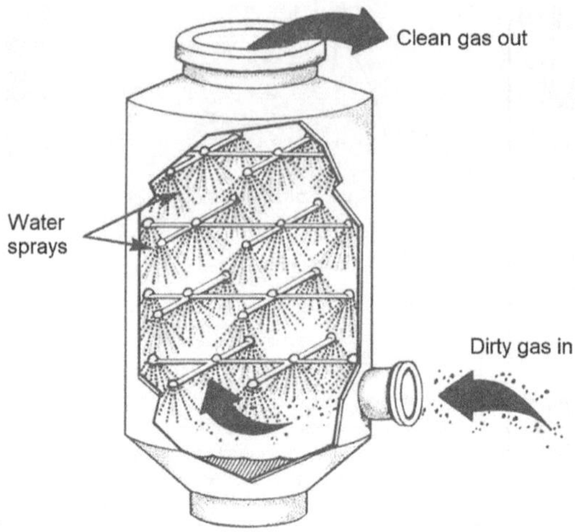

FIGURE 6.31 Schematic of a spray tower scrubber (Modified from USEPA, 2012).

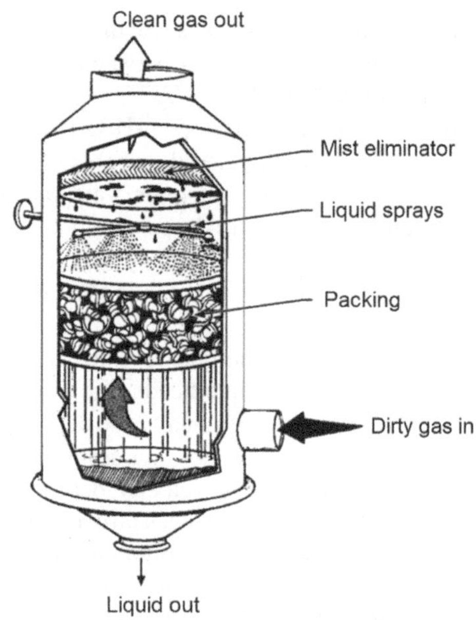

FIGURE 6.32 Schematic of a packed-bed scrubber (Modified from USEPA, 2012).

efficiency. Removal efficiencies for fine particles (<3 μ) are very low. Too high a particulate concentration may cause plugging in the packing materials (USEPA, 2012).

An *impingement-plate scrubber* is a vertical chamber with several plates/trays mounted horizontally inside. It is another example of a medium-energy wet scrubber. Water and air flow counter-currently, with water flowing downward. The plates are metallic plates with openings of ~3/16 inches (5 mm) in diameter. Small baffle

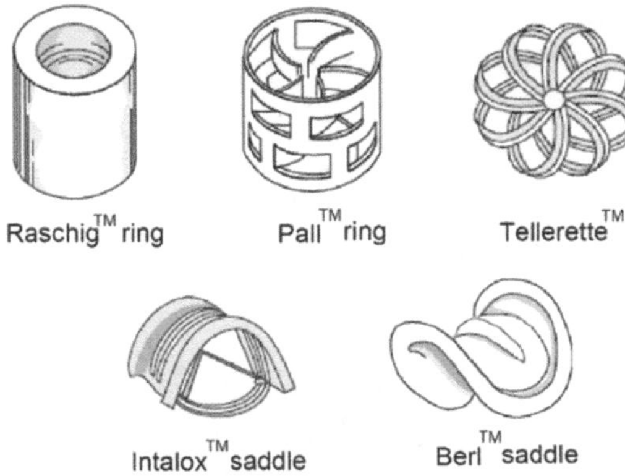

FIGURE 6.33 Common types of packing (Modified from USEPA, 2012).

plates are mounted directly above each of the holes. The openings allow the gas to pass through and create a bubbling action so that the particles can be collected by impact. The liquid depth in each tray is set by overflow weirs on each tray. The liquid overflows downward to the next tray through downcomers (Figure 6.34). They have limited removal efficiencies for particles smaller than 1 μ, due to the limitation of the gas stream velocity through the openings of the trays (USEPA, 2012).

The *Venturi scrubber* is an example of a high-energy wet scrubber. The fixed-throat Venturi scrubber (Figure 6.35) is one of the most common designs. The gas stream

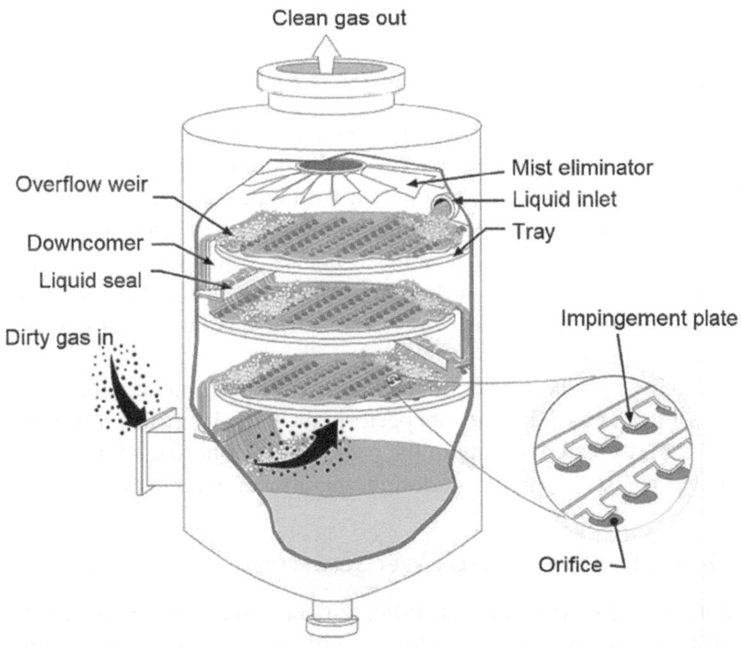

FIGURE 6.34 Schematic of an impingement-plate scrubber (Modified from USEPA, 2012).

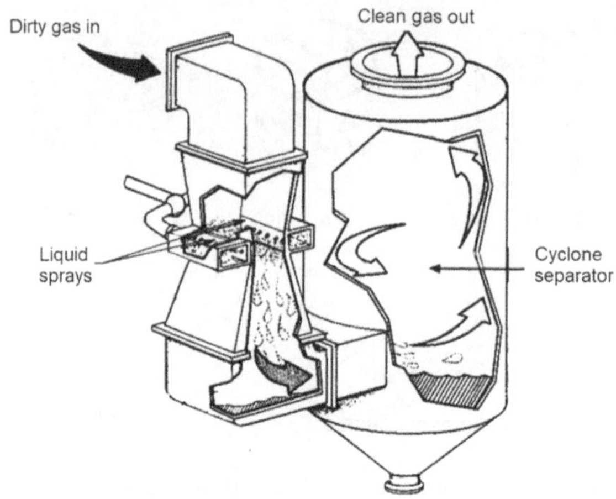

FIGURE 6.35 Schematic of a Venturi scrubber (Modified from USEPA, 2012).

entering the converging section of the Venturi is accelerated to a velocity between 200 and 600 ft/s (60 and 180 m/s) at the inlet of the throat. The liquid is injected into the throat and atomized into fine droplets to enhance the gas–liquid contact. The size of the droplets depends on the throat gas velocity and the liquid-to-gas ratio, and the typical mean size of the fine liquid droplets is 50 to 75 μm. The effectiveness of a Venturi scrubber depends on the relative velocity between the gas stream and liquid droplets in the throat. They can remove sub-micron particles.

There are other types of wet scrubbers available (see USEPA, 2012).

6.7.3 Factors Affecting the Performance of Wet Scrubbers

The liquid flow rate is essential for any wet scrubber's removal performance. The liquid flow rate is often expressed in terms of the liquid-to-gas ratio. Most wet scrubber systems for particulate removal operate with liquid-to-gas ratios between 4 and 20 gallon/1,000 ft^3, equivalent to a dimensionless ratio of about 0.0005 to 0.0025. Higher ratios do not usually enhance performance. Instead, they may exert a slight adverse impact due to changes in the size distribution of the droplets. On the other hand, low liquid-to-gas ratios can have severe adverse effects because of insufficient collection targets present.

Although a higher relative gas velocity across the packing bed, the tray openings, or the Venturi throat will enhance its performance, it will incur higher power costs. In addition, it may create flooding in the packing bed. The size of water drops in the spray scrubber is also an important design/operational parameter.

6.7.4 Design Calculations for Wet Scrubbers

Many parameters (e.g., liquid-to-gas ratio, temperature, dimensions of the scrubber, particle size, and types of packings/trays) will affect the capacity/performance of a scrubber system. Two general approaches can be used: (1) pilot-scale tests and

(2) empirical equations based on previously installed scrubbers for similar sources. Below are some empirical equations mentioned in USEPA (2012). It should be noted that the inertial impact number (Ψ_I) is the most influential factor for removal, and the liquid-to-gas ratio is also important because it significantly affects the droplet sizes and their distribution [Note: water droplets are the collection targets in scrubbers].

Spray tower wet scrubbers. For counter-current spray tower scrubbers, collection efficiency due to impaction by a single droplet (η_I) can be estimated by (USEPA, 2012; Calvert et al., 1972; Calvert, 1977; Engineering Science, 1979):

$$\eta_I = \left(\frac{\Psi_I}{\Psi_I + 0.35}\right)^2 \tag{6.65}$$

where Ψ_I is the inertia impact number, as shown in Eq. 6.29.

The removal efficiency of the tower can be estimated by:

$$\eta_{\text{spray tower}} = 1 - \exp\left\{-\left[\frac{1.5 u_t \eta_I z}{D_d\left(u_t - v_g\right)}\right]\left(\frac{L}{G}\right)\right\} \tag{6.66}$$

where u_t is the terminal velocity of the droplets, z is the height of the scrubber, D_d is the water droplet diameter, v_g is the gas velocity, and (L/G) is the liquid-to-gas flow rate ratio (dimensionless).

Example 6.24: Performance of a Spray Tower Scrubber

Estimate the removal efficiency of particles with D_p=4μ (ρ_p=1,200 kg/m³) for a gas stream (150 m³/min) by a counter-current spray tower (height=3 m, T=293 K). The liquid flow rate is 120 L/min (droplet diameter=500μ), and the gas velocity inside the scrubber is 1 m/s.

Solution:

a. Use Eq. 6.15 to find the terminal velocity of the water droplet:

$$u_t = (1.0)\left[\frac{(9.81)(1{,}000 - 1.21)\left(500\times10^{-6}\right)^2}{(18)\left(1.81\times10^{-5}\right)}\right] = 7.52\ \text{m/s}$$

b. Use Eq. 6.19 to find the Cunningham slip correction factor (C_c) for the 4-μ particles:

$$C_c = 1 + \frac{6.21\times10^{-4}T}{D_p} = 1 + \frac{\left(6.21\times10^{-4}\right)(293)}{4} = 1.05$$

c. Use Eq. 6.29 to find the inertial impact number (Ψ_I):

$$\Psi_I = \left[\frac{(1.05)(4\times10^{-6})^2(1,200)(7.52-1.0)}{18(1.81\times10^{-5})}\right] \div (500\times10^{-6}) = 0.807$$

d. Use Eq. 6.65 to find the removal efficiency of a single droplet (η_I):

$$\eta_I = \left(\frac{\Psi_I}{\Psi_I + 0.35}\right)^2 = \left(\frac{0.807}{0.907+0.35}\right)^2 = 0.486$$

e. Use Eq. 6.66 to find the removal efficiency by the scrubber:

$$\eta_{\text{spray tower}} = 1 - \exp\left\{-\left[\frac{1.5(7.52)(0.486)(3)}{(500\times10^{-6})(7.52-1.0)}\right]\left(\frac{0.12}{150}\right)\right\} = 98.2\%$$

Discussion:

1. In part (a), a value of unity was used for the Cunningham slip factor because the droplet size is much larger than a few microns, and 1,000 kg/m^3 (the density of water) was used as the density of the water droplets.
2. In part (b), C_c for the 4-μ particles was found to be 1.05, which is very close to unity.
3. In part (c), the relative velocity (the difference between the droplet velocity and the gas velocity) should be used, and a value of 6.52 m/s (= 7.52 – 1.0) is used.
4. The removal efficiency of 4-μ particles by this spray scrubber is relatively high, at 98.2%.

Packed-bed wet scrubbers. For packed-bed scrubbers, the removal efficiency can be estimated (USEPA, 2012; Calvert et al., 1972; Calvert, 1977; Engineering Science, 1979):

$$\eta_{\text{packed bed}} = 1 - \exp\left\{-\left[\frac{\pi z \Psi_I}{D_c \varepsilon (j - j^2)}\right]\right\} \tag{6.67}$$

where z is the scrubber height, j is the channel width as a fraction of packing diameter (dimensionless), ε is the bed porosity, and D_c is the diameter of packings.

The packing diameter should be used instead of the water droplet diameter to calculate the inertial impact number. The channel width as a fraction of packing diameter (j) depends on the type of packing used and typically ranges from 0.165 to 0.192. The bed porosity (ε) depends on the type of packing used and typically ranges from 0.57 to 0.94. The values of these two parameters should be obtained from the manufacturers.

Example 6.25: Performance of a Packed-Bed Scrubber

Referring to Example 6.24, a packed bed of 3 m will be used instead (packing = 5 cm Raschig™ rings). Assume $j = 0.165$ and porosity = 0.75.

Solution:

a. Use Eq. 6.29 to find the inertial impact number (Ψ_I):

$$\Psi_I = \left[\frac{(1.05)(4\times10^{-6})^2(1,200)(1.0)}{18(1.81\times10^{-5})}\right] \div (0.05) = 1.24\times10^{-3}$$

b. Use Eq. 6.67 to find the removal efficiency by the packed-bed scrubber:

$$\eta_{\text{packed bed}} = 1-\exp\left\{-\left[\frac{\pi z \Psi_I}{D_c \varepsilon (j - j^2)}\right]\right\} = 1-\exp\left\{-\left[\frac{\pi(3)(1.24\times10^{-3})}{(0.05)(0.75)(0.165-0.165^2)}\right]\right\}$$
$$= 89.6\%$$

Discussion: In part (a), the target diameter is the size of the packing (0.05 m), and the velocity is the gas velocity (1 m/s).

Impingement-plate wet scrubbers. For tray scrubbers, the removal efficiency can be estimated by (USEPA, 2012; Calvert et al., 1972; Calvert, 1977; Engineering Science, 1979):

$$\eta_{\text{tray}} = 1-\left[\exp\left(-80F^2\Psi_I\right)\right]^n \tag{6.68}$$

where F is the foam density fraction (dimensionless) and n is the number of trays.

The gas velocity through the holes in the plate and the diameter of the holes, instead of the diameter of the water drops, should be used to calculate the inertial impact number. The foam density fraction in the scrubbers refers to the proportion of the tray volume occupied by foam, which is the mixture of liquid and gas. The foam density fraction typically ranges from 0.38 to 0.65.

Example 6.26: Performance of a Tray Scrubber

Referring to Example 6.24, a tray scrubber will be used instead (3 trays with 10 mm holes; the gas velocity through the holes is 18.0 m/s). Assume the foam density fraction (F) = 0.50.

Solution:

a. Use Eq. 6.29 to find the inertial impact number (Ψ_I):

$$\Psi_I = \left[\frac{(1.05)(4\times10^{-6})^2(1,200)(18.0)}{18(1.81\times10^{-5})}\right] \div (0.01) = 0.106$$

b. Use Eq. 6.68 to find the removal efficiency by the packed-bed scrubber:

$$\eta_{\text{tray}} = 1 - \left[\exp\left(-80F^2\Psi_t\right)\right]^n = 1 - \left[\exp\left(-80(0.50)^2(0.106)\right)\right]^3 = 98.8\%$$

Discussion: In part (a), the target diameter is the size of the hole (0.01 m), while the velocity is the gas velocity through the hole (18.0 m/s), which is much larger than the gas velocity through the bulk of the scrubber (1 m/s).

Venturi scrubbers. Most pressure-drop equations for Venturi scrubbers are of the following form (USEPA, 2002):

$$\Delta P = k \times V_t^2 \times \rho_g \times (L/G) \tag{6.69}$$

where ΔP is the pressure drop across the Venturi scrubber, k is the correlation factor for a specific scrubber design, V_t is the throat velocity, ρ_g is the gas density, and (L/G) is the liquid-to-gas ratio, respectively. One of the more widely accepted equations was published by Calvert (1977), and it is:

$$\Delta P = \left(5.4 \times 10^{-4}\right) \times V_t^2 \times \rho_g \times (L/G) \tag{6.70}$$

where ΔP is in in-H_2O, V_t in ft/s, ρ_g in lb/ft^3, and (L/G) in gal/1,000 ft^3.

Eq. 6.70 predicts pressure drop reasonably well at moderate liquid-to-gas ratios. The equation performed well at ratios between 3 and 10 gal/1000 ft^3; the equation overpredicts the pressure drop by 80% or more at or above 12 gal/1000 ft^3 (USEPA, 2002).

In NCEES (2025), the equations for pressure drop across a Venturi scrubber are given as (Corbitt, 1989):

$$\Delta P = \left(10^{-6}\right) \times V_t^2 \times (L/G) \tag{6.71a}$$

where ΔP is in cm-H_2O, V_t in cm/s, and (L/G) in L/m^3.

$$\Delta P = \left(5.0 \times 10^{-5}\right) \times V_t^2 \times (L/G) \tag{6.71b}$$

where ΔP is in in-H_2O, V_t in ft/s, and (L/G) in gal/1,000 ft^3.

Example 6.27: Pressure Drop across a Venturi Scrubber

A Venturi scrubber (area of the throat = 0.04 m^2) removes particulates from an air stream (Q = 10,000 m^3/h) and the liquid flow rate = 9,000 L/h. Estimate the pressure drop across the scrubber.

Solution:

a. Throat velocity = 10,000 m^3/h ÷ 0.04 m^2 = 250,000 m/h = 69.4 m/s = 228 ft/s
b. The L/G ratio = 9,000 L/h ÷ 10,000 m^3/h = 0.9 L/m^3 (= 6.73 gal/1,000 ft^3)
c. Using Eq. 6.70 and density of air = 1.21 kg/m^3 (= 0.075 lb/ft^3),

$$\Delta P = \left(5.4 \times 10^{-4}\right) \times V_t^2 \times \rho_g \times (L/G) = \left(5.4 \times 10^{-4}\right) \times (228)^2 \times 0.075 \times 6.73 = 14.2 \text{ in-H}_2\text{O}$$

Discussion:

1. Eq. 6.70 is an empirical equation, and the units used should be as specified.
2. The air density is that of air at T = 20°C and P = 1 atm (see Example 2.2).

Example 6.28: Pressure Drop across a Venturi Scrubber

A Venturi scrubber (area of the throat = 0.04 m^2) removes particulates from an air stream (Q = 10,000 m^3/h) and the liquid flow rate = 9,000 L/h. Use the equations in NCEES (2025) to estimate the pressure drop across the scrubber.

Solution:

a. Using Eq. 6.71a with V_t = 6,940 cm/s and L/G = 0.9 L/m^3,

$$\Delta P = \left(10^{-6}\right) \times V_t^2 \times (L/G) = \left(10^{-4}\right) \times (6{,}940)^2 \times (0.9) = 43.3 \text{ cm-H}_2\text{O}$$

b. Using Eq. 6.71b with V_t = 228 ft/s and L/G = 6.73 gal/1,000 ft^3,

$$\Delta P = \left(5 \times 10^{-5}\right) \times V_t^2 \times (L/G) = \left(5 \times 10^{-5}\right) \times (228)^2 \times (6.73) = 17.5 \text{ in-H}_2\text{O}$$

Discussion:

1. Eqs. 6.71a and 6.71b are empirical equations, and the units used should be as specified.
2. The results of part (a) and part (b) are essentially the same.
3. The air density is included in Eq. 6.70, but not in Eqs. 6.71a and 6.71b. This may cause the difference in the pressure drop estimates between Examples 6.27 and 6.28.

For a Venturi scrubber, the gas stream atomizes the liquid into droplets. The droplet diameter depends on the gas velocity through the throat (V_t) and the liquid-to-gas ratio. The Sauter mean diameter is most representative of the diameter of the water droplet in Venturi scrubbing. The *Sauter mean diameter* is the diameter of a droplet/particle having the same volume/area ratio as the entire size distribution. For a gas–liquid system, this droplet diameter can be estimated as (USEPA, 2012; Calvert et al., 1972; Calvert, 1977; Engineering Science, 1979):

$$D_{\text{Sauter}} = \frac{16{,}400}{V_t} + 1.45\left(\frac{L}{G}\right)^{1.5} \tag{6.72}$$

where D_{Sauter} is the mean droplet diameter (μ), V_t is the gas velocity at the throat (ft/s), and (L/G) is the liquid-to-gas ratio (gal/1,000 ft³).

For Venturi scrubbers, the removal efficiency can be estimated by:

$$\eta_{venturi} = 1 - \exp\left\{-k\sqrt{\Psi_I}\left(\frac{L}{G}\right)\right\} \tag{6.73}$$

where k is an empirical constant (in 1,000 ft³gal), typically 0.1 to 0.2 1,000 ft³/gal (i.e., 100 to 200 ft³/gal), and (L/G) is the liquid-to-gas ratio (in gal/1,000 ft³).

Eqs. 6.72 and 6.73 are in the U.S. customary units. For use of SI units, they can be modified to:

$$D_{Sauter} = \frac{5,000}{V_t} + 938,000\left(\frac{L}{G}\right)^{1.5} \tag{6.74}$$

where D_{Sauter} is the mean droplet diameter (μ), V_t is the gas velocity at the throat (m/s), and (L/G) is the liquid-to-gas ratio (dimensionless).

$$\eta_{venturi} = 1 - \exp\left\{-k\sqrt{\Psi_I}\left(\frac{L}{G}\right)\right\} \tag{6.75}$$

where k is an empirical constant, typically 750 to 1,500 and (L/G) is the liquid-to-gas ratio (dimensionless).

Example 6.29: Performance of a Venturi Scrubber (the U.S. Customary Units)

Estimate the removal efficiency of particles with D_p = 1 μ (ρ_P = 1,500 kg/m³) by a Venturi scrubber having a throat velocity of 328 ft/s (100 m/s) and a liquid-to-gas ratio of 8.0 gallon/1,000 ft³ (= 1.07×10^{-3} dimensionless). Assume T = 68°F (20°C) and k = 150 ft³/gal (= 0.15 1,000 ft³/gal = 1,122 dimensionless).

Solution:

a. Use Eq. 6.72 to calculate the mean droplet diameter (D_{Sauter}):

$$D_{Sauter} = \frac{16,400}{328} + 1.45(8.0)^{1.5} = 82.8\,\mu$$

b. Use Eq. 6.19 to find the Cunningham slip correction factor (C_c) for the 1-μ particles:

$$C_c = 1 + \frac{6.21 \times 10^{-4} T}{D_p} = 1 + \frac{(6.21 \times 10^{-4})(293)}{1} = 1.18$$

c. Use Eq. 6.29 to calculate the inertial impact number (Ψ_I):

$$\Psi_I = \left[\frac{(1.18)(1\times10^{-6})^2(1,500)(100)}{18(1.81\times10^{-5})}\right] \div (82.8\times10^{-6}) = 6.56$$

d. Use Eq. 6.73 to find the removal efficiency by the Venturi scrubber:

$$\eta_{\text{venturi}} = 1 - \exp\left\{-0.15\sqrt{6.56}\,(8.0)\right\} = 0.954 = 95.4\%$$

Discussion: In part (b), the target diameter is the mean size of the liquid droplets (82.8 μ).

Example 6.30: Performance of a Venturi Scrubber (the SI Units)

Referring to Example 6.29, use the values in SI units and Eqs. 6.74 and 6.75 to estimate the removal efficiency of particles with D_p =1 μ.

Solution:

a. Use Eq. 6.74 to calculate the mean droplet diameter (D_{Sauter}):

$$D_{\text{Sauter}} = \frac{5,000}{100} + (938,000)(1.07\times10^{-3})^{1.5} = 82.8$$

b. Use Eq. 6.29 to calculate the inertial impact number (Ψ_I):

$$\Psi_I = \left[\frac{(1.18)(1\times10^{-6})^2(1,500)(100)}{18(1.81\times10^{-5})}\right] \div (82.8\times10^{-6}) = 6.56$$

c. Use Eq. 6.75 to find the removal efficiency by the Venturi scrubber:

$$\eta_{\text{venturi}} = 1 - \exp\left\{-(1,122)\sqrt{6.56}\,(1.07\times10^{-3})\right\} = 0.954 = 95.4\%$$

Discussion: 1,000 ft^3 = 7,480 gallons. The dimensionless k value for 0.15 1,000 ft^3/gal is 1,122 (= 0.15 × 7,480).

6.8 CONSIDERATIONS IN SELECTION OF TYPES OF REMOVAL DEVICES

Selecting a device to remove/collect particles from a gas stream, characteristics of the particles (e.g., shape, density, stickiness, oiliness, and acidity), characteristics of the gas stream (e.g., temperature, humidity, and presence of acidic gas), gas stream flow rate, and mass concentrations and sizes of the particles to be collected need to be taken into consideration.

Below are some general suggestions:

1. Post-treatment of the collected dust/liquids should be considered part of the treatment process.
2. Tolerance of construction materials to high temperatures should be considered. For example, materials in fabric filters should have proper temperature ratings.
3. Reducing the temperature of the gas stream would reduce its flow rate. This may be beneficial because it will reduce the pressure drop across the device (i.e., for a lower energy cost) and increase the residence time in a control device (i.e., for better treatment performance). However, it may cause condensation, which can be troublesome if acidic gas components are present.
4. Sticky and greasy particles may create problems on the collection surfaces of the cyclones, fabric filters, or ESPs.
5. Particles in the gas stream may be abrasive to the control devices.
6. Wet scrubbers would generally be a better choice for explosive gases than the other types of control devices.
7. The cut size of gravity settling chambers is about 50 μ and that of cyclones is about 5 μ. If finer particles also need to be removed, gravity settling chambers and/or cyclones can serve as pretreatment for fabric filters, ESPs, or wet scrubbers, which can remove finer particles.

6.9 CONCLUSION

1. Sizes of particles in gas streams are typically log-normally distributed. Fine particles, with diameters <10 μ, are of greater concern about air pollution.
2. Major particle removal mechanisms in common control devices include gravitation, centrifugation, electrostatic attraction, inertial impaction, and Brownian motion.
3. Common particulate removal devices include gravity settling chambers, cyclones, ESPs, fabric filters, and wet scrubbers.
4. In gravity settling chambers, cyclones, and ESPs, particles are driven to a collection surface mainly by gravitational, centrifugal, and electrostatic forces. The design equations are relatively similar.
5. In fabric filters and wet scrubbers, the gas stream is divided by the collection targets (i.e., fabric in the filters and water droplets in the wet scrubbers). The collection targets will collect the particulates, while the gas stream moves around them. The design equations are different.
6. The cut size of gravity settling chambers is about 50 μ and that of cyclones is about 5 μ. If fine particles also need to be removed, gravity settling chambers and/or cyclones can serve as pretreatment to fabric filters, ESPs, or wet scrubbers, which can remove finer particles.

EXERCISE QUESTIONS

1. To determine the diameter of a small particle (density = 2.5 g/cm^3), one lets it settle in still air (T = 20°C) in the field of view of a microscope. The settling/terminal velocity was 0.005 m/s. Use Stokes' law to calculate the particle diameter (in microns).
2. The separation number for a 2-μ particle (density = 2.0 g/cm^3) in an air stream against a sphere was calculated to be 1.0. Estimate the target efficiency for a 1-μ particle (density = 2.5 g/cm^3) in the same air stream.
3. To determine the diameter of a small particle (density = 2.5 g/cm^3), one lets it settle in still air (T = 20°C) in the field of view of a microscope. The settling/terminal velocity was 0.005 m/s, and the particle diameter was calculated to be 8 μ. What is the Reynolds number of this particle traveling at the stated velocity?
4. A spherical particle with a diameter of 5 μ and density of 3,000 kg/m^3 is settling in still air (T = 20°C; P = 1 atm; viscosity = 1.81×10^{-5} kg/m·s) and the Cunningham correction factor = 1.0.
   a. Determine its terminal velocity in this air (in m/s).
   b. Determine its Reynolds number at this settling speed.
   c. What is its aerodynamic diameter?
   d. What would be its Stokes stopping distance if it is ejected at 10 m/s?
   e. Determine the separation number if this particle moves at 5 m/s against a stationary 1-cm sphere.
5. A spherical particle with a diameter of 2 μ and density of 2,000 kg/m^3 is settling in still air (T = 20°C and P = 1 atm). Its terminal velocity is 2.4×10^{-4} m/s; the Reynolds number at this velocity is 3.2×10^{-5}, and diffusivity in this air is 1.2×10^{-11} m^2/s. For a spherical particle with a diameter of 4 μ and a density of 3,000 kg/m^3 in the same air, determine:
   a. its terminal velocity in this air (in m/s),
   b. the Reynolds number at the settling speed, and
   c. its diffusivity in this air (in m^2/s).
6. A log-normally distributed particle mass has a geometric mean diameter of 5 μ, and the diameter of particles with 84.13% by weight of all particles in this particle mass is smaller than 25 μ. Determine:
   a. the geometric standard deviation (in μ),
   b. the standard deviation (in μ), and
   c. the diameter of particles with 15.87% by weight of all particles smaller (in μ).
7. A cyclone separator has an inlet width of 0.5 m treating an air stream (viscosity = 2.1×10^{-5} kg/m·s) with an inlet gas velocity of 14 m/s, and the number of effective turns within the cyclone is 5. The density of all particles is 1,500 kg/m^3. Determine:
   a. the cut diameter of this cyclone, using the block flow model (in μ), and
   b. the removal efficiency for particles of 10 μ (in %).

8. A cyclone separator was operated at 500 scfm with $D_{cut}=5$ μ. However, the flow has recently been decreased to 250 scfm. Without any modifications to this cyclone,
   a. what is the new cut size?
   b. will the cyclone provide higher removal efficiencies at this reduced flow rate?
9. A cyclone separator was operated at 800 scfm with $D_{cut}=5$μ. However, the flow has been increased to 1,000 scfm due to the expansion in manufacturing activities.
   a. What is the new cut size without modifying this cyclone?
   b. What is the new removal efficiency for 5-μ particles?
10. A gas stream contains particles of two sizes: 20 μ (30% by weight) and 2μ (70% by weight). The air is to be treated by a cyclone ($D_{cut}=4$μ). Estimate the collection efficiency (% by weight) of all particles.
11. A cyclone has an outside diameter of 1 m and a rectangular inlet (Height = 0.4 m and Width = 0.2 m). The gas flow rate is 2.4 m^3/s (85 ft^3/s), carrying particles with a density of 3,000 kg/m^3.
    a. Demonstrate that the gas flow velocity inside the cyclone equals 30 m/s.
    b. Determine the cut diameter with $N=5$ and viscosity of air $=1.8 \times 10^{-5}$ kg/m·s.
    c. Estimate its removal efficiency for particles with a diameter = 2 μ.
12. A cyclone separator is operated at 1,000 cfm with $D_{cut}=5$ μ. With an improvement, the air stream flow rate will be reduced to 600 cfm.
    a. What is the new cut size without modifying this cyclone?
    b. Will the reduction in the inlet flow rate improve the particle removal efficiency of the cyclone?
13. Estimate the head loss through a cyclone if the gas velocity is 25 ft/s (7.6 m/s) and $T=68$°F (20°C).
14. Referring to the cyclones in Question #13, estimate the pressure drop using SI units.
15. Referring to the cyclone in Questions #13 & #14, estimate the annual electricity cost if it operates continuously and the blower efficiency is 85%. The flow rate is 20 m^3/s (706 ft^3/s), and the electricity cost is \$0.20 per kilowatt hour (\$0.20/kWh).
16. A bag house has 10 compartments (200 bags per compartment with a bag length of 12 ft and a diameter of 6 inches) and controls particulate emission from an air stream ($Q=60{,}000$ ft^3/min). When two compartments are taken offline for bag cleaning and maintenance, calculate the air-to-cloth ratio (ft^3/min/ft^2).
17. A 500-MW coal-fired plant produces 1 million acfm of stack gas to be treated by a shaker bag house. From a design handbook, the engineer found that this type of application's typical air/cloth ratio of 4 cfm/ft^2 is appropriate. Estimate the number of bags (40 ft long and 1 ft in diameter) needed.
18. A bag house will be constructed to control particulate emissions from an air stream ($Q=30$ m^3/s). The filter bags measure 0.5 m in diameter and 4.0 m in

length. The engineer decided to use the air/cloth ratio (3 $ft^3/min/ft^2$) found in a design handbook for the design. Determine the minimum number of bags needed for the bag house.

19. A shake-deflate baghouse has eight compartments. During normal operation, six compartments are in service, while the other one is offline for cake removal or maintenance. Each compartment contains 120 bags that are 0.4 m in diameter and 5 m long. The air-to-cloth ratio is 1.2 m/min.
    a. Calculate the available surface area of each bag.
    b. What does this baghouse treat the airflow rate?
20. Assume the Deutsch-Andersen equation applies. An ESP treats a particle-laden air stream ($T=600°F$), collecting 90% of the particles at an (A/Q) ratio of 0.2 min/ft.
    a. Determine the drift velocity of these particles.
    b. Assuming the drift velocity stays the same if the air stream is cooled down to 300°F before entering the ESP, will the removal efficiency increase or decrease?
    c. What would the new removal efficiency be for these particles?
21. An ESP system is designed to treat an air stream of 500 m^3/min ($T=500°C$) with 90% particulate removal efficiency. However, the recent regulatory requirement requires a 98% removal. The owner decided to lower the temperature of the air to 200°C. Will this approach meet the new requirement (assume drift velocity is a constant and use the Deutsch–Anderson equation)?
22. A gas stream contains particles of two sizes: 2 and 4 μ. The particle density is the same for both sizes. Assume the Deutsch–Anderson equation applies and the drift velocity is proportional to the diameter. If the collection efficiency for the 4 particles is 99%, estimate the collection efficiency for the 2-μ particles.
23. An ESP system is designed to treat an air stream of 500 m^3/min. The system will provide a 95% removal efficiency for particles with a drift velocity of 0.2 m/s. The size of each plate is 5 m tall and 2 m long. Calculate the minimum number of plates needed using the Deutsch–Anderson equation.
24. An electrostatic precipitator treats an air stream of 500 m^3/min. The system is to remove particles with a drift velocity of 0.2 m/s. Using the Deutsch–Anderson equation to determine
    a. the plate area required for 90% removal efficiency (in m^2),
    b. the plate area required for 99% removal efficiency (in m^2), and
    c. the plate area required for 99.9% removal efficiency (in m^2).
25. Estimate the removal efficiency of particles with $D_p=5$ μ ($\rho_p=1{,}500\,kg/m^3$) for a gas stream (160 m^3/min) by a counter-current spray tower (height = 4 m, $T=293$ K). The liquid flow rate is 150 L/min (droplet diameter = 400μ), and the gas velocity inside the scrubber is 1.2 m/s.
26. Referring to Question #25, a packed bed will be used instead (packing = 5 cm Raschig™ rings). Assume $j=0.165$ and porosity = 0.75. Determine its removal efficiency.

27. Referring to Question #25, a tray scrubber will be used instead (3 trays with 10 mm holes; gas velocity through the holes is 18.0 m/s). Assume the foam density fraction (F) = 0.50. Determine its removal efficiency.
28. Estimate the removal efficiency of particles with D_p = 0.8 (ρ = 2,000 kg/m^3) by a venture scrubber with a throat velocity of 328 ft/s and a liquid-to-gas ratio of 9.0 gallon/1,000 ft^3. Assume T = 68°F and k = 0.16 1,000 ft^3/gal.
29. Estimate the removal efficiency of particles with D_p = 0.8 (ρ = 2,000 kg/m^3) by a venture scrubber with a throat velocity of 100 m/s and a liquid-to-gas ratio of 0.0012 (dimensionless). Assume T = 20°C and k = 1,200.

BIBLIOGRAPHY

Calvert, S. (1977). "*How to Choose a Particulate Scrubber*," *Chem. Eng.* 54–68, August 29, 1977.

Calvert, S., Goldschmid, J., Leith, D., Mehta, D. (1972). *Scrubber Handbook. Vol.* I, *Wet Scrubber System Study*, EPA-R2-72-118a, U.S. Environmental Protection Agency.

Corbitt, R.A. (1989). *Standard Handbook of Environmental Engineering*, 1st edition, McGraw-Hill.

de Nevers, N. (2000). *Air Pollution Control Engineering*, 2nd edition, McGraw-Hill Companies, Inc.

Engineering Science. (1979). *Scrubber Emissions Correlation*, Final Report, U.S. EPA Contract 68-01-4146, Task Order 49, U.S. Environmental Protection Agency.

NCEES. (2025). *PE/Environmental Reference Handbook (Version 1.6)*, National Council of Examiners for Engineering and Surveying (NCEES), Greenville, SC.

USEPA. (2002). *EPA Air Pollution Control Cost Manual*, 6th edition, EPA/452/B-02-001, Office of Air Quality Planning and Standards, United States Environmental Protection Agency, Research Triangle Park, NC.

USEPA. (2003). *APTI 445: Inspection of Particle Control Devices - Student Manual*, 2nd revision, Air Pollution Training Institute (APTI), Environmental Research Center, United States Environmental Protection Agency, Research Triangle Park, NC.

USEPA. (2012). *APTI 413: Control of Particulate Matter Emissions - Student Manual*, 5th edition, prepared by Tidewater Operations Center of C² Technologies for Air Pollution Training Institute, United States Environmental Protection Agency, Research Triangle Park, NC.

USEPA. (2013). *APTI 450: Source Sampling for Pollutants – Student Guide*, Air Pollution Training Institute, United States Environmental Protection Agency, Research Triangle Park, NC.

USEPA. (2025). *Air Pollutant Emissions Trend Data*, https://www.epa.gov/air-emissions-inventories/air-pollutant-emissions-trends-data, last updated on 04/28/2025.

7 Control of Gaseous Organic Pollutant Emissions

7.1 INTRODUCTION

Gaseous pollutants in ambient air can be primary or secondary. Unless for indoor air, it is uncommon for us to draw ambient air into a treatment unit to remove air pollutants. Instead, eliminating and reducing anthropogenic pollution emissions is essential for good ambient air quality. This chapter focuses on reductions of primary gaseous contaminants from point sources. These reductions are typically achieved by applying an end-of-the-pipe control system (i.e., add-on control devices).

Primary gaseous pollutants can be inorganic or organic. Organic gaseous pollutants include volatile organic compounds (VOCs) and other organic compounds. Inorganic gaseous contaminants of concern include SO_x, NO_x, CO, GHGs, acidic gases (e.g., HCl and HF), as well as H_2S and other reduced sulfur compounds. Emission reductions of SO_x, NO_x, CO, and GHGs are covered in separate chapters in this book. Many treatment processes for organic compounds (e.g., adsorption, absorption, thermal, and biological processes) also apply to acidic gases and reduced sulfur compounds. Consequently, this chapter focuses on the end-of-the-pipe treatment of organic gaseous pollutants.

7.1.1 Sources and Emissions of VOCs

Table 7.1 shows sources and emissions of VOCs in the U.S. in 2024. The total VOC emission was about 12.8 million tons in 2024. Contributions to VOC emissions are 54.4%, 25.9%, 14.8%, and 4.9% from the industrial and other processes, miscellaneous, transportation, and stationary fuel combustion sectors, respectively. Petroleum and related industries are the most significant industrial sources, accounting for 22.7% of the total emissions, while solvent utilization is the second largest industrial source (21.7%) (USEPA, 2025).

Figure 7.1 depicts trends of annual VOC emissions in the U.S. since 1970. As shown, the emissions have been decreasing, except for the miscellaneous sector. Most of the reductions come from the transportation sector, and some from the industry sector.

7.1.2 Types of Emission Reduction Technologies

End-of-the-pipe control processes for gaseous pollutants can be classified into recovery and destruction. *Recovery processes* refer to transferring compounds of concern

DOI: 10.1201/9781003532705-7

TABLE 7.1
VOC Emission Rates and Source Categories in the U.S. in 2024

Source Category	(1,000 tons)	(%)
Stationary fuel combustion	**629**	**4.9**
Electric utility	27	0.2
Industrial	116	0.9
Others	486	3.8
Industrial and other processes	**6,950**	**54.4**
Chemical & allied product manufacturing	69	0.5
Metals processing	22	0.2
Petroleum & related industries	2,899	22.7
Other industrial processes	386	3.0
Solvent utilization	2,777	21.7
Storage & transport	605	4.7
Waste disposal & recycling	192	1.5
Transportation	**1.886**	**14.8**
Highway vehicles	896	7.0
Off-Highway	990	7.7
Miscellaneous	**3.317**	**25.9**
Wildfires	974	7.6
Others	2,343	18.3
Total	**12,782**	**100.0**

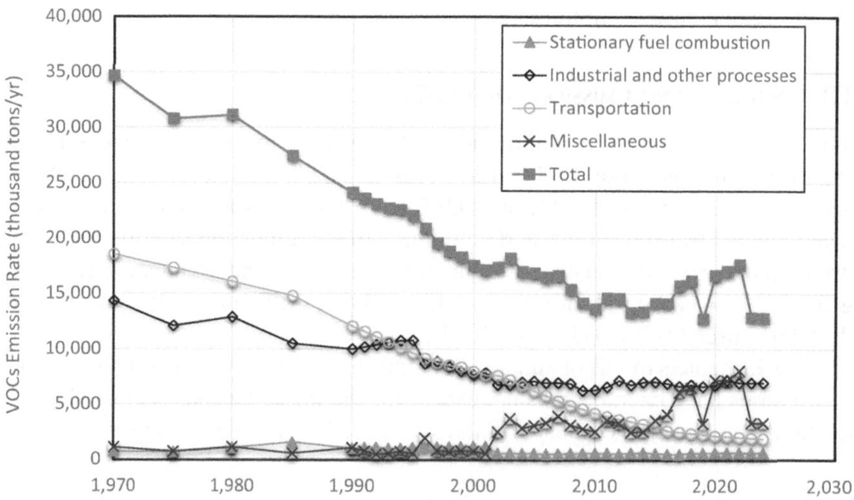

FIGURE 7.1 Annual trends of VOC emissions from major source categories in the U.S. (1990–2024).

(COCs) in a gas stream to collection media, which can be solid or liquid. Standard recovery processes include adsorption, absorption, and condensation. *Destruction processes* include thermal oxidation (direct or catalytic incineration), chemical oxidation, and biodegradation, which convert COCs into more acceptable end products. Although chemical oxidation has reportedly been used for destructing gaseous organic pollutants, it is not commonly practiced in the field.

7.1.3 Important Gas Stream Characteristics

As mentioned, many technologies are available for the emission reduction of gaseous pollutants. In addition to economic considerations, the selection and design of a treatment system would depend significantly on the characteristics of the gas stream. They include flow rate, temperature, pressure, pollutant concentrations, heating values, oxygen level, moisture content, and particulate concentrations.

Flow rate, temperature, and pressure data are needed to size a treatment device to meet the design operational conditions. They can also be used to determine the flow velocity through the treatment unit, which is another significant operating parameter. From the flow rate and the pollutant concentration, the mass loading to the control device, another critical operating parameter, can be determined. Heating values are of concern for thermal processes. Oxygen level and moisture content are parameters of concern for both thermal and biological processes. For adsorption, moisture content and particulates are of concern because water molecules may compete with COCs for the limited adsorption sites, and particulates may mask those sites.

7.2 ADSORPTION

Adsorption processes have been widely used to remove gaseous organic and inorganic (e.g., H_2S) pollutants. Adsorption is a surface phenomenon. For air pollution control, *adsorption* removes undesirable COCs from a gas stream by having them attached/adhered to the surface of a solid. The participants of this process include *adsorbent* (the solid), *adsorbate* (the COCs), and *carrier gas* (the inert portion of the gas stream, typically air).

7.2.1 Principles of Adsorption

Adsorption is mass transfer from the bulk gas phase to a solid surface. The COCs travel through the bulk gas phase to the adsorbent, enter the pores of the adsorbent (mainly due to diffusion caused by the concentration gradient), accumulate in the pores, and get adsorbed onto the surface of the adsorbent (Figure 7.2).

Adsorption can be a physical or chemical process. A chemical bond is formed between the adsorbent and the COC molecule in *chemical adsorption*. In contrast, in *physical adsorption*, the solid surface only retains the COC molecule through intermolecular cohesion. In most industrial applications, physical adsorption is the preferred mode since it is readily reversible (i.e., desorption can readily occur when the operational conditions change). *Desorption* is an approach to regenerate spent adsorbent, and *regeneration* is to refresh the adsorbent. It should be noted that the

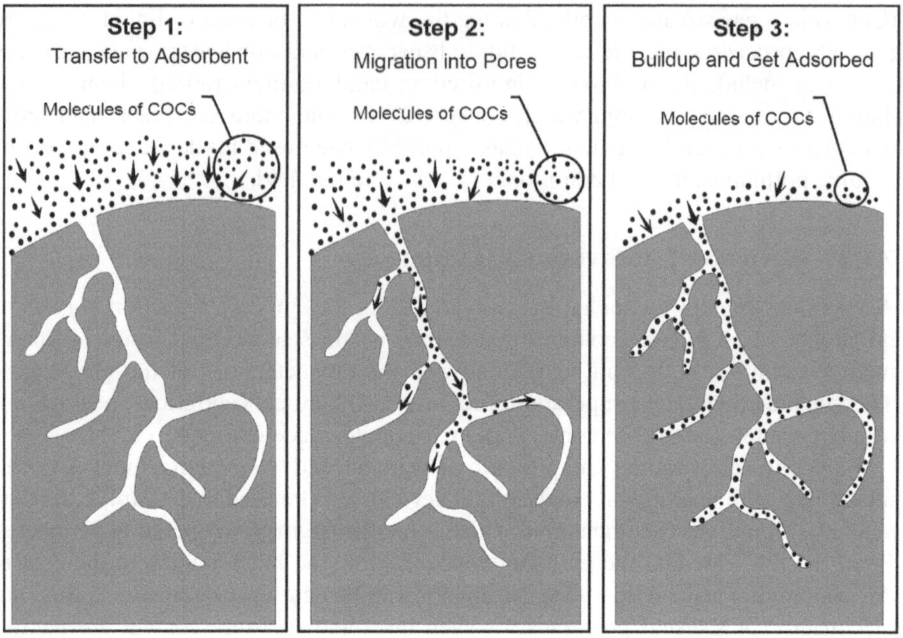

FIGURE 7.2 Adsorption steps (Modified from USEPA, 2000).

adsorption process is *exothermic* (i.e., heat would be released due to adsorption). Consequently, lower temperatures favor adsorption, while higher temperatures are better for desorption.

7.2.2 Adsorption Systems and Their Components

The main component of an adsorption system is the adsorbent. An effective adsorbent should have a large specific surface area with a good affinity for COCs. The *specific surface area* can be defined as the total surface area per unit mass (e.g., m^2/g). Common adsorbents are activated carbon, zeolites, synthetic polymers, alumina, and silica gel.

Activated carbon is the most commonly used adsorbent for air pollution control. It can be produced from different raw carbonaceous materials. The raw material first goes through pyrolysis to drive off all volatile materials [Note: *pyrolysis* is a process of thermal decomposition of organic material in the absence of oxygen]. The remaining carbon is then "activated" by steam, air, or CO_2 to have a porous and amorphous structure with a large specific surface area. The activated carbon can be in the form of granules or powders. Using granular activated carbon (GAC) is more common for air pollution control. For example, the size of GAC for an application is 20×40 mesh (i.e., 90% or more of the GAC will pass through a 20-mesh sieve (0.810 mm opening) and be retained by a 40-mesh sieve (0.420 mm opening). Due to its non-polar surface, GAC can adsorb various organic and inorganic gases. The typical internal porosity, specific surface area, dry bulk density, and mean pore diameter of GAC are 55%–75%, 600–1,600 m^2/g, 0.35–0.50 g/cm^3, and 1,500–2,000 Å [Note: 1 Å= 10^{-10} m], respectively (EPA, 2012a).

Zeolites (molecular sieves) refer to natural or synthetic microporous aluminosilicates with a crystalline structure. They have been used to remove moisture and NO_x from air pollution sources. Synthetic polymeric adsorbents have various functional groups and high adsorption capacities for selected organic compounds, such as ketones, aldehydes, and reactive compounds, which may have unwanted reactions on the surface of GAC. They are also used for gas streams with a relative humidity (RH) greater than 50% because they have less affinity to water vapor than GAC. Silica gel and activated alumina mainly remove moisture in a gas stream (USEPA, 2012a).

Each type of adsorbent has its maximum capacity, or a limited number of active sites, for adsorption. The removal efficiency will drop significantly once the adsorbed COCs occupy the most available sites. If the operation continues beyond this point, the breakthrough point (defined later) will be reached, and the effluent concentration will rise sharply. Eventually, the adsorbent in service would be *saturated*, *exhausted*, or *spent* when most of its sites are occupied.

Adsorbers have been employed for a wide range of COC concentrations. For odor control and low COC concentration applications (<10 ppm), the adsorbent bed is often discarded when the adsorbent approaches saturation with the COCs. These are *non-regenerative systems* because the adsorbent is not regenerated for reuse. The non-regenerative adsorbers usually consist of thin adsorbent beds (Figure 7.3a as an example), with thickness ranging from 1.3 to 10 cm (0.25 to 4 inches). The pressure drop across a bed is relatively low (<2.5 in-H_2O, or 0.62 kPa) with a typical superficial gas velocity of 6 to 20 m/min (20 to 60 ft/min) through the bed. The *superficial gas velocity* can be determined by dividing the gas flow rate by the cross-sectional area of the adsorbent bed. The design service lives of these adsorbent beds range from six months for heavy odor concentrations to up to two years for trace concentrations or

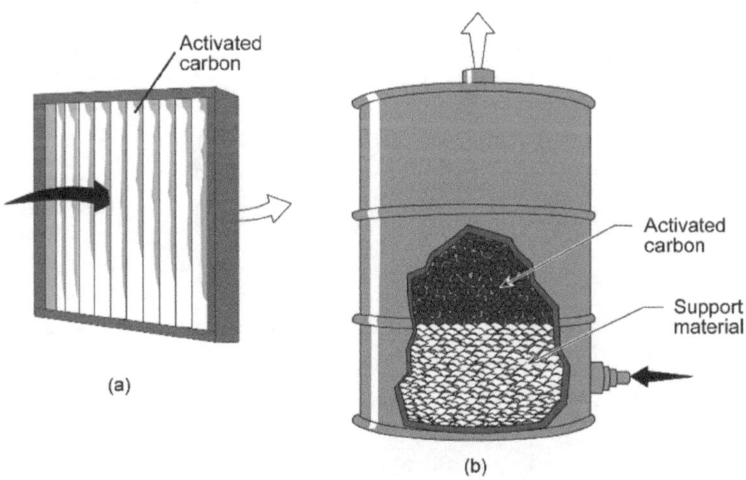

FIGURE 7.3 Non-regenerative adsorbers: (a) pleated thin bed and (b) drum adsorber (Modified from USEPA, 2000.

intermittent operations. Figure 7.2b illustrates a thick-bed non-regenerative adsorber with GAC housed in a 55-gallon (0.2 m^3) drum (USEPA, 2012a).

For applications with high COC concentrations (up to 10,000 ppm), it would become uneconomical to discard the exhausted adsorbent after it reaches its first saturation with the COCs. Instead, the adsorbent is regenerated to be refreshed for the next cycle. In the meantime, the previously adsorbed COCs may also be recovered for potential recycling/reuse. Figure 7.4 illustrates a 3-bed fixed-bed regenerative adsorption system. In the operation mode, one or more beds treat the influent gas stream, while the remaining ones are being regenerated or cooled down. The regeneration is often done by low-pressure steam (or hot air), in which the adsorbed COCs are desorbed from the adsorbent and move into the steam. The steam and COCs mixture is then chilled and condensed for separation into the COCs and water. As shown, there is a particulate filter to remove particles that may mask the active adsorption sites, and a heat exchanger to lower the gas temperature to a range of 15 to 38°C (60 to 100°F) for optimal adsorption efficiency and adsorbent service life (USEPA, 2012a).

7.2.3 Factors Affecting Adsorption

In addition to the characteristics of the adsorbent (e.g., grain size, pore size, specific surface area, water vapor affinity, and polarity), several operating parameters would affect the performance of an adsorption system. Two pretreatment processes are often employed on the influent gas stream to enhance the treatment performance. The first is cooling, while the second is dehumidification. Adsorption of COCs is exothermic, which is favored by lower temperatures. In addition, at higher temperatures, the

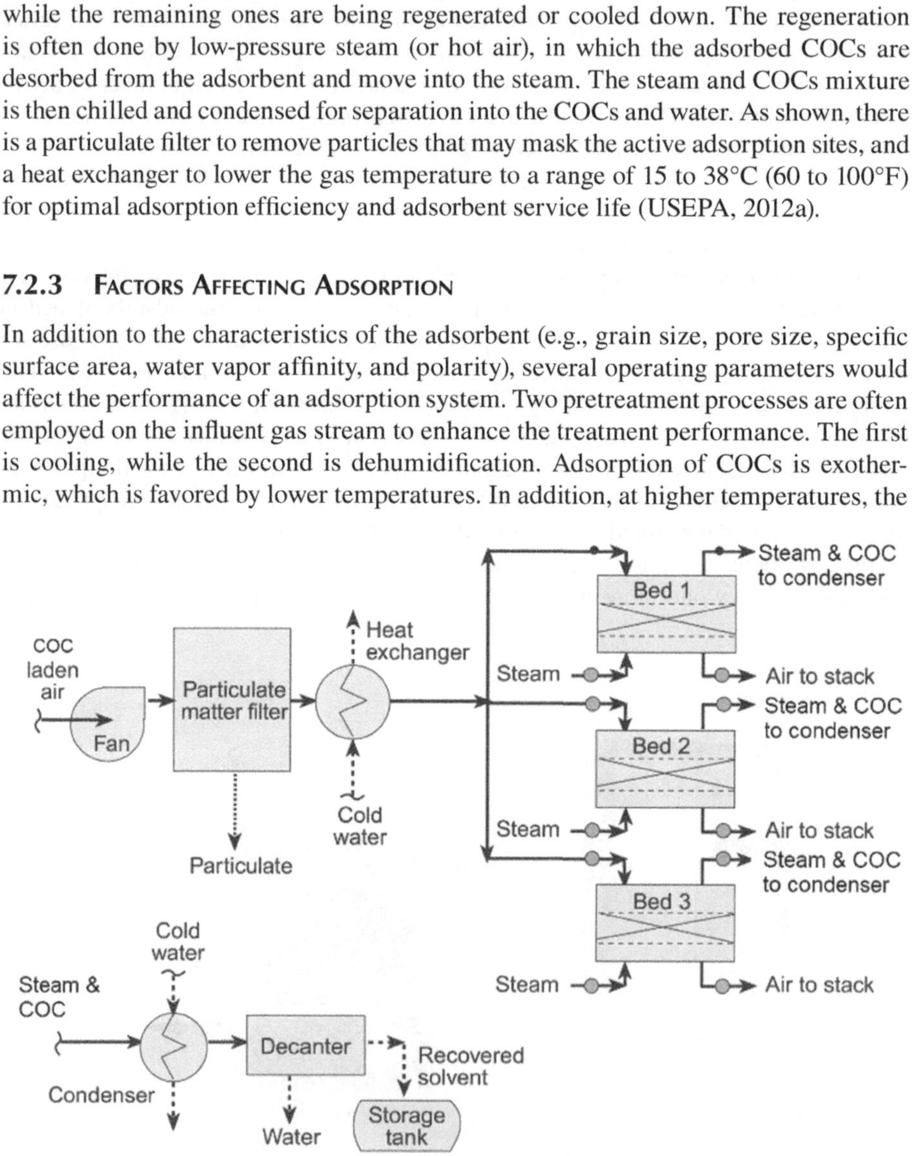

FIGURE 7.4 A 3-bed fixed-bed regenerative adsorption system (Modified from USEPA, 2000).

vapor pressures of COCs would be higher so that the adsorbed COCs may have sufficient energy to overcome the attraction force between them and the adsorbent. The gas stream to be treated must be cooled to <130°F (54°C). On the other hand, water vapor molecules in the gas stream will compete with COCs for available adsorption sites. The influent gas stream's RH should generally be reduced to <50%. For example, the off-gas stream from an air stripper is typically saturated with water. The gas stream needs to be cooled down (e.g., using chiller water) to condense the moisture and then heated (e.g., using an electrical heater) to some extent to raise its RH before entering a GAC system.

7.2.4 Design Considerations for Adsorbers

Using GAC systems as an example, the sizing of a treatment system depends primarily on the following parameters:

- Volumetric flow rate of the gas stream
- Concentration or mass loading of VOCs
- Adsorption capacity of the GAC
- Design GAC regeneration or change-out frequency

The design flow rate affects sizing, the cross-sectional area of the adsorbent bed, the fan and motor, and the ductwork. The other three parameters (i.e., mass loading, GAC adsorption capacity, and regeneration frequency) affect the size and the number of adsorption units, as well as the amount of GAC required for a specific project.

The adsorption capacity of GAC depends on the characteristics of GAC and COCs, COC concentrations, the presence of other species competing for adsorption, and system temperature and pressure. At a given T, an equilibrium relationship exists between the mass of COC adsorbed per unit mass of GAC and the COC concentration (or the partial pressure) in the gas stream; the relationship is often referred to as an *adsorption isotherm*. For many COCs, their adsorption isotherms can be fitted well by a power curve, also known as the *Freundlich isotherm*, as discussed in Chapter 3:

$$q = a\left(P_{\text{COC}}\right)^{m} \tag{7.1}$$

where q is the equilibrium/saturation adsorption capacity (in kg COC/kg GAC, or lb COC/lb GAC), P_{COC} is the partial pressure of VOC in the gas stream (in atm, or psi), and a and m are empirical constants. The empirical constants of the Freundlich isotherms for selected VOCs can be obtained from vendors. It should be noted that the values of these empirical constants are different, depending on the types of GAC and COC, as well as the concentration range and operating temperature.

The *saturation (equilibrium) capacity*, determined from the adsorption isotherm, is the maximum amount of COC the GAC can adsorb. The adsorption capacity in the field applications should be lower than the saturation capacity. Typically, design engineers take 25 to 50% of the equilibrium value as the *design/working capacity* (q_{design}) as a safety factor.

The maximum amount of COCs that can be removed or held (M_{removal}) by a given amount of GAC (M_{GAC}) can be determined as:

$$M_{\text{removal}} = q_{\text{design}} \times M_{\text{GAC}} = q_{\text{design}} \times \left[V_{\text{GAC}} \times \rho_b\right] \tag{7.2}$$

where V_{GAC} is the volume of the GAC and ρ_b is the bulk density of the GAC.

Example 7.1: Adsorption Capacity of a GAC Adsorber

Off-gases from a soil venting project are to be treated by GAC adsorbers. The m-xylene (C_8H_{10}) concentration in the off-gas is 800 ppm. The gas flow rate is 6 m³/min (210 ft³/min) @T=20°C. Two 500-kg (1,100-lb) GAC adsorbers are to be used. Determine the maximum amount of m-xylene that can be held by one GAC adsorber before it is exhausted. The adsorption isotherm (T=20°C) for m-xylene concentrations ranging from 0.0001 to 0.05 psi is shown below, where q is in kg/kg and P in psi:

$$q = 0.527 \times P^{0.0703}$$

Solution:

a. Convert the concentration from ppm to psi as:

$$P = 800 \text{ ppm} = 800 \times 10^{-6} \text{atm} = 0.0118 \text{ psi} \ [\text{Note}: 1 \text{ atm} = 14.696 \text{ psi}]$$

b. Saturation capacity (q_{sat}) = (0.527) × (0.0118)$^{0.0703}$ = 0.386 kg xylene/kg GAC.

c. Assuming design capacity = 40% of the saturation capacity, then

$$q_{\text{design}} = (40\%) \times (q_{\text{sat}}) = (40\%) \times (0.386) = 0.154 \text{ kg/kg}$$

d. Amount of xylene retained by the GAC before being exhausted

$$= (M_{\text{GAC}}) \times (q_{\text{design}})$$
$$= (500 \text{ kg/unit}) \times (0.154 \text{ kg xylene/kg GAC}) = 77 \text{ kg xylene/unit.}$$

Discussion:

1. The saturation capacity of vapor-phase GAC is typically less than 40% of its mass (i.e., <0.4 kg COC/kg GAC), so the typical design capacity is <0.2 kg COC/kg GAC. For comparison, the adsorption capacity of liquid-phase GAC (used for water/wastewater treatment) is much smaller, typically in the order of 0.01 kg COC/kg GAC, due to the presence of other compounds that are also adsorbable.
2. Care should be taken to ensure that the units of two concentrations in an adsorption isotherm go with the given empirical constants. In addition, the influent pollutant concentration should be within the applicable range.

3. The influent COC concentration in the gas stream, not the effluent concentration, should be used in an adsorption isotherm to determine the adsorption capacity.

To achieve efficient adsorption, the gas flow rate through the GAC adsorber should be kept low to allow for a sufficient contact time. The design gas velocity (v_{gas}) is often 18 m/min (60 ft/min), and 30 m/min (100 ft/min) is considered to be the maximum. A minimal design velocity of 6 m/min (20 ft/min) is to avoid flow problems such as channeling (USEPA, 2012a).

The required cross-sectional area of the GAC adsorbers (A_{GAC}) can be readily determined as:

$$A_{GAC} = \frac{Q_{Gas}}{v_{gas}} \tag{7.3}$$

where Q_{gas} is the volumetric influent gas flow rate.

For practical design considerations, there are minimum and maximum depths for the adsorbent bed. Figure 7.5 illustrates the *mass transfer zone* (MTZ), the zone of the bed where mass transfer occurs. When the leading edge of the MTZ reaches the end of the bed, the effluent COC concentration starts to rise rapidly, referred to as the *breakthrough point*. The adsorbent bed's design depth is usually 0.6 m (2 ft) or more, to accommodate the MTZ. The actual depth would also depend on the design change-out rate.

The allowable pressure drop across the bed limits the maximum practical bed depth. Figure 7.6 shows that the pressure drop per unit depth increases with the superficial gas velocity and decreases with the size of GAC. With a gas velocity of 30 cm/s (18 m/min

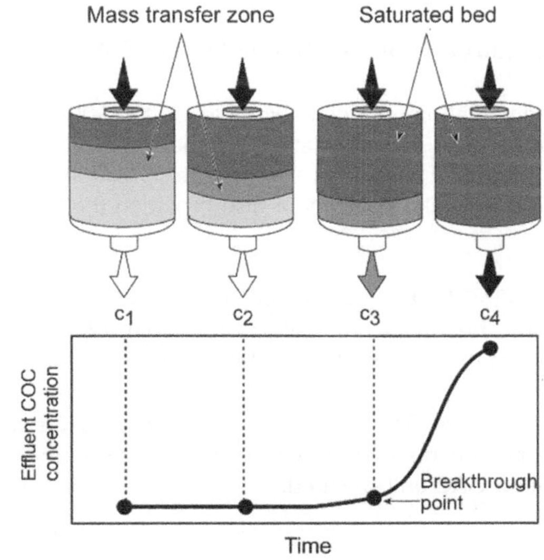

FIGURE 7.5 Mass transfer zone (Modified from USEPA, 2000).

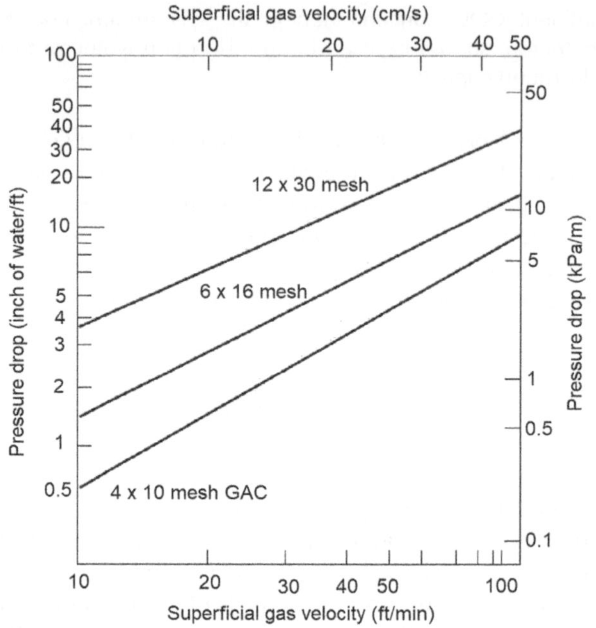

FIGURE 7.6 Pressure drop vs. superficial gas velocity through deep-bed GAC adsorbers (Modified from USEPA, 2012a).

or 60 ft/min), the pressure drop will be about 5.7 kPa/m of media depth (7 in-H_2O/ft) using 6×16 mesh GAC (the middle line). If the total pressure drop across the bed is limited to <6 kPa (24 in-H_2O), the bed length should be kept <1 m (3 ft).

Example 7.2: Required Cross-Sectional Area of GAC Adsorbers

Referring to the remediation project in Example 7.1, the 500-kg GAC units are out of stock. To avoid a delay in remediation, off-the-shelf 55-gallon GAC units are proposed in the interim. The type of GAC in the 55-gallon (0.21 m^3) units is the same as in the 500-kg units. The vendor also provided the following information about the units:

- Diameter of GAC bed in each 55-gallon drum = 1.5 ft (0.5 m).
- Depth of GAC bed in each 55-gallon drum = 3 ft (1.0 m).
- Bulk density of the GAC = 28 lb/ft^3 (449 kg/m^3).

Determine (a) the mass of GAC in each 55-gallon unit, (b) the amount of xylene that each GAC unit can remove before being exhausted, and (c) the minimum number of 55-gallon GAC units needed.

Solution:

a. Volume of GAC inside a 55-gallon drum

$$= \pi\left(r^2\right)(h) = \pi\left[(1.5/2)^2\right](3) = 5.3\ \text{ft}^3\left(= 0.15\ \text{m}^3\right)$$

Mass of the GAC inside a 55-gallon drum

$$= \left(V_{\text{GAC}}\right)(\rho_b) = \left(5.3\ \text{ft}^3\right) \times \left(28\ \text{lb/ft}^3\right) = 148\ \text{lb}\left(= 67\ \text{kg}\right)$$

b. The mass of xylene can be retained before the GAC becomes exhausted

$$= \left(M_{\text{GAC}}\right)\left(q_{\text{design}}\right) = (148\ \text{lb/drum}) \times (0.154\ \text{lb xylene/lb GAC})$$
$$= 22.8\ \text{lb xylene/drum}\left(= 10.4\ \text{kg xylene/drum}\right)$$

c. Assuming a design gas flow velocity of 60 ft/min (18 m/min), the required cross-sectional area of the GAC adsorbers can be found using Eq. 7.3 as:

$$A_{\text{GAC}} = \frac{Q_{\text{Gas}}}{v_{\text{gas}}} = \frac{210}{60} = 3.5\ \text{ft}^2\ \left(0.325\ \text{m}^2\right)$$

If the adsorption system is tailor-made, a system with a cross-sectional area of 3.5 ft^2 will do the job. However, the off-the-shelf 55-gallon drums are to be used. We must determine the number of drums providing the required cross-sectional area.
Area of the GAC bed inside a 55-gallon drum

$$= \pi\left(r^2\right) = (\pi)\left[(1.5/2)^2\right] = 1.77\ \text{ft}^2/\text{drum}\left(= 0.164\ \text{m}^2/\text{drum}\right)$$

Number of drums in parallel to meet the design gas flow velocity

$$= \left(3.5\ \text{ft}^2\right) \div \left(1.77\ \text{ft}^2/\text{drum}\right) = 1.98\ \text{drums}$$

Using two drums in parallel would provide the required cross-sectional area to meet the design superficial gas velocity of 60 ft/min (18 m/min).

Discussion:

1. The bulk density of vapor-phase GAC is typically 30 lb/ft^3 (480 kg/m^3) or less. The amount of GAC in a 55-gallon (0.21 m^3) drum is approximately 150 pounds (68 kg).
2. The minimum number of 55-gallon drums needed for this project is two to meet the gas flow velocity requirement. The number of drums in the treatment system should be increased to meet the monitoring requirements and the desired change-out frequency. If multiple GAC adsorbers are used, the adsorbers are often arranged in series and parallel. If two adsorbers are arranged in series, the monitoring point can be located at the effluent end of the first adsorber. In contrast, the effluent end of the second adsorber should serve as the sampling point for regulatory

compliance. A high effluent concentration from the first adsorber indicates that it is reaching its capacity. The first adsorber should be taken offline, and the second should be shifted forward to be the first. Consequently, the capacity of both adsorbers can be fully utilized, and compliance can also be met.

For this sample project, there should be a minimum of four GAC drums: two parallel trains, and each train has two GAC drums in series. In addition, three parallel trains should be considered because one train can always be taken offline for regeneration or maintenance to operate the system continuously at the design gas flow rate.

The COC removal rate by a GAC adsorber ($R_{removal}$) can be calculated by using the following formula:

$$R_{removal} = \left(G_{in} - G_{out}\right) \times \left(Q_{Gas}\right) \tag{7.4}$$

In practical applications, the effluent concentration (G_{out}) is kept below the discharge limit, which is often very low. Therefore, the term G_{out} can be deleted from Eq. 7.4 as a safety factor or as a conservative approach in design. The mass removal rate is then the same as the mass loading rate ($R_{loading}$):

$$R_{removal} \sim R_{loading} = \left(G_{in}\right) \times \left(Q_{Gas}\right) \tag{7.5}$$

The mass loading rate is nothing but the product of the gas flow rate and the mass concentration of the COC.

Example 7.3: Determine Mass Removal Rate by GAC Adsorbers

Referring to the remediation project described in Example 7.2, the discharge limit for xylene is 100 ppb. Determine the mass removal rate by the GAC system.

Solution:

a. MW of xylene (C_8H_{10}) = (12)(8) + (1)(10) = 106

At 20°C (77°F), 1 ppm = (106/385) $\times 10^{-6}$ lb/ft³ = 2.75×10^{-7} lb/ft³

$$= (106/24.05)\text{mg/m}^3 = 4.41\text{mg/m}^3$$

$$800 \text{ ppm} = (800) \times \left(2.75 \times 10^{-7}\text{lb/ft}^3\right) = 2.2 \times 10 -^4 \text{ lb/ft}^3 \left(= 3.53 \text{ g/m}^3\right)$$

b. $R_{removal} \sim R_{loading} = (2.2 \times 10^{-4} \text{ lb/ft}^3) \times (210 \text{ ft}^3/\text{min}) = \underline{0.046 \text{ lb/min}}$

$$= \left(3.53 \text{ mg/m}^3\right) \times \left(6 \text{ m}^3/\text{min}\right) = \underline{0.021 \text{ kg/min}}$$

Once the GAC in an adsorber reaches its capacity, it should be regenerated or adequately disposed of. The service time between two regenerations (or the service life of a fresh batch of GAC without regeneration), service (t_{service}), can be found by dividing the capacity of the GAC adsorber by R_{removal} as:

$$t_{service} = \frac{M_{capacity}}{R_{removal}} = \frac{(M_{GAC})(q_{design})}{R_{removal}} \tag{7.6}$$

Example 7.4: Determine the Change-Out Frequency of GAC Adsorbers

Referring to the remediation project described in Example 7.3, the discharge limit for xylene is 100 ppb. Determine the service life of two 55-gallon GAC units.

Solution:

As shown in Example 7.2, each unit can retain 22.8 lb (10.4 kg) of xylene before exhaustion. The total service life of these two units is:

$$t_{\text{service}} = \frac{M_{\text{capacity}}}{R_{\text{removal}}} = \frac{(2)(22.8\,\text{lb/unit})}{(0.046\,\text{lb/min})} = 991\,\text{min} < 1\,\text{day}$$

Discussion:

1. Although two drums in parallel can provide a proper cross-sectional area, the relatively high COC concentration makes the service life of two 55-gallon GAC drums very short.
2. A 55-gallon GAC drum typically costs several hundred dollars or more. In this example, two drums last less than one day. With the labor and disposal costs, this alternative will be cost-prohibitive. A GAC system with on-site regeneration or other treatment alternatives should be considered.

If the COC concentration of the gas stream is high, a GAC adsorption system with on-site regeneration would become a more attractive option. The amount of GAC required for on-site regeneration depends on the mass loading rate, the GAC's adsorption capacity, the design service time between two regenerations, and the ratio between the number of GAC units in regeneration mode and those in adsorption mode. The total amount of GAC needed (M_{GAC}) can be determined by using the following formula (USEPA, 1991):

$$M_{\text{GAC}} = \frac{(R_{\text{removal}})(t_{\text{service}})}{q_{\text{design}}}\left[1 + \frac{N_{\text{des}}}{N_{\text{ads}}}\right] \tag{7.7}$$

where M_{GAC} is the total amount of GAC required, t_{service} is the adsorption time between two consecutive regenerations, N_{ads} is the number of GAC beds in the adsorption mode, and N_{des} is the number of GAC beds in the regeneration (desorption) mode.

Example 7.5: Amount of GAC Needed for On-site Regeneration

Referring to the remediation project described in Example 7.3, an on-site regeneration GAC system is proposed to deal with the high COC mass loading. The system consists of three GAC adsorbers. Two will be in the adsorption mode, while the other will be in the regeneration mode. The adsorption cycle time is 24 hours. Determine the amount of GAC required for this system.

Solution:

The total amount of GAC required in all three adsorbers:

$$M_{\text{GAC}} = \frac{(0.046\ \text{lb/min})(1440\ \text{min})}{0.154\ \text{lb/lb}}\left[1+\frac{1}{2}\right] = 645\ \text{lb}\,(293\ \text{kg})$$

A total of 645 lb (293 kg) of GAC is needed (i.e., 215 lb (98 kg) in each bed.

7.3 ABSORPTION

For air pollution control, *absorption* refers to removing undesirable gaseous components from a gas stream by dissolving them into a liquid. It is a mass transfer process, and the primary removal mechanism is dissolution. The participants of this process include (a) *absorbent*, the liquid, usually water, into which the COC is absorbed/retained; (b) *absorbate*, the COC that is being absorbed into the absorbent; and (c) *carrier gas*, the inert part of the gas stream, typically air, from which the COCs are to be removed.

7.3.1 Principles of Absorption

What is the main difference between a<u>d</u>sorption and a<u>b</u>sorption? A<u>d</u>sorption is a surface phenomenon, and the COCs are retained on the surface of the solid adsorbent. In a<u>b</u>sorption, the COCs in the gas phase go through the air–liquid interface, are retained and distributed throughout the liquid absorbent. A *two-film theory* is commonly used to describe the mass transfer of a compound across the interfacial boundary between the gas and the liquid. In this theory, there is a gas film on one side of the boundary and a liquid film on the other. They are stagnant films and hypothetical (Figure 7.7). The gas pollutant concentration in its bulk phase is G, while the pollutant concentration in the bulk liquid phase is C. The gas concentration (G) is greater than the concentration in equilibrium with the liquid concentration (C). Therefore, the gaseous compound tends to move into the liquid phase. Consequently, the gas concentration would decrease to G_i at the interface to be in equilibrium with C_i, the liquid concentration at the interface. The concentration gradients between the bulk phases and the interface are the driving forces for the mass transfer.

The dissolution of the COCs into the liquid is the primary removal mechanism. Solubility is the most critical criterion for determining the applicability of the

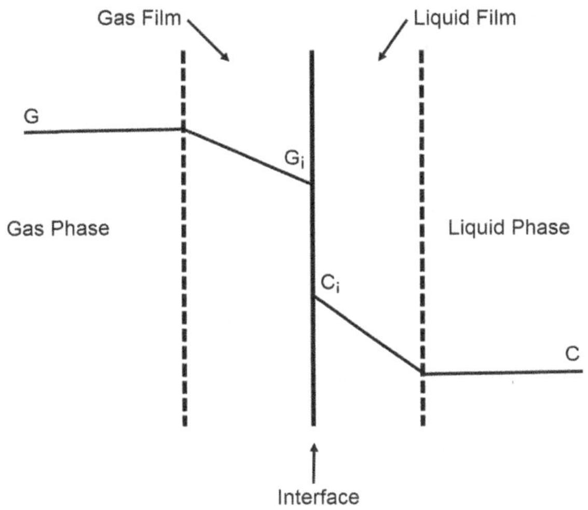

FIGURE 7.7 Two-film theory for mass transfer in absorption.

absorption process. Thus, adsorbers apply to relatively soluble gaseous organic and many inorganic (e.g., H_2S, HCl, and NH_3) pollutants.

The solubility limit can be overcome by changing the pH or adding chemicals to the liquid to react with the dissolved gas pollutants. For example, lowering the liquid pH would improve the effectiveness of an adsorber for the removal of NH_3; adding an oxidant such as hypochlorite (OCl^-) for the removal of reduced compounds; and adding lime, $Ca(OH)_2$, to react irreversibly with SO_2 to form $CaSO_4$ as in flue gas desulfurization (FGD).

7.3.2 Absorption Systems and Their Components

Absorbers in air pollution control use a liquid to remove/scrub COCs from a gas stream and operate under the absorption principle. Since it is a mass transfer process, a large air–liquid interfacial area and contact time available for mass transfer are the critical factors affecting the performance of absorbers. As mentioned, wet scrubbers/absorbers can simultaneously remove gaseous pollutants and particulates from a gas stream. However, the primary mechanisms for particulate removal and gaseous pollutants are different. The primary mechanism for particulate removal is inertial impaction, and the treatment systems are often called (wet) scrubbers. Dissolution is the primary mechanism for removing gaseous pollutants, and the removal systems are often called absorbers.

The main component of an absorption system is a device that enhances the close contact between the gas stream and the liquid. The subsequent discussion is focused on operations based on dissolution.

To provide enough contact time, an absorber is often a tall column where the air flows counter-currently with the liquid. The standard configurations include a spray tower, packed column, and tray column. *Packed columns* are the most common configuration for absorbers and are the focus here. In addition to packing, the system consists of a liquid pump, an air blower, a liquid distributor (or sprayers) to distribute the liquid, and a mist eliminator to minimize the carry-over of liquid droplets exiting at the top of the column. There are a packing support plate at the bottom and a restrainer at the top to keep the packing in place. A sump at the bottom of the column is also standard for liquid storage, and the liquid can be recirculated or added back as part of the scrubbing liquid (Figure 7.8).

Packing materials provide a large specific surface area on which the liquid will form a thin layer for mass transfer between the absorbent and the gas. A particular packing is described by its trade name and overall size (e.g., 2-inch Jaegar Tri-Packs®). Some common types of packing have been shown in Figure 6.32. Their dimensions typically range from 2.5 to 10cm (1 to 4 inches). They are made of plastics (e.g., polyethylene (PE), polypropylene (PP)), ceramic, or metal. However, plastic packing materials are common for air pollution control applications. The primary considerations in selecting the packing materials include size, temperature rating, chemical resistance, bulk density, and cost. Figure 7.9 illustrates the different sizes of Jaeger Tri-Packs®.

Crossflow type absorbers (vs. counter-flow type described above) are also used. For this configuration, the air stream flows horizontally through the packing materials, and the scrubbing liquid flows down through the packing.

An air stripper operates on the same principle as an absorber, based on the mass transfer of COCs between the liquid and air. Air strippers use clean air to treat dirty liquid, while absorbers use clean water to treat dirty air.

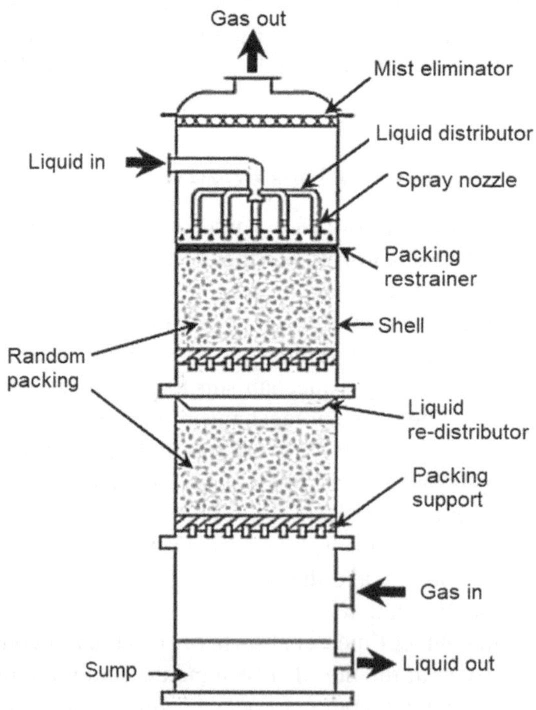

FIGURE 7.8 Packed tower for gas absorption (Modified from USEPA, 2002).

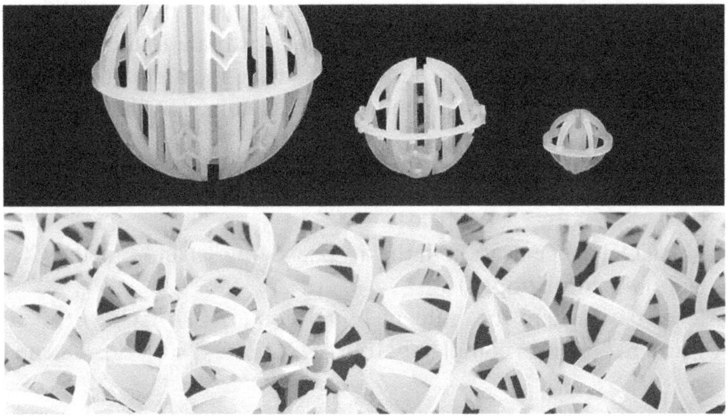

FIGURE 7.9 Different sizes of Jaeger Tri-Packs® (http://www.jaeger.com/)

7.3.3 Factors Affecting Absorption

Henry's constant of the COC is the most critical characteristic affecting its removal by absorption. A COC with a large Henry's constant implies it prefers staying in the air. Therefore, the absorption will work better for compounds with smaller Henry's constants. In addition, Henry's constants of most VOCs increase with temperature. Thus, the extent of absorption of VOCs will be greater at lower temperatures.

The type and size of packing used will affect the mass transfer between the liquid and gas streams. For a given type of packing, for example, a smaller packing size should provide a better mass transfer since it has a larger specific surface area. However, the trade-off is a higher pressure drop. A thicker packing bed should yield better removal, but has higher capital and operating costs (i.e., a larger head loss). The liquid-to-air ratio is another significant operating parameter.

7.3.4 Design Calculations

Absorption is also a widespread process in chemical engineering practices. Design examples for absorbers in most, if not all, of the environmental engineering literature used the classical chemical engineering approach in which the concentrations are expressed in mole fraction and graphical methods are used. Let us take a different approach here, using the mass concentrations and no graphics. It is more straightforward and comprehensible, in my opinion.

Sizing an absorber depends primarily on the volumetric flow rate of the gas stream, the influent COC concentration, the required removal efficiency, and the solubility of the COC in the liquid (actually, Henry's constant will be a better indicator). The volumetric gas flow rate (Q_G) determines the cross-sectional area of the column (A) as well as the fan and motor size. The concentration, the required removal, and Henry's constant determine the residence time (which, in turn, determines the height of the column) as well as the liquid flow rate (Q_L) needed. Types of packing materials would impact the mass transfer coefficient (consequently, the required residence time for a specific removal) and the head loss across the column. The packing provides a large surface area for COCs to migrate from the air stream to the liquid stream.

A mass balance equation can be derived by letting the amount of COCs removed from the gas stream be equal to the amount of COCs retained by the liquid (Figure 7.10):

$$Q_G\left(G_{\text{in}} - G_{\text{out}}\right) = Q_L\left(C_{\text{out}} - C_{\text{in}}\right) \tag{7.8}$$

where G is the COC concentration in the air phase (mg/L) and C is the COC concentration in the liquid phase (mg/L).

For an ideal case where the influent liquid contains no COCs ($C_{\text{in}}=0$) and the effluent gas stream becomes completely free of COC ($G_{\text{out}}=0$), Eq. 7.8 can be simplified into:

$$(Q_G)\left(G_{\text{in}}\right) = (Q_L)\left(C_{\text{out}}\right) \tag{7.9}$$

Assuming Henry's law applies, and the influent gas stream is in equilibrium with the effluent liquid, then:

$$G_{\text{in}} = H^* \times C_{\text{out}} \tag{7.10}$$

where H^* is Henry's constant of the COC in a dimensionless form. Combining the two equations above, the following relationship can be derived:

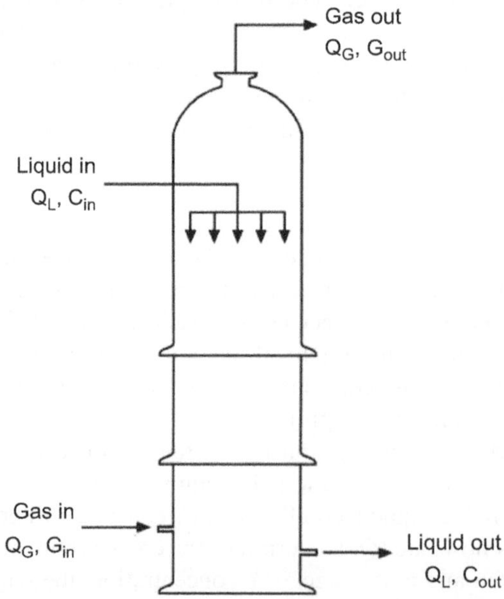

FIGURE 7.10 Mass balance around an absorber (Modified from USEPA, 2003a).

$$\left(\frac{1}{H^*}\right)\left[\frac{Q_L}{Q_G}\right]_{\text{min}} = 1 \tag{7.11}$$

where $(Q_L/Q_G)_{\text{min}}$ is the minimum liquid-to-air ratio (in vol/vol) for this ideal/theoretical case. It would require a relatively large packing height to reach the equilibrium and to achieve 100% removal. The liquid-to-air ratio in real-life applications $(Q_L/Q_G)_{\text{actual}}$ should be larger than $(Q_L/Q_G)_{\text{min}}$.

The *absorption factor* (AF), which is the product of $(1/H^*)$ and $(Q_L/Q_G)_{\text{actual}}$, is commonly used in design:

$$\text{AF} = \left(\frac{1}{H^*}\right)\left[\frac{Q_L}{Q_G}\right]_{\text{actual}} \tag{7.12}$$

For field applications, the values of AF should be greater than one. The practical values of AF range from 1.5 to 2.

Example 7.6: Liquid-to-Air Ratio for an Absorber

A packed-column absorber is designed to remove ammonia (NH_3) from a gas stream (Q = 85 m^3/min = 3,000 ft^3/min). Determine (a) the minimum liquid-to-air ratio, (b) the design liquid-to-air ratio, and (c) the design liquid flow rate. Use the following information in calculations:

- Henry's constant = 0.78 mole fraction in air/mole fraction in water.
- Absorption factor (AF) = 1.5.
- T = 298 K and P = 1 atm.

Solution:

a. Convert the given Henry's constant into the dimensionless value (see Table 2.3):

$$H^* \text{ (dimensionless)}$$
$$= H[\text{in (mole frac. in air)/(mole frac. in liq.)}] \times W \div [R \times T \times \gamma \times 1{,}000] \times P$$
$$= (0.78) \times (18) \div [(0.082) \times (298) \times (1) \times (1{,}000)] \times 1 = 5.75 \times 10^{-4}$$

$$\left(\frac{1}{H^*}\right)\left[\frac{Q_L}{Q_G}\right]_{\text{min}} = 1 = \left(\frac{1}{5.75 \times 10^{-4}}\right)\left[\frac{Q_L}{Q_G}\right]_{\text{min}}$$

So, $(Q_L/Q_G)_{\text{min}} = 5.75 \times 10^{-4}$

b. $(Q_L/Q_G)_{\text{actual}} = (Q_L/Q_G)_{\text{min}} \times \text{AF} = (5.75 \times 10^{-4}) \times (1.5) = 8.62 \times 10^{-4}$

c. $Q_L = (85)(8.62 \times 10^{-4}) = 7.33 \times 10^{-2}$ m^3/min = 19.4 gal/min

Discussion: An absorption factor value of 1.5 means the $(Q_L/Q_G)_{\text{actual}}$ ratio is 1.5 times the $(Q_L/Q_G)_{\text{min}}$ ratio. The $(Q_L/Q_G)_{\text{actual}}$ can be obtained by multiplying the $(Q_L/Q_G)_{\text{min}}$ by the AF value.

One of the critical components in sizing an absorber is determining the diameter of the column. The diameter depends mainly on the gas flow rate. The larger the volumetric gas flow rate, the larger the column diameter. For a given cross-sectional area and a specific liquid flow rate, if the gas flow velocity increases to a level, the liquid may start to be held up in the void of the packing bed. Due to reduced available channels for gas travel through the bed, the pressure drop across the bed will increase, and the mass transfer between the gas and the liquid will adversely decrease.

If the gas velocity increases further, the liquid stops flowing and collects on the top of the packing bed. This condition is called "*flooding*"; the relevant gas velocity is called the *flooding velocity*. The flooding velocity depends mainly on the density of the gas stream, the density and viscosity of the liquid, and the type of packing. Getting specific monographs from the vendors of packing materials is advisable to derive the site-specific flooding velocity. The design gas velocity (v_G) is typically 50 to 75% of the flooding velocity. This velocity is then used to determine the required cross-sectional area of the absorber column (A_{absorber}):

$$A_{\text{absorber}} = \frac{Q_G}{v_G} \tag{7.13}$$

Another critical design component is the required depth of the packing column (Z) for a specific removal efficiency. A taller packing column would be necessary for a larger removal efficiency. The packing height can be determined using the classical chemical engineering transfer unit concept:

$$Z = \text{HTU} \times \text{NTU} \tag{7.14}$$

where HTU is the height of the transfer unit and NTU is the number of transfer units.

The HTU value depends on the gas flow velocity and the *overall gas-phase mass transfer coefficient*, $K_G a$, in which K_G is the mass transfer rate constant (m/s) and a is the specific surface area (m^2/m^3). $K_G a$ has a unit of 1/time. The $K_G a$ value for a particular application should be best determined from pilot testing. There are also empirical equations available to estimate $K_G a$ values. HTU has a unit of length and can be calculated by:

$$\text{HTU} = \frac{v_G}{K_G a} \tag{7.15}$$

The NTU value can be determined by using the following formula:

$$\text{NTU} = \left(\frac{\text{AF}}{\text{AF}-1}\right)\ln\left\{\left(\frac{G_{\text{in}} - H^* \times C_{\text{in}}}{G_{\text{out}} - H^* \times C_{\text{in}}}\right)\left(\frac{\text{AF}-1}{\text{AF}}\right) + \frac{1}{\text{AF}}\right\} \tag{7.16}$$

The equation can be simplified if the influent liquid is free of COC by having $C_{\text{in}} = 0$, as follows:

$$\mathrm{NTU} = \left(\frac{\mathrm{AF}}{\mathrm{AF}-1}\right)\ln\left\{\left(\frac{G_{\mathrm{in}}}{G_{\mathrm{out}}}\right)\left(\frac{\mathrm{AF}-1}{\mathrm{AF}}\right)+\frac{1}{\mathrm{AF}}\right\} \tag{7.17}$$

The equation can be further simplified for exceptional cases where a chemical reaction occurs once the COC enters the liquid or if the COC is highly soluble. The COC will exhibit insignificant vapor pressure for these two cases, so Henry's constant is essentially zero. With that,

$$\mathrm{NTU} = \ln\left(\frac{G_{\mathrm{in}}}{G_{\mathrm{out}}}\right) \tag{7.18}$$

For this case, the NTU for a 90% reduction in the influent COC concentration would equal ln10 (= 2.303), in which $G_{\mathrm{out}} = 10\%$ of G_{in}.

Example 7.7: Sizing an Absorber

For the absorber in Example 7.6, if the influent ammonia concentration is 3%, the required removal is 90%. The diameter of the column is 0.75 m, and $K_G a = 6.5/\mathrm{s}$. Determine the packing height needed.

Solution:

a. Cross-sectional area $= \pi(0.75/2)^2 = 0.442\ \mathrm{m}^2$
b. Using Eq. 7.13,

$$v_G = \frac{85}{0.442} = 192.4\ \mathrm{m/min} = 3.21\ \mathrm{m/s}$$

c. Using Eq. 7.15,

$$\mathrm{HTU} = \frac{3.21\ \mathrm{m/s}}{6.5/\mathrm{s}} = 0.49\ \mathrm{m}$$

d. Using Eq. 7.17,

$$\mathrm{NTU} = \left(\frac{1.5}{1.5-1}\right)\ln\left\{\left(\frac{0.03}{0.003}\right)\left(\frac{1.5-1}{1.5}\right)+\frac{1}{1.5}\right\} = 4.16$$

e. Using Eq. 7.14,

$$Z = \mathrm{HTU} \times \mathrm{NTU} = (0.49\ \mathrm{m}) \times (4.16) = 2.05\ \mathrm{m}$$

7.4 THERMAL OXIDATION

Combustion is the process of burning. For a more scientific definition, *combustion* is a chemical reaction process in which a substance reacts rapidly with oxygen at an elevated temperature. The substance is more reduced and often called the *fuel*, while the source of oxygen is called the *oxidizer*. The fuel is typically an organic compound that possesses a *heating value*. Combustion is a destructive process in which the fuel is destroyed, and the process is exothermic (i.e., energy is released from the reaction). It is often called a *thermal oxidation* process. We burn fossil fuels, such as coal, natural gas, and gasoline, to generate energy for many daily activities. *Incineration* is a waste treatment process that destroys organic substances in a waste solid, liquid, or gas stream.

Regarding air pollution control, thermal oxidation processes commonly treat gas streams containing VOCs and toxic organic compounds. Thermal oxidation is also applicable to some inorganic compounds, such as H_2S. They are often grouped into (direct) thermal oxidation and catalytic oxidation. Both groups operate under the same principle, which employs elevated temperatures to oxidize/destruct organic compounds, except that catalysts are used in *catalytic oxidation*. The first section will focus on (direct) thermal oxidation, while the next section is on catalytic oxidation.

7.4.1 Principle of Thermal Oxidation

Most fuels used for combustion are composed essentially of carbon and hydrogen. A simplified reaction for thermal oxidation of hydrocarbons can be expressed as:

$$C_xH_y + \left(x + \frac{y}{4}\right)O_2 \rightarrow x\ CO_2 + \left(\frac{y}{2}\right)H_2O + \text{Heat} \tag{7.19a}$$

Similarly, the reaction for thermal oxidation of carbohydrates ($C_xH_yO_z$) can be expressed as:

$$C_xH_yO_z + \left(x + \frac{y}{4} - \frac{z}{2}\right)O_2 \rightarrow x\ CO_2 + \left(\frac{y}{2}\right)H_2O + \text{Heat} \tag{7.19b}$$

If the organic compounds contain sulfur ($C_xH_yO_zS_s$), it can be expressed as:

$$C_xH_yO_zS_s + \left(x + \frac{y}{4} - \frac{z}{2} + s\right)O_2 \rightarrow x\ CO_2 + \left(\frac{y}{2}\right)H_2O + s\ SO_2 + \text{Heat} \tag{7.19c}$$

If combustion happens in air instead of pure O_2, nitrogen (N_2) would appear as a "spectator" on both sides of the equation at an amount of "79/21 × the amount of O_2". However, some N_2 in the air may react with O_2 to form unwanted nitrogen oxides during combustion. In addition, some nitrogen oxides can also be formed if the organic compounds contain nitrogen (see Chapter 8).

The reaction products from combusting hydrocarbons are carbon dioxide and water. For methane, as an example,

$$CH_4 + 2O_2 \rightarrow CO_2 + 2H_2O \tag{7.20}$$

It requires two moles of oxygen to completely combust one mole of methane. The oxygen needed for complete oxidation is called the *stoichiometric* (or *theoretical*) *amount*. The *stoichiometric ratio* of O_2:CH_4 in the combustion of methane is 2:1. If air, instead of pure oxygen, is used in combustion, the stoichiometric ratio of air:CH_4 would be approximately 9.52:1, assuming the air contains 21% oxygen (i.e., $9.52 = 2 \div 21\%$). The air–fuel mixture is (fuel) *rich* if the amount of oxygen is less than the stoichiometric amount to burn all the fuel. This type of combustion is often called *rich burning*. When the amount of oxygen is more than stoichiometric, the air–fuel mixture will be *lean*, and combustion is called *lean burning*.

The minimum concentration of a combustible vapor to support its combustion is defined as its *Lower Explosive Limit* (*LEL*). In other words, the LEL is the lowest concentration of gas/vapor in air that can ignite and cause an explosion if an ignition source is present. Below this level, the mixture is "too lean" to burn. The maximum concentration of a compound that will burn in the air is defined as its *Upper Explosive Limit* (*UEL*). Above this level, the mixture is "too rich" to burn. The range between the LEL and UEL is the *flammable range* for that compound. Table 7.2 shows some common compounds' LEL, UEL, and stoichiometric mixture (USEPA, 2012b).

The LEL and UEL values in Table 7.2 and those in the literature are often different and should be used cautiously. It is because they are measured values, and the data might be derived at different *T*, *P*, and oxygen concentrations. Because of uncertainty in the LEL values and the accuracy of real-time monitoring, influent concentrations to the gas control system are usually kept less than 25% of the LEL to provide some margin of protection from fires and explosions within the system. The LEL and UEL values of a gas mixture, without having actual laboratory measurements, can be estimated as the weighted average of each combustible constituent in the gas stream:

$$\text{LEL}_{\text{mixture}} = \frac{1}{\left[\sum \left(y_i/\text{LEL}_i\right)\right]}; \ \text{UEL}_{\text{mixture}} = \frac{1}{\left[\sum \left(y_i/\text{UEL}_i\right)\right]} \tag{7.21}$$

where y_i is the mole fraction of combustible component *i* of all the explosive components. For a more conservative approach, one can use the lowest LEL of these combustible compounds as the LEL of the mixture (EPA, 2012b).

Most gas compositions in combustion calculations are expressed in % by volume; however, a mass/weight ratio is sometimes used, such as the "*air/fuel ratio*" (*A*/*F*) for internal combustion (IC) engines. The *A*/*F* ratio for an air–fuel mixture can be found as:

$$A/F = \frac{n_{\text{air}} \times \text{MW}_{\text{air}}}{n_{\text{fuel}} \times \text{MW}_{\text{fuel}}} \tag{7.22}$$

where n_{air} is the number of moles of air needed to stoichiometrically combust a specific number of moles of the fuel (n_{fuel}), and MW is the molecular weight.

Example 7.8: LEL of a Gas Mixture

A gas stream contains 1,250 ppm of benzene, 750 ppm of toluene, and 500 ppm of xylenes. Does it exceed 25% of the LEL for the entire gas stream?

Solution:

a. From Table 7.2, LELs of benzene, toluene, and xylenes are 1.2%, 1.1%, and 0.9%, respectively.
b. With a total concentration of 2,500 ppm (1,250+750+500), the mole fractions of benzene, toluene, and xylenes are 0.5, 0.3, and 0.2, respectively.

$$LEL_{mixture} = \frac{1}{\left[\frac{0.5}{1.2\%} + \frac{0.3}{1.1\%} + \frac{0.2}{0.9\%}\right]} = 1.10\% = 11,000 \text{ ppmV}$$

$$25\% \text{ of the } LEL_{mixture} = (25\%) \times (11,000) = 2,750 \text{ ppm}$$

The total concentration (i.e., 2,500 ppm) is less than the 25% LEL value (i.e., 2,750 ppm).

c. For a more conservative approach:
The lowest *LEL* of these three compounds is 0.9% (xylenes),

$$25\% \text{ of } 9,000 \text{ ppm} = (25\%) \times (9,000) = 2,250 \text{ ppm}$$

The total concentration (i.e., 2,500 ppm) exceeds the 25% LEL value (i.e., 2,250 ppm).

Discussion: 1% by volume = (1/100) by volume = (10,000/1,000,000) by volume = 10,000 ppm

TABLE 7.2
LEL, UEL, and Stoichiometric Mixture (% by Volume)

Compound	LEL	UEL	Stoichiometric Mixture
Acetone	2.5	12.8	4.97
Benzene	1.2	7.8	2.71
Ethane	2.8	15.3	5.64
Hydrogen	2.0	80.0	29.5
Methane	4.4	15.5	9.48
Octane	1.0	6.5	1.65
Pentane	1.4	9.2	2.55
Propane	2.1	9.5	4.02
Toluene	1.1	7.1	2.27
Xylenes	0.9	7.0	1.96
Gasoline	1.4	7.6	Varies with/Compositions

Example 7.9: Stoichiometric Mixture and A/F Ratio

Determine the methane concentration (in % by volume) in a stoichiometric mixture for combustion and the corresponding *A/F* ratio.

Solution:

a. As shown in Table 7.2, the stoichiometric ratio of air to methane is 9.48.

$$\text{Concentration of CH}_4 = \frac{n_{\text{methane}}}{n_{\text{methane}} + n_{\text{air}}} = \frac{1}{1+9.48} = 9.5\%$$

b.

$$\frac{A}{F} = \frac{n_{\text{air}} \times \text{MW}_{\text{air}}}{n_{\text{fuel}} \times \text{MW}_{\text{fuel}}} = \frac{(1-9.5\%)(29)}{(9.5\%)(16)} = 17.3 \frac{\text{kg air}}{\text{kg methane}} = 17.3 \frac{\text{lb air}}{\text{lb methane}}$$

Discussion: The LEL, UEL, and stoichiometric mixture concentrations (in % by volume) vary considerably with the fuel types; however, they differ much less on a weight basis (e.g., the *A/F* ratio).

7.4.2 Thermal Oxidation Systems and Their Components

A thermal oxidizer has a chamber that has gas- or oil-fired burners. The burners heat the gas stream to exceed the auto-ignition temperatures of the COCs. *Auto-ignition temperature* is the lowest temperature at which a compound spontaneously ignites without an external ignition source in a standard atmosphere, while the *flash point* is the lowest temperature at which the vapor of a volatile compound can be ignited in the presence of an ignition source. *Auxiliary fuel* refers to extra or secondary fuel used to support the main fuel system, while *auxiliary air* is the additional air supplied to help the primary combustion process. The thermal oxidation reaction will occur in the combustion chamber, which is lined with refractory insulation materials. Although a heat recovery system is not mandatory, a heat recovery device is commonly employed to recover some heat/energy from the flue gas before it is discharged into the atmosphere (Figure 7.11).

The heat from the hot flue process can be recovered by preheating the influent stream to the thermal oxidizer (i.e., *primary heat recovery*) or by heating other methods at the site (i.e., *secondary heat recovery*). Although directly recycling the flue gas into the combustion chamber is theoretically feasible, it would increase the total flow rate to the combustion chamber to reduce the residence time.

Recuperative and regenerative devices refer to the type of heat exchangers used to enhance heat recovery efficiency. Figure 7.12 illustrates a thermal oxidation system with a recuperative heat exchanger. In this system, the COC-laden gas (T=120°C=248°F) is heated up to 351°C (664°F), while the flue gas from the combustion zone (T=700°C=1,292°F) is cooled down to 315°C (599°F) before moving to the stack for discharge.

A *recuperative heat exchanger* typically consists of tubes of good heat transfer properties (e.g., copper) so that the heat of the hotter gas on one side of the tube can be readily transferred to the colder gas on the other side (see Figure 7.13 for a

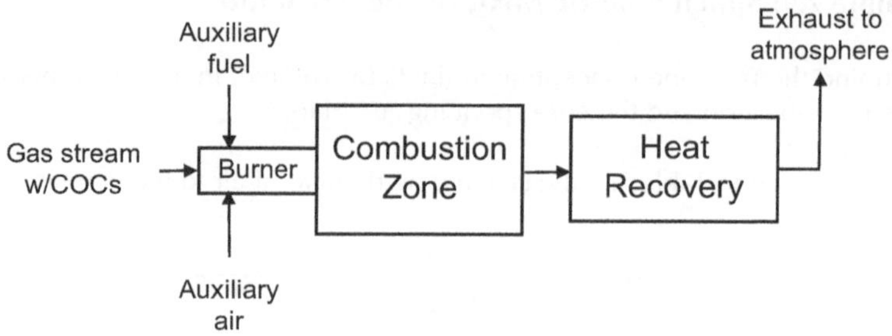

FIGURE 7.11 Schematic of thermal oxidation system.

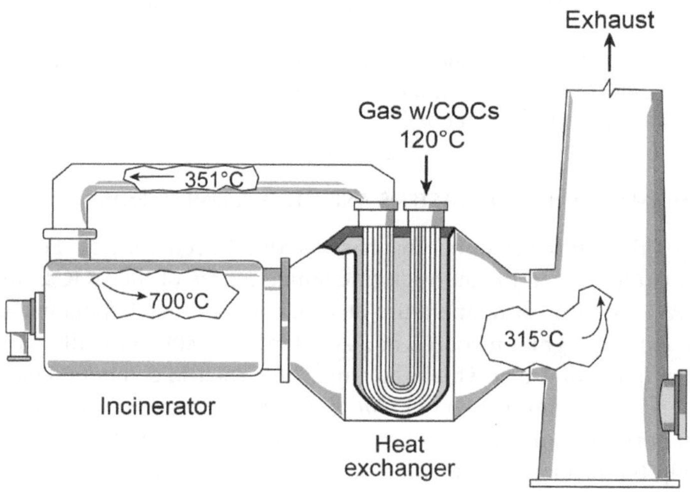

FIGURE 7.12 Recuperative thermal oxidizer (Modified from USEPA, 2000).

shell-tube heat exchanger); there is no mixing between the hot fluid and the cold fluid. *Regenerative heat recovery* devices use heat-absorbing materials to store heat from a hot gas stream, which will be released to a cooler stream later. It should be noted that there are at least two limitations to maximum heat recovery. First, the temperature of the hot gas should not be cooled below the dew point of any gases because the condensate may be corrosive. Second, the temperature of the cooler stream should not be raised above its auto-ignition temperature.

7.4.3 Factors Affecting Thermal Oxidation

In addition to sufficient oxygen concentrations, an efficient combustion process requires three T's: a sufficiently high <u>temperature</u>, an adequate residence <u>time</u>, and good contact between the fuel and the oxidizer (i.e., good <u>turbulence</u>) in the combustion chamber. These three parameters govern the speed and completeness of thermal oxidation. It should be noted that they are independent variables. The temperature

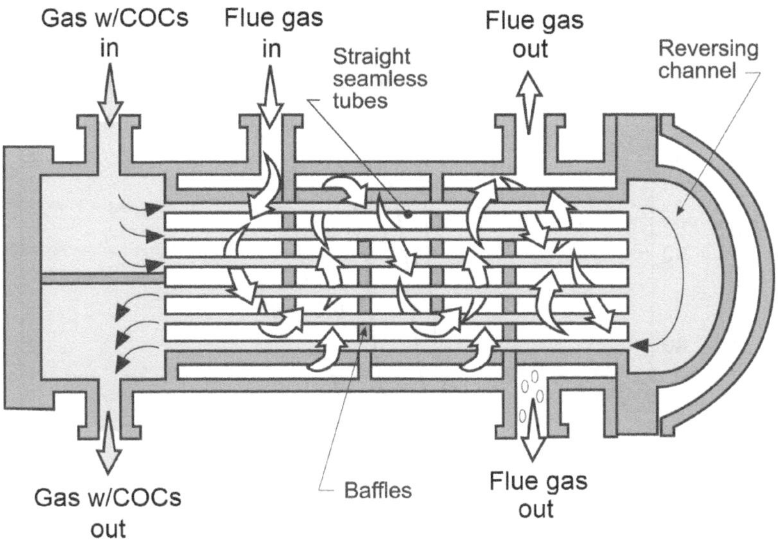

FIGURE 7.13 Schematic of a recuperative shell-tube heat exchanger (Modified from USEPA, 2000).

greatly influences the rate of COC oxidation in the combustion chamber. The higher the temperature, the faster the oxidation reaction proceeds. Figure 7.14 illustrates the effects of combustion temperature and residence time on the destruction of a hypothetical COC. For example, the same destruction efficiency can be achieved at a higher temperature with a shorter residence time or at a lower temperature with a longer residence time. In addition, a higher temperature or a longer residence time would yield a higher destruction efficiency.

Table 7.3 shows the auto-ignition temperatures of several organic compounds. They range from 869 to 1,245°F (465 to 674°C). The combustion temperatures for efficient oxidation are generally 200 to 500°F (~110 to 280°C) above the auto-ignition temperature of the COC, which is most challenging to oxidize in the gas stream. The minimum thermal oxidation temperature is ~1,300°F (~700°C), and most organic vapors can be oxidized at this temperature. However, the equilibrium conditions among CO, O_2, and CO_2 would increasingly favor CO, instead of CO_2, as the combustion product if the temperature is below 1,300°F (~700°C). Consequently, most thermal oxidizers operate at temperatures ranging from 1,300 to 1,800°F (~700 to 1,000°C) (USEPA, 2012a).

If the waste air stream has a low oxygen content (below 13 to 16%), auxiliary air would be needed to raise the oxygen level. This ensures the flame stability of the burner and a low concentration of CO in the flue gas [Note: CO is one of the six criteria air pollutants].

The residence time needed to achieve a specific destruction efficiency in the combustion chamber depends partly on the rate of COC oxidation at the prevailing temperature and partly on the degree of mixing between the gas stream and the auxiliary air in the combustion zone. To promote mixing, baffles can be installed within

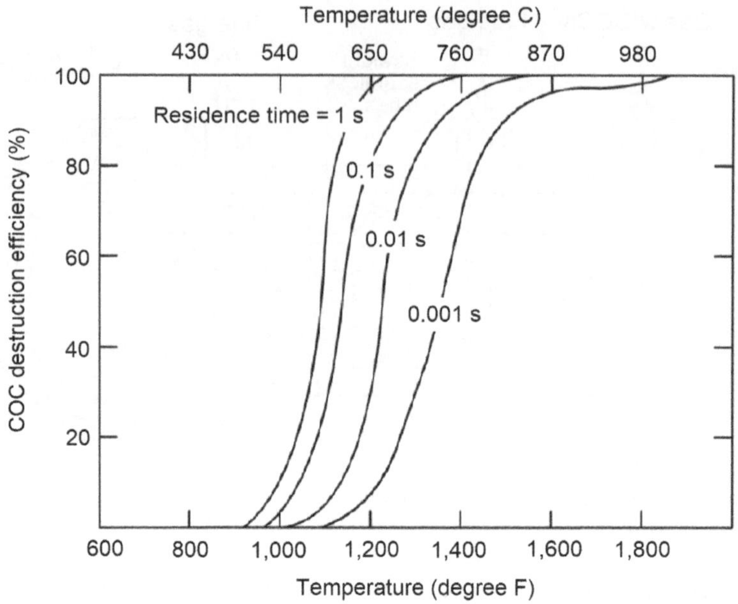

FIGURE 7.14 Effects of temperature and residence time on COC destruction (Modified from USEPA, 2012a).

TABLE 7.3
Auto-ignition Temperatures

	Temperature	
Compound	**(°F)**	**(°C)**
Acetone	869	465
Benzene	1,097	592
Chlorobenzene	1,245	674
Methanol	878	470
Methyl ethyl ketone (MEK)	960	516
Toluene	997	536
Xylene	924	496

Source: Modified from USEPA (2012a).

the combustion chamber to increase the turbulence. The residence times for typical thermal oxidizers are between 0.3 and 2 seconds (USEPA, 2002).

For incineration of hazardous waste containing *principal organic hazardous constituent* (POHC), *destruction and removal efficiency* (DRE) is used as an indicator of removal efficiency:

$$\text{DRE } (\%) = \frac{W_{\text{in}} - W_{\text{out}}}{W_{\text{in}}} \times 100 \tag{7.23}$$

where W_{in} is the influent mass flow rate and W_{out} is the effluent mass flow rate of a particular POHC, in mass/time.

In addition, *combustion efficiency* (CE) is calculated by using volume concentrations of CO and CO_2 (i.e., [CO] & [CO_2]) in the flue gas as:

$$\text{CE }(\%) = \frac{[\text{CO}_2]}{[\text{CO}_2]+[\text{CO}]} \times 100 \quad (7.24)$$

Higher DRE and CE values are usually more desirable.

7.4.4 Design Calculations for Thermal Oxidizers

The design of a thermal oxidation system depends on the concentrations and types of COC in the influent gas stream, as well as the required destruction efficiencies or the required effluent COC concentrations. Achieving a specific destruction efficiency is generally more accessible if the influent COC concentrations are higher.

Higher COC concentrations in the influent gas stream would help to raise the combustion temperature because of the higher heat content. That is one of the reasons why thermal incineration is a favorable alternative for treating a gas stream with high organic concentrations. However, for gas streams containing hazardous air pollutants, influent concentrations of flammable vapors to a thermal oxidizer are generally limited to 25% of the LEL, imposed by insurance companies for safety concerns. Vapor concentrations up to 40 to 50% of the LEL may be permissible if online monitoring of VOC concentrations and automatic process control and shutdown are employed. *Flares* are used for gas stream combustion with vapor concentrations >100% of the UEL (e.g., waste gases and vented gas in oil refineries, chemical plants) (USEPA, 2012b); they are operated in a less controlled *T*, *P*, and air–fuel ratio, just enough to keep the flame stable. Flares consist of a stack or chimney where waste gases are directed and burned at the tip. There are two types of flares: (i) *open flares*, featuring an open flame at the stack tip, visible from a distance, and (ii) *enclosed flares*, containing the flame within a combustion chamber.

The volumetric gas flow rate is commonly expressed in ft^3/min in the U.S. customary system, which is cubic feet per minute (cfm). Since the volumetric flow rate of an air stream is a function of temperature and the gas stream undergoes zones of different temperatures in a thermal process, the airflow rates are shown in actual cfm (acfm) or standard cfm (scfm). The acfm refers to the volumetric flow rate under the actual temperature and pressure, while the scfm is the flow rate under standard conditions. The standard conditions are the basis for comparison. Unfortunately, as mentioned earlier, the definition of the standard conditions is not universal. For USEPA's documents related to combustion, the standard conditions are usually $T = 77°F$ (25°C) and $P = 1$ atm. Consequently, this standard temperature is used for combustion calculations here.

Conversions between an actual flow rate and its corresponding standard flow rate can be readily done using the formula below. It assumes that the ideal gas law is valid, and the pressure stays relatively constant:

$$\frac{Q_{\text{actual}}}{Q_{\text{standard}}} = \frac{273 + T_{\text{actual}}\,(\text{in °C})}{273 + 25} = \frac{460 + T_{\text{actual}}\,(\text{in °F})}{460 + 77} \tag{7.25}$$

Example 7.10: Conversion between Actual and Standard Flow Rates

A thermal oxidizer is used to treat off-gas from an industrial process. The oxidizer operates at 700°C (1,292°F) to achieve removal efficiency. A heat exchanger is employed to recuperate heat (see Figure 7.12), and the temperature of the stack is 315°C (599°F). The actual flow rate at the exit of the thermal oxidizer is 1,000 actual m^3/min (35,300 acfm).

a. What would the flow rate be at the exit of the thermal oxidizer under standard conditions?
b. Determine the exit velocity from the stack (diameter = 1 m = 3.28 ft).

Solution:

a. Use Eq. 7.25 to convert the actual flow rate at the exit of the thermal oxidizer to its standard flow rate:

$$\frac{1{,}000}{Q_{\text{standard}}} = \frac{273 + 700}{273 + 25} = \frac{460 + 1{,}292}{460 + 77}$$

$$Q_{\text{standard}} = 306 \text{ standard m}^3\text{/min } (10{,}810 \text{ scfm})$$

b. Use Eq. 7.25 again to convert the standard flow rate to the actual flow rate at the stack:

$$\frac{Q_{\text{actual}}}{306} = \frac{273 + 315}{273 + 25} = \frac{460 + 599}{460 + 77}$$

$$Q_{\text{actual}} = 604 \text{ actual m}^3\text{/min}(21{,}330 \text{ acfm})$$

$$\text{Exit velocity} = Q/A = \left(604 \text{ m}^3\text{/min}\right) \div [\pi (1)^2 / 4] = 770 \text{ m/min} = 2{,}520 \text{ ft/min}$$

Discussion: If the actual flow rate at one temperature is known, the exact flow rate at another temperature can be directly determined by using:

$$\frac{Q_{\text{actual}} @ T_1}{Q_{\text{actual}} @ T_2} = \frac{273 + T_1(\text{in °C})}{273 + T_2(\text{in °C})} = \frac{460 + T_1(\text{in °R})}{460 + T_2(\text{in °R})} \tag{7.26}$$

The reported heating value of a COC is often based on mass. If the data are unavailable, Dulong's formula can be used (Eqs. 2.23 and 2.24). The heat content of a gas stream is the sum of the heat contents of all the COCs contained:

$$\text{Heat content of a gas stream}\left(\frac{\text{kJ}}{\text{m}^3}\text{ or }\frac{\text{Btu}}{\text{scf}}\right)= \tag{7.27}$$

$$\sum\left\{\text{Heat content of COC}\left(\frac{\text{kJ}}{\text{kg}}\text{ or }\frac{\text{Btu}}{\text{lb}}\right)\times\text{COC conc.}\left(\frac{\text{kg}}{\text{m}^3}\text{ or }\frac{\text{lb}}{\text{scf}}\right)\right\}$$

The heat content of a gas stream in Btu/lb or kJ/kg can be converted from the corresponding value in Btu/ft^3 or kJ/m^3 by using its density:

$$\text{Heat content of a gas stream}\left(\frac{\text{kJ}}{\text{kg}}\text{ or }\frac{\text{Btu}}{\text{lb}}\right)=\text{Its heat content}\left(\frac{\text{kJ}}{\text{m}^3}\text{ or }\frac{\text{Btu}}{\text{ft}^3}\right)\div \tag{7.28}$$

$$\text{its density}\left(\frac{\text{kg}}{\text{m}^3}\text{ or }\frac{\text{lb}}{\text{ft}^3}\right)$$

Example 7.11: Estimate the Heating Value of a Gas Stream

Referring to the project described in Example 7.3, a thermal oxidizer is also considered an alternative to treat the off-gas. Estimate the heat content of the gas stream that contains 800 ppm of xylene.

Solution:

a. Use Dulong's formula (Eq. 2.23) to estimate the heating value of xylene:

$$\text{MW of xylene}\left(C_8H_{10}\right)=12\times8+1\times10=106$$

$$\text{Weight percentage of C}=(12\times8)\div106=90.57\%$$

$$\text{Weight percentage of H}=(1\times10)\div106=9.43\%$$

$$\text{Heat Content}\left(\text{in }\frac{\text{Btu}}{\text{lb}}\right)=(145.4)(90.57)+(620)(9.43)=19{,}015$$

b. To determine the heating value of the gas stream containing 800 ppm xylene, we need to determine the mass concentration of xylene, which has been previously determined in Example 7.3:

$$800\text{ ppm of xylene}=2.2\times10^{-4}\text{lb/ft}^3=3.53\text{ g/m}^3$$

$$\text{Heat content}=(19{,}015\text{ Btu/lb})\left(2.2\times10^{-4}\text{lb/ft}^3\right)=4.11\text{ Btu/ft}^3$$

c. Use Eq. 7.28 to convert the heat content from Btu/ft^3 to Btu/lb:

$$\text{Heat content}=\left(4.11\text{ Btu/ft}^3\right)\div\left(0.0739\text{ lb/ft}^3\right)=55.6\text{ Btu/lb}$$

Discussion:

1. The heat content of xylenes calculated from Dulong's formula, 19,015 Btu/lb, is similar to that found in literature, 18,650 Btu/lb.
2. Using Dulong's formula in SI, one can easily find the corresponding values of 44,190 kJ/kg, 153 kJ/m^3, and 129 kJ/kg for (a), (b), and (c).
3. The density of the gas stream was assumed to be the same as that of the air. At 77°F (25°C), the density of air = (29 lb)/(392 ft^3) = (29 g)/(24.46 L).

When a gas stream has a vapor concentration more significant than 25% of the LEL, dilution air must be used to lower the COC concentrations to <25% of its LEL before combustion. The 25% LEL corresponds to a heat content of 176 Btu/lb or 13 Btu/scf (409 kJ/kg or 484 kJ/m^3) in most cases (USEPA, 1991).

Example 7.12: Heat Content of a Gas Stream at 25% of Its LEL

A gas stream contains an elevated level of toluene (C_7H_8). The heat content of toluene is 18,500 Btu/lb (43,000 kJ/kg). Estimate the heat content of this gas stream, which corresponds to 25% of its LEL.

Solution:

a. From Table 7.2, the LEL of toluene in air is 1.1% by volume.

$$\text{Then, } 25\% \text{ of the LEL} = (25\%)\times(1.1\%) = 0.275\% \text{ by volume} = 2{,}750 \text{ ppm}$$

$$\text{MW of toluene } (C_7H_8) = 12\times 7 + 1\times 8 = 92$$

$$1 \text{ ppm} = (92/392)\times 10^{-6} = 0.235\times 10^{-6} \text{ lb/ft}^3$$

$$= (92/24.46) = 3.755 \text{ mg/m}^3 \text{ @ } T = 77°\text{F } (25°\text{C})$$

$$2{,}750 \text{ ppm} = (2{,}750)(0.235\times 10^{-6}) = 6.46\times 10^{-4} \text{ lb/ft}^3 \ (= 10.33 \text{ g/m}^3)$$

b. Heat content = (18,500 Btu/lb)(6.46 × 10^{-4} lb/ft^3) = 11.95 Btu/ft^3

$$= (43{,}000 \text{ kJ/kg})(10.33\times 10^{-3} \text{kg/m}^3) = 444.2 \text{ kJ/m}^3$$

c. Heat content = (11.95 Btu/ft^3) ÷ (0.0739 lb/ft^3) = 161.7 Btu/lb

$$= (444.2 \text{ kJ/m}^3) \div (1.184 \text{ kg/m}^3) = 375.3 \text{ kJ/kg}$$

Discussion: The calculated heating values are in line with those of typical VOCs at their 25% LELs.

If the heat content of the waste gas to be treated is too high, dilution air will be added to the combustion device as part of the influent stream. Assuming the dilution air has negligible heat content, the volumetric flow rate of the dilution air (containing no heat content) can be found as (USEPA, 1991):

$$Q_w \times H_w = (Q_{\text{dilution}} + Q_w) \times H_i \tag{7.29a}$$

$$Q_{\text{dilution}} = [(H_w/H_i) - 1]Q_w \tag{7.29b}$$

where Q_{dilution} is the required flow rate of dilution air, Q_w is the flow rate of the waste gas stream to be treated, H_w is the heat content of the waste gas stream, and H_i is the heat content of the desired influent entering the combustion chamber.

Example 7.13: Determine the Amount of Dilution Air

Direct thermal oxidation will treat the waste gas stream described in Example 7.12. The heat content of the gas stream was determined to be 700 kJ/kg (300 Btu/lb). The insurance policy limits the COC concentration in the influent to the combustion chamber to ≤25% of its LEL. The flow rate of the gas stream is 20 standard m³/min (706 scfm). Estimate the required dilution air flow rate.

Solution:

Use 375 kJ/kg (162 Btu/lb), from Example 7.12, as the heat content corresponding to 25% LEL. Use Eq. 7.29:

$$Q_{\text{dilution}} = [(700/375) - 1](20) = 17.3 \text{ standard m}^3/\text{min} (= 610 \text{ scfm})$$

If the waste air to be treated has a low oxygen content (<16%), supplemental air may be needed to raise the oxygen level to ensure burning flame stability. If the exact composition of the waste gas stream is known, one can determine the stoichiometric amount of air (oxygen) for complete combustion. In addition, excess air is added to ensure complete combustion in general practice. The following example illustrates how the stoichiometric amount of air and excess air for combusting a landfill gas are determined.

Example 7.14: Stoichiometric and Supplemental Air for Combusting Landfill Gas

A landfill gas stream (Q = 100 standard m³/min = 3,530 scfm) is to be treated by a thermal oxidizer. The landfill gas comprises 60% CH_4 volume and 40% CO_2. The gas will be burned with 20% excess air at 950°C (1,742°F). Determine:

a. the stoichiometric amount of air required,
b. the supplemental air required,
c. the total influent flow rate to the thermal oxidizer, and
d. the total effluent flow rate from the thermal oxidizer.

Solution:

a. Influent Q of CH_4 = (60%) × (100) = 60 standard m³/min
   Influent Q of CO_2 = (40%) × (100) = 40 standard m³/min
   The reaction for the complete combustion of methane is:

$$CH_4 + 2O_2 \rightarrow CO_2 + 2H_2O$$

   The stoichiometric amount of O_2 = (60) × (2/1) = 120 standard m³/min [Note: CH_4:O_2 = 1:2 in methane combustion]
   The stoichiometric amount of air = (120) ÷ (21%) = 570 standard m³/min

b. Q of O_2 in the influent = (1 + 20%) × (120) = 144 standard m³/min
   Q of the influent air = (1 + 20%) × (570) = 684 standard m³/min
   Q of N_2 in the influent air = (79%) × (684) = 540 standard m³/min

c. Total influent Q = 60 (CH_4) + 40 (CO_2) + 684 (total air)
   = 784 standard m³/min

d. Q of O_2 in the effluent = (20%) × (120) = 24 standard m³/min (20% excess)
   Q of N_2 in the effluent = Q of N_2 in the influent = 540 standard m³/min
   Q of CO_2 in the effluent = CO_2 in the landfill gas + CO_2 from combustion
   = 40 + 60 = 100 standard m³/min [Note : CH_4 : CO_2 = 1 : 1 in methane combustion]

   Q of water vapor in the effluent = water vapor produced from combustion = (2/1) × (60) = 120 standard m³/min [Note: CH_4:H_2O = 1:2 in methane combustion]

$$\text{Total effluent } Q = 24 + 540 + 100 + 120 = 784 \text{ standard m}^3/\text{min}$$

Discussion:

1. The following table summarizes the flow rate of each component in this process:

	CH_4	O_2	N_2	CO_2	H_2O
Influent	60	[(2)(60)](1.2) = 144	540	40	0
Effluent	0	144 − 120 = 24	540	40 + 60	120

2. Flow rates of the influent and the effluent are the same at 784 standard m³/min under the standard conditions.
3. The CO_2 concentration in the effluent = 110/784 = 12.7%.

If COC concentrations in the waste gas stream are low, supplementary/auxiliary fuel (e.g., methane gas) is needed to raise the combustion temperature to a desirable range. The following equation can be used to determine the amount of supplementary fuel required (USEPA, 1991):

$$Q_{sf} = \frac{\rho_{wg} \times Q_{wg}\left[C_p\left(1.1 \times T_{com} - T_{he} - 0.1T_{ref}\right) - H_{wg}\right]}{\rho_{sf}\left[H_{sf} - 1.1 \times C_p\left(T_{comb} - T_{ref}\right)\right]} \quad (7.30)$$

where Q_{sf} is the flow rate of the supplementary fuel (scfm or standard m^3/min); ρ_{wg} is the density of the waste gas stream (typically 0.0739 lb/ft^3 or 1.18 kg/m^3); ρ_{sf} is the density of the supplementary fuel (0.0408 lb/ft^3 or 0.653 kg/m^3 for CH_4); T_{comb} is the combustion temperature (°F or °C); T_{he} is the temperature of the waste gas stream after the heat exchanger (°F or °C); T_{ref} is the reference temperature (77°F or 25°C); C_p is the average specific heat capacity of air between T_{comb} and T_{ref} (Btu/lb-°F or kJ/kg·K), and the specific heat capacity value can be extrapolated from data in Table 2.2; H_{wg} is the heat content of the waste gas stream (Btu/lb, kJ/kg); and H_{sf} is the heating value of the supplementary fuel (21,600 Btu/lb or 50,200 kJ/kg for methane).

If the temperature of the waste gas stream after the heat exchanger (T_{he}) is not specified, use the following equation to calculate T_{he}:

$$T_{he} = \left(\frac{HR}{100}\right)T_{comb} + \left[1 - \frac{HR}{100}\right]T_{wg} \tag{7.31}$$

where HR is the heat recovery in the heat exchanger, % (If no other information is available, a value of 70% may be assumed). If no heat exchangers are employed to recuperate the heat, then $T_{he} = T_{wg}$.

Example 7.15: Heat Recovery Efficiency of a Heat Exchanger

Referring to the recuperative heat exchanger in Figure 7.12, estimate the heat recovery efficiency of that heat exchanger.

Solution:

As shown, $T_{wg} = 120°C$, $T_{comb} = 700°C$, and $T_{he} = 351°C$, so

$$624 = \left(\frac{HR}{100}\right)(973) + \left[1 - \frac{HR}{100}\right](393)$$

$$\text{Heat recovery (HR)} = \underline{39.8\%}.$$

Discussion: The heat recovery efficiency of this system is 39.8%, which is less than 70%. Using temperatures in K or °C will achieve the same result.

Example 7.16: Supplemental Fuel Requirement

Referring to the remediation project described in Example 7.11, a thermal oxidizer with a recuperative heat exchanger will treat an off-gas stream (Q = 210 scfm = 6 standard m^3/min) containing 800 ppm of xylene. The combustion temperature is set at 982°C (1,800°F). Determine the flow rate of methane as the supplementary fuel, if required.

Solution:

a. Assuming heat recovery = 70% and T of the waste gas stream = 25°C (77°F), T_{he} can be found as:

$$T_{he} = \left(\frac{70}{100}\right)(982) + \left[1 - \frac{70}{100}\right](25) = 695°C(= 1,283°F)$$

b. The average specific heat can be derived from Table 2.3 as 1.07 kJ/kg·K (0.256 Btu/lb-°F) for the interval between 25°C and 982°C (77°F and 1,800°F).
c. The heat content of the waste gas is 129 kJ/kg (55.6 Btu/lb), as determined in Example 7.11.
d. The supplementary fuel required can be estimated by using Eq. 7.30 as:

$$Q_{sf} = \frac{(1.18)(6)\{(1.07)[1.1(982) - 695 - 0.1(25)] - 129\}}{0.653[50,200 - 1.1(1.07)(982 - 25)]} = 0.062\ m^3/min(= 2.2\ scfm)$$

The flow rate of the influent (Q_{inf}) to the combustion chamber is the sum of those of the waste gas (Q_{wg}), auxiliary (dilution or supplemental) air (Q_{aux}), and the supplementary/auxiliary fuel (Q_{sf}) as:

$$Q_{inf} = Q_{wg} + Q_{aux} + Q_{sf} \tag{7.32}$$

In most cases, one can assume that the flow rate of the combined gas stream (Q_{inf}) entering the combustion chamber under the standard conditions is approximately equal to that of the flue gas leaving the combustion chamber (Q_{fg}) under the standard conditions, because the volume change across the incineration chamber is insignificant due to the combustion of COCs and the supplementary fuel is insignificant.

The flue gas flow rate under actual conditions (Q_{actual}) can be determined from that under the standard conditions:

$$Q_{actual} = Q_{standard}\left[\frac{T_{actual}}{T_{standard}}\right] \tag{7.33}$$

Just a reminder, the combustion and standard temperatures in Eq. 7.33 should be absolute temperatures.

The volume of the combustion chamber (V_{comb}) can be readily determined from the residence time (in seconds) and Q_{actual}, as follows:

$$V_{comb} = \left[\left(\frac{Q_{actual}}{60}\right)\tau\right] \times 1.05 \tag{7.34}$$

TABLE 7.4
Typical Design Values for Thermal Oxidizers

Required Destruction Efficiency (%)	Non-Halogenated Compounds Combustion Temperature	Non-Halogenated Compounds Residence Time (s)	Halogenated Compounds Combustion Temperature	Halogenated Compounds Residence Time (s)
98	1,600°F (538°C)	0.75	2,000°F (1,093°C)	1.0
99	1,800°F (982°C)	0.75	2,200°F (1,204°C)	1.0

Source: USEPA (1991).

The equation is nothing but "residence time = volume ÷ flow rate". The factor of 1.05 is a safety factor, which is an industrial practice to account for minor fluctuations in the flow rate. Table 7.4 shows typical thermal oxidizer system design values for toxic air pollutants.

Example 7.17: Sizing a Thermal Oxidizer

Referring to the remediation project described in Example 7.16, a thermal oxidizer with a recuperative heat exchanger will treat an off-gas stream (Q = 210 scfm = 6 standard m³/min) containing 800 ppm of xylene. The combustion temperature is set at 982°C (1,800°F) to achieve a destruction efficiency of 99%. Determine the size of the thermal oxidizer.

Solution:

a. The flue gas flow rate under standard conditions:

$$Q_{fg} \sim Q_{inf} = Q_{wg} + Q_{aux} + Q_{sf} = 6.0 + 0 + 0.062 = 6.06 \text{ m}^3/\text{min} (= 214 \text{ scfm})$$

b. The flue gas flow rate at actual combustion conditions:

$$Q_{actual} = 6.06 \times [(273 + 982) \div (273 + 25)] = 25.5 \text{ m}^3/\text{min} (= 900 \text{ ft}^3/\text{min})$$

c. The required residence time is 1 second for 99% destruction (Table 7.4):

$$V_{comb} = \left[\left(\frac{25.5}{60}\right)(1)\right] \times 1.05 = 0.44 \text{ m}^3 = 15.6 \text{ ft}^3$$

Thermal oxidation of halogenated organic compounds would generate acidic gases such as hydrochloric acid (HCl) and hydrofluoric acid (HF). This type of compound is corrosive, and discharge limits are often imposed. Consequently, the emission rates of these compounds need to be calculated to determine if a removal system, such as an absorber, is required after the thermal oxidizer.

Example 7.18: Emission Rate and Concentration of Acid Gas

A thermal oxidizer is designed to treat a 100 standard m^3/min (3,530 scfm) gas stream containing 400 ppm benzene (C_6H_6) and 500 ppm carbon tetrachloride (CCl_4). The oxidizer uses 6 and 20 standard m^3/min (219 and 706 scfm) of natural gas and supplemental air, respectively. Estimate the emission rate and concentration of HCl in the flue gas.

Solution:

a. MW of $CCl_4 = (12) \times (1) + (35.5) \times (4) = 154$
   Mass concentration of CCl_4 in the gas stream

$$= (500) \times (154/24.46) = 3{,}142 \text{ mg/m}^3 = 3.14 \text{ g/m}^3$$

   The mass flow rate of CCl_4 in the influent gas stream

$$= \left(3.14 \text{ g/m}^3\right) \times \left(100 \text{ m}^3/\text{min}\right) = 314 \text{ g/min}$$

   The molar flow rate of CCl_4 in the influent gas stream

$$= 314 \text{ g/min} \div 154 \text{ g/mole} = 2.04 \text{ mole/min}$$

b. Since one mole of CCl_4 will generate four moles of HCl, the molar flow rate of HCl in the flue gas = (2.04 mole/min) × (4/1) = 8.16 mole/min

$$\text{MW of HCl} = (1) \times (1) + (35.5) \times (1) = 36.5$$

   The mass flow rate of HCl in the flue gas

$$= (8.16 \text{ mole/min}) \times \left(36.5 \text{ g/mole}\right) = 298 \text{ g/min}$$
$$= 17.9 \text{ kg/h} (= 39.4 \text{ lb/h})$$

c. The flue gas flow rate under standard conditions:

$$Q_{fg} \sim Q_{inf} = Q_{wg} + Q_{aux} + Q_{sf} = 100 + 20 + 6 = 126 \text{ m}^3/\text{min} (= 4{,}470 \text{ scfm}$$

   The mass concentration of HCl in the flue gas under the standard conditions

$$= \left(298 \text{ g/min}\right) \div \left(126 \text{ m}^3/\text{min}\right) = 2.365 \text{ g/m}^3 = 2{,}365 \text{ mg/m}^3$$

   The volume concentration of HCl in the flue gas

$$= (2{,}365) \div \left[(36.5)/(24.05)\right]$$
$$= 1{,}560 \text{ ppm} \left[\text{Note} : 1 \text{ ppm HCl} = (36.5/24.05) \text{mg/m}^3 \text{ @ } 20°\text{C}\right].$$

7.5 CATALYTIC INCINERATION

Similar to direct thermal oxidation, catalytic incineration employs elevated temperatures to destroy combustible COCs in gas streams. However, its combustion temperatures are lower than those of thermal oxidation for a given destruction efficiency due to the participation of catalysts in the reaction. The lower combustion temperature means less fuel usage.

7.5.1 Principles of Catalytic Incineration

A *catalyst* is a substance that increases the rate of a chemical reaction by lowering its activation energy without being used up or going through permanent chemical changes. Typical catalysts used in catalytic incineration are precious metal oxides of platinum (Pt), palladium (Pd), or rhodium (Rh). Some other metal oxides, such as vanadium pentoxide (V_2O_5), titanium dioxide (TiO_2), or manganese oxide (MgO), are also being used.

7.5.2 Catalytic Incineration Systems and Their Components

A *catalytic thermal oxidizer* is very similar to a thermal oxidizer, except that a catalyst bed is incorporated. Passing through a zone with burners, the waste gas stream is heated to the required temperature and then passes a bed containing the catalyst. The catalysts are deposited as a thin layer on a material with a high specific surface area, such as alumina, which is bonded to a support structure. Typical configurations of support structures are honeycombs, grids, and mesh pads so that they can provide large surface areas for reactions and facilitate uniform flow with smaller head losses.

A recuperative or regenerative heat exchanger is often employed. Figure 7.15 illustrates a recuperative catalytic thermal oxidizer.

7.5.3 Factors Affecting Catalytic Incineration

The three T's (temperature, residence time, and turbulence) are also critical design parameters for catalytic oxidation. In addition, the type of catalyst has a significant impact on the system's performance and cost.

Most organic COCs have catalytic auto-ignition temperatures of about 400 to 500°F (~200 to 260°C). It should be noted that they are smaller than their corresponding auto-ignition temperatures (see Table 7.3). The inlet temperature entering the catalyst bed is generally 100°F (55.6°C) above the catalytic auto-ignition temperature of the most difficult-to-destroy COC in the gas stream, so that the minimum inlet temperature to the bed is around 500 to 600°F (260 to 315°C).

The temperature of the gas stream would increase by about 27°F (15°C) for each 1% of the LEL oxidized in the catalyst bed. For a gas stream at its 25% LEL, its temperature can go up by 675°F (375°C), which will bring the temperatures of the catalyst bed and the gas stream at its outlet to about 1,175 to 1,275°F (635 to 690°C). This is very close to the temperature that can deactivate/damage most catalysts (LADCO,

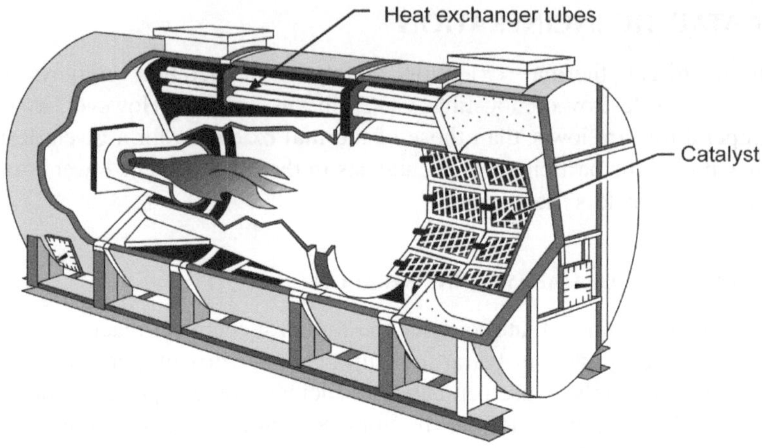

FIGURE 7.15 Recuperative catalytic oxidizer (USEPA, 2012a).

2009). This is why organic vapor concentrations entering catalytic thermal oxidizers are usually kept at less than 20% or 25% of the LEL.

7.5.4 Design Calculations for Catalytic Thermal Oxidizers

The concentration of flammable vapors in a catalytic incinerator is generally limited to 10 Btu/scf (372.4 kJ/m^3) or 135 Btu/lb (314 kJ/kg), which is equivalent to 20% of the LEL for most VOCs (USEPA, 1991). When dilution is required, the volumetric flow rate of the dilution air can be found in Eq. 7.29, as shown earlier.

Example 7.19: Dilution Air for a Catalytic Thermal Oxidizer

Referring to Example 7.12, the heat content of the gas stream was 11.95 Btu/ft^3 (445 kJ/m^3) at 25% of the LEL. The flow rate of the waste gas stream is 10 standard m^3/min (353 scfm). Determine the flow rate of dilution air, if needed, if the gas stream is to be treated by catalytic incineration.
Solution:

The heating content of the waste gas stream at 25% of its LEL exceeds the 10 Btu/scf (372.4 kJ/m^3) limit. Therefore, air dilution is required to bring the heat content of the influent stream to 10 Btu/ft^3. The flow rate of the dilution air can be found by using Eq. 7.29 as:

$$Q_{\text{dilution}} = \left[(11.95/10) - 1\right](10) = 1.95\ \text{m}^3/\text{min}\left(= 65.3\ \text{scfm}\right)$$

Discussion: For this waste gas stream, dilution air may not be needed for thermal oxidation, but it would be necessary for catalytic incineration.

For catalytic incineration, supplementary heat can be provided by electrical heaters. If natural gas is used, use Eq. 7.30 to determine the required flow rate of the supplementary fuel. Before calculating the supplementary heat requirement, the following two equations should be used to estimate the flue gas temperature (T_{out}), which can achieve the desired destruction efficiency without damaging the catalyst. T_{out} can be estimated with the temperature of the waste gas after the heat exchanger but before the catalyst bed (T_{in}) and the heat content of the gas:

$$T_{out}\ (°F) = T_{in}(°F) + 50 \times H_w\ (\text{in Btu/scf}) \tag{7.35}$$

$$T_{out}\ (°C) = T_{in}(°C) + 0.746 \times H_w\ (\text{in kJ/m}^3) \tag{7.36}$$

where H_w is the heat content of the waste gas stream.

These two equations assume a 50°F temperature increase for every Btu/scf of heat content (or 0.746°C temperature increase for every kJ/m^3) in the influent to the catalyst bed. These equations can also determine the required influent temperature to achieve the desired temperature in the catalyst bed.

Example 7.20: Estimate the Temperature of a Catalyst Bed

Referring to the remediation project described in Example 7.3, an off-gas stream (Q=6 m^3/min=210 scfm) containing 800 ppm of xylenes is to be treated by a catalytic incinerator with a recuperative heat exchanger. After the heat exchanger, the temperature of the diluted waste gas is 550°F (288°C). Estimate the temperature of the catalyst bed.

Solution:
After the air dilution, the heat content of the waste gas is 10 Btu/scf (372.4 kJ/m^3).

$$T_{out} = 550 + 50 \times 10 = 1{,}050\ °F$$

or

$$T_{out} = 288 + 0.746 \times 372.4 = 566°C$$

Discussion: The calculated flue gas temperature (T_{out}), 1,050°F (566°C), falls within the typical temperature range for catalyst beds (1,000 to 1,200°F or 540 to 650°C).

Similar to thermal oxidation, the flow rate of the influent to a catalyst bed is the sum of the waste gas, auxiliary (dilution or supplemental) air, and supplementary fuel. Because of the short residence time in the catalyst bed, space velocity is commonly used to relate the volumetric influent flow rate and the volume of the catalyst bed. The space velocity (SV) is defined as the volumetric flow rate of the influent entering the catalyst bed divided by the volume of the catalyst bed. It is actually the inverse of the residence time. Table 7.5 provides the typical design parameters for catalytic incinerators. It should be noted that the flow rate used in the space velocity

TABLE 7.5
Typical Design Parameters for Catalytic Thermal Oxidizers

Desired Destruction Efficiency (%)	*T* at the Bed Inlet	*T* at the Bed Outlet	Space Velocity (hr^{-1})	
			Base Metal	Precious Metal
95	600°F (315°C)	1,000 to 1,200°F (540 to 650°C)	10,000–15,000	30,000–40,000

calculation is the influent gas flow rate at standard conditions, not that of the catalyst bed or the bed effluent.

The size of the catalyst can be determined by:

$$V_{cat} = \frac{60 \times Q_{inf}}{SV} \tag{7.37}$$

where V_{cat} is the volume of the catalyst bed (in m^3 or ft^3), Q_{inf} is the influent flow rate to the catalyst bed (in standard m^3/min or scfm), and SV is the space velocity (in hr^{-1}).

Example 7.21: Sizing a Catalytic Thermal Oxidizer

Referring to the remediation project described in Example 7.20, a catalytic thermal oxidizer with a recuperative heat exchanger will treat an off-gas stream (Q=210 scfm=6 standard m^3/min) containing 800 ppm of xylene. The design space velocity is 12,000 hr^{-1}. Determine the size of the catalyst bed.

Solution:

a. The flue gas flow rate at standard conditions:

$$Q_{fg} \sim Q_{inf} = Q_{wg} + Q_{aux} + Q_{sf} = 6.0 + 0 + 1.95 = 7.95 \text{ m}^3/\text{min} (= 281 \text{ scfm})$$

b. With a space velocity of 12,000 hr^{-1}, the size of the catalyst bed:

$$V_{cat} = \frac{60 \times 7.95}{12,000} = 0.04 \text{ m}^3 \left(= 1.4 \text{ ft}^3\right)$$

Discussion: The size of the catalyst, 0.04 m^3 (1.4 ft^3), is much smaller than the volume of the combustion chamber needed for thermal oxidation, 0.44 m^3 (15.6 ft^3).

7.6 CONDENSERS

Using condensation for air pollution control, COCs are removed from the gas stream by causing them to change from a gaseous state to a liquid state. It can be done by reducing the system temperature, increasing the system pressure, or combining both. However, most condensers for air pollution control use temperature reduction because it is less expensive than compressing a gas stream. Compared to other air pollution control technologies, condensation is more favorable for cases of high COC concentrations, and the condensed products have good recycling/reuse potential.

7.6.1 Principles of Condensation

As a COC-laden gas stream is cooled down, the partial pressure of COC in the gas stream will decrease. The condensation will start when the gas stream is cooled to a temperature below the dew point of the COC, causing it to condense into liquid droplets. With further reduction in the system temperature, the vapor pressure of the COC in the liquid state decreases, so its concentration in the gas stream gets lower.

A *coolant* is needed to absorb the heat/energy from the gas stream, and, in turn, its temperature will increase. It is essentially a heat transfer process. The condensation process is typically applied to cases where the organic COCs can be economically recovered. It is also limited to the cases where only one or two organic compounds are in the gas stream to be recovered because the subsequent separation of multiple organic compounds in the condensate could be difficult. In addition, particulates in gas streams can cause fouling problems in most heat transfer systems; this process is typically limited to gas streams with low particulate concentrations (USEPA, 2002).

7.6.2 Condensation Systems and Their Components

A condensation system must facilitate heat transfer between the gas stream and the coolant. Three types of condensers are used for air pollution control: (i) conventional, (ii) refrigeration, and (iii) cryogenic systems.

Conventional condensation systems are relatively simple and employ a coolant (usually chilled water) to reduce the temperature of the gas stream to as low as 40°F (4.4°C). Using brine as the coolant can cool the gas stream to 0°F (−17.8°C). There are two types of conventional systems: one uses direct-contact condensers and the other uses surface condensers. A *direct-contact condenser* is physically similar to an absorber/wet scrubber, except that the liquid has a much lower temperature to condense the COC in the gas stream and absorption, if the COC is water soluble. The coolant and the condensed/absorbed COC leave the condenser unit as a single stream needing further treatment. The condenser units can be spray towers, tray towers, or ejectors (Figure 7.16). *Ejectors* employ high-pressure spray nozzles to create suction that moves the inlet gas stream. The coolant is sprayed into the throat of the ejector, and COC vapors condense while passing through. Direct-contact condensers are inexpensive and straightforward. However, a wastewater stream containing the spent coolant and COCs must often be treated.

A typical *surface condenser* is a shell-and-tube heat exchanger, similar to the exchanger used in a recuperative thermal oxidizer (Figure 7.13). The coolant flows through the tubes, while the waste gas flows through the shell outside the tubes. The COCs will condense on the outside surface of the tubes and then be collected. The tubes are made of materials that are conductive to heat, such as copper.

Refrigeration systems have lower operating temperatures than conventional condensers and use refrigerants and compressors. As shown in Figure 7.17, the COC-laden gas stream is drawn into a pre-condenser where it is cooled to ~40°F

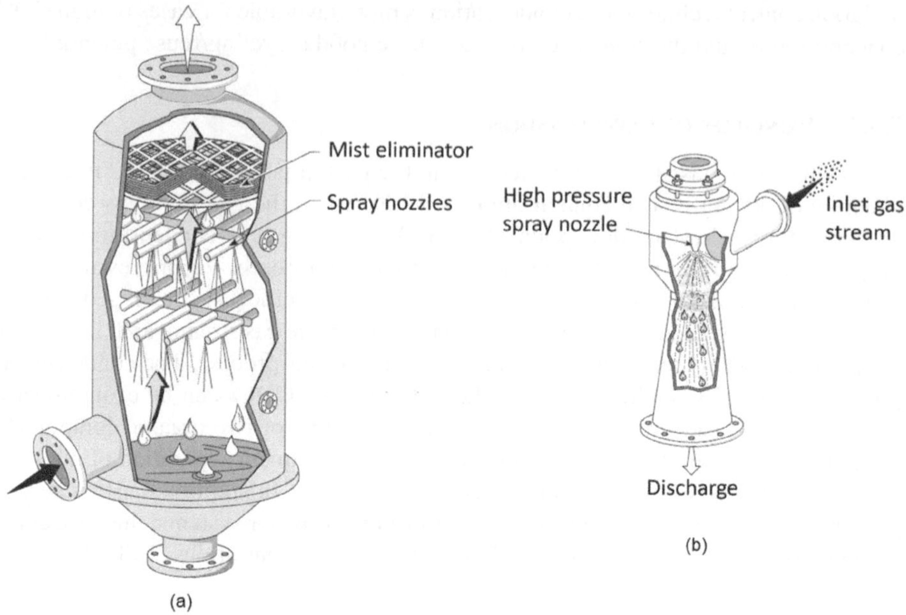

FIGURE 7.16 Direct-contact condensers: (a) spray tower and (b) ejector (USEPA, 2000).

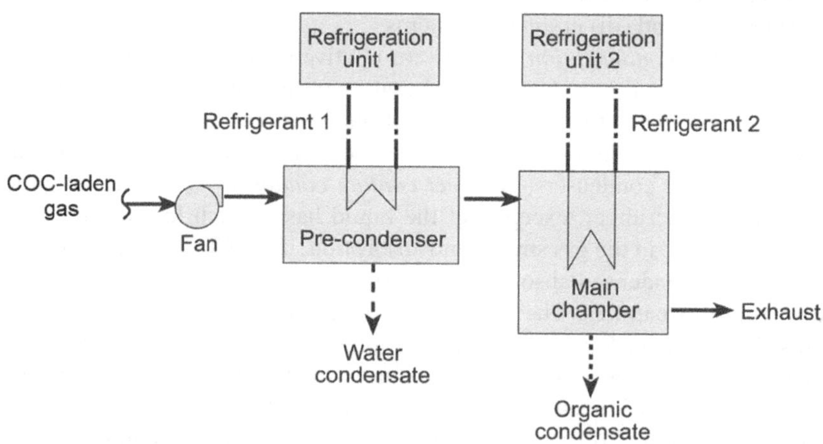

FIGURE 7.17 Process flow diagram of a refrigeration system (Modified from USEPA, 2000).

(4.4°C) to reduce water content to <0.02 lb H_2O/lb air (or kg H_2O/kg air). It minimizes the build-up of frost on top of the heat exchange surfaces of the main chamber, which operates at much lower temperatures, ranging from −50 to −150°F (−43 to −101°C) (USEPA, 2012a).

Cryogenic systems use liquefied nitrogen or carbon dioxide, with operating temperatures ranging from −100 to −320°F (−73 to −160°C) to cool waste gas

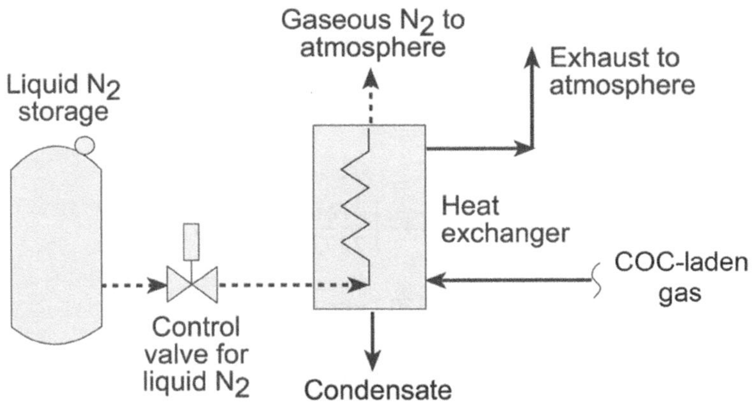

FIGURE 7.18 Process flow diagram of a cryogenic system (Modified from USEPA, 2000).

streams to the freezing point of the COCs (USEPA, 2012a). Figure 7.18 illustrates a simplified process flow diagram of a cryogenic system.

7.6.3 Factors Affecting Condensation

The effectiveness of a condensation system depends on the characteristics of the COCs and the coolant, the configuration of the system, and the operation parameters. As mentioned, condensation for air pollution control is essentially a heat transfer process. Heat must be removed from the waste stream, and the coolant picks it up. Consequently, condensate is formed, and the temperatures of the condensate and the effluent stream are lower than the influent waste gas stream, while the temperature of the coolant will increase.

7.6.4 Design Calculations for Condensers

A COC in the treated gas stream leaving a condensation system will have a reduced partial pressure. Assuming the condensate and the effluent gas are in equilibrium, the partial pressure of the COC in the gas stream (P_{COC}) will be equal to the vapor pressure of the condensate ($\mathrm{VP}_{\mathrm{condensate}}$). The maximum removal efficiency ($\eta_{\max}$), based on the assumption of an equilibrium, can be estimated by:

$$\eta_{\max}\ (\%) = \frac{P_{\mathrm{COC,\ influent\ gas}} - \mathrm{VP}_{\mathrm{condensate}}}{P_{\mathrm{COC,\ influent\ gas}}} \times 100 \tag{7.38}$$

Example 7.22: Removal Efficiency of a Condenser

What would be the maximum possible removal efficiency for acetone (C_3H_6O) in an indirect-contact condenser operating at 0°C (32°F) at 1 atm? The acetone concentration in the influent gas stream is 300,000 ppm. The Antonine equation for acetone ($T = -12.9$ to 55.3°C) is shown below (USEPA, 2012):

$$\log_{10}(P \text{ in mm-Hg}) = 7.11714 - \left[\frac{1210.595}{229.664 + T \text{ (in °C)}}\right]$$

Solution:

a. At $T = 0°C$,

$$\log_{10}(P) = 7.11714 - \left[\frac{1210.595}{229.664 + 0}\right] = 1.846$$

$$P = 70.1 \text{ mm-Hg}$$

$$y_{\text{acetone}} = P^{*}_{\text{actone}} / P_{\text{total}}$$

$$= (70.1 \text{ mm-Hg}) \div (760 \text{ mm-Hg}) = 0.0923 = 92{,}300 \text{ ppm}$$

b. The maximum possible removal efficiency:

$$\eta_{\max}(\%) = \frac{300{,}000 - 92{,}300}{300{,}000} \times 100 = 69.2\%$$

Discussion: Vapor concentrations in ppm, instead of partial pressure, were used to determine the maximum removal efficiency.

Often, the influent gas stream is above its dew point (superheated); it needs to be cooled to its dew point so condensation can begin. In addition, the condensate needs to be subcooled so that it will not re-vaporize. This is typically done by maintaining a condensate level that covers the tubes at the bottom of the last heat exchanger, a condition known as flooding. The total heat that needs to be taken away by the coolant includes (i) the sensible heat change of the waste gas, (ii) the latent heat (heat of condensation or vaporization) of the condensate, and (iii) the sensible heat change of the condensate from the dew point to the subcooling temperature. The total rate of heat transfer (q) can be determined as:

$$q = (m_g)(C_{p,g})(T_{g,i} - T_{g,f}) + (m_{\text{con}})(\Delta H_v) + (m_{\text{con}})(C_{p,\text{con}})(\Delta T_{\text{supercooling}}) \quad (7.39)$$
$$= (m_L)(C_{p,L})(T_{L,f} - T_{L,i})$$

where m_g, m_{con}, and m_L are the mass flow rates of the gas, the condensate, and the coolant, respectively; $C_{p,g}$, $C_{\text{p,con}}$, and $C_{p,L}$ are the specific heat capacities of the gas, the condensate, and the coolant, respectively; $T_{g,i}$, $T_{g,f}$, $T_{L,i}$ and $T_{L,f}$ are the initial and final temperatures of the gas and the coolant, respectively; ΔH_v is the heat of vaporization of the condensate; and $\Delta T_{\text{supercooling}}$ is the temperature drop of the condensate in the supercooling stage.

For an indirect condenser, the rate of heat transfer across the tube surface depends on the overall heat transfer coefficient (U), the heat transfer surface area (A), and the temperature difference between the gas and liquid across the tube (ΔT) as:

$$q = U \times A \times \Delta T \tag{7.40}$$

The *overall heat transfer coefficient* (U) measures the total resistance to heat transfer from the hot side to the cool side across the tube. Typical values of overall heat transfer constant, for the cases where the gas stream is organic solvent vapor with high percentages of non-condensable gases and water is the coolant, are 20 to 60 Btu/°F-ft^2-hr, or 410 to 1,230 kJ/h·m^2·K (USEPA, 2012). [Note that 1 Btu/°F-ft^2-hr = 20.44 kJ/h·m^2·K = 0.34 kJ/min·m^2·K].

In an indirect-contact heat exchanger, the gas stream and the coolant typically flow concurrently or counter-currently. In either case, the temperature difference between the two fluids varies throughout the entire length of the heat exchanger. To account for the variations, the log-mean of the temperature differences ($\Delta T_{\text{log mean}}$) is usually used, and it can be found by:

$$\Delta T_{\text{logmin}} = \frac{\left(\Delta T_{\text{inlet}} - \Delta T_{\text{outlet}}\right)}{\ln\left(\Delta T_{\text{inlet}} / \Delta T_{\text{outlet}}\right)} \tag{7.41}$$

In almost all condensation applications used for air pollution control, the waste gas to be treated will contain one or more condensable COCs with non-condensable components. The non-condensable components would offer resistance to heat and mass transfer, which needs to be considered in the design. The following two examples are for a simplified case (pure solvent) adapted from USEPA (2012a). The primary purpose is to illustrate calculations related to heat transfer.

Example 7.23: Required Mass Flow Rate of Coolant for Condensation

A single-pass, counter-current flow, indirect-contact condenser is used for a gas stream of pure acetone (C_3H_6O; MW = 58) at T = 120°C and P = 1 atm. The boiling point of acetone is 56.2°C (133.2°F). The feed rate of acetone is 600 kg/h. Water is the coolant that enters the condenser at 20°C and exits at 40°C, and the condensate is subcooled to 30°C. Use the following information to determine the coolant flow rate:

- Heat capacity of acetone vapor = 0.084 kJ/g-mole·K.
- Heat capacity of acetone liquid = 0.13 kJ/g-mole·K.
- Heat capacity of water = 4.182 kJ/kg·K = 0.0753 kJ/g-mole·K.
- Heat of vaporization of acetone = 25.1 kJ/g-mole·K.

Solution:

a. Since the given heat capacity values are on the molar basis,

$$\text{Molar flow rate of acetone} = (600 \text{ kg/h})(1{,}000 \text{ g/kg}) \div (58 \text{ g/g-mole})$$
$$= 10{,}340 \text{ g-mole/h} = 172 \text{ g-mole/min}$$

b. Total heat needs to be removed (q) = (cooling the superheat vapor from 120°C to its boiling point of 56.2°C) + (isothermal condensation of acetone at 56.2°C) + (subcooling the condensate from 56.2°C to 30°C)

$$q = (m_g)(C_{p,g})(T_{g,i} - T_{g,f}) + (m_{con})(\Delta H_v) + (m_{con})(C_{p,con})(\Delta T_{supercooling})$$

$$= (172)\{(0.084)(120 - 56.2) + 25.1 + (0.13)(56.2 - 30)\} = 5,820 \text{ kJ/min}$$

c. Total heat taken away by the coolant (q):

$$q = 5,820 \text{ kJ/min} = (m_L)(C_{p,L})(T_{L,f} - T_{L,i}) = (m_L)(4.182)(40 - 20)$$

$$\text{Coolant flow rate } (m_L) = \underline{69.6 \text{ kg/min}}$$

Discussion:

1. Since the gas stream is pure acetone, all of it condensed, thus $m_{gas} = m_{condensate}$.
2. We assumed that all the heat capacity values are constant. It is justifiable because all the temperature ranges are relatively narrow.

Example 7.24: Surface Area of Heat Exchangers for Condensation

Referring to Example 7.23, use the overall heat transfer coefficient (U) values given below to determine the required surface area of heat exchangers:

- $U = 40$ Btu/°F-ft²-h (cooling superheated acetone vapor).
- $U = 100$ Btu/°F-ft²-h (condensing acetone).
- $U = 50$ Btu/°F-ft²-h (subcooling acetone liquid).

[Note: 1 Btu/°F-ft²-h = 20.44 kJ/h·m²·K = 0.34 kJ/min·m²·K]

Solution:

a. For the stage of cooling superheated acetone vapor:

$$\text{Removal of superheat } (q) = (m_g)(C_{p,g})(\Delta T)$$
$$= (172) \times (0.084) \times (120 - 56.2) = 922 \text{ kJ/min}$$

$$\text{Temperature change of the coolant } (\Delta T) = q \div [(m_L)(C_{p,L})]$$
$$= 922 \div [(69.6) \times (4.182)] = 3.2 \text{ K}$$

Temperature of the coolant entering this stage = 40 − 3.2 = 36.8°C

$$\Delta T_{\text{logmin}} = \frac{(\Delta T_{\text{inlet}} - \Delta T_{\text{outlet}})}{\ln(\Delta T_{\text{inlet}}/\Delta T_{\text{outlet}})} = \frac{[(120-40)-(56.2-36.8]}{\ln[(120-40)/(56.2-36.8)]} = 42.8$$

$$\text{Required heat transfer area}(A_1) = q \div [(U)(\Delta T_{\text{log mean}})]$$
$$= 922 \div \{[(40)\times(0.34)]\times(42.8)\} = 1.6\ \text{m}^2$$

b. For the stage of acetone condensation:

$$\text{Heat of condensation}(q) = (m_{\text{con}})\times(\Delta H_v) = (172)\times(25.1) = 4{,}320\ \text{kJ/min}$$

$$\text{Temperature change of the coolant }(\Delta T) = q \div [(m_L)(C_{p,L})]$$
$$= 4{,}320 \div [(69.6)\times(4.182)] = 14.8\ \text{K}$$

$$\text{Temperature of the coolant entering this stage} = 36.8 - 14.8 = 22.0°\text{C}$$

$$\Delta T_{\text{logmin}} = \frac{(\Delta T_{\text{inlet}} - \Delta T_{\text{outlet}})}{\ln(\Delta T_{\text{inlet}}/\Delta T_{\text{outlet}})} = \frac{[(56.2-36.8)-(56.2-22]}{\ln[(56.2-36.8)/(56.2-22)]} = 26.1\text{K}$$

$$\text{Required heat transfer area}(A_2) = q \div [(U)\times(\Delta T_{\text{log mean}})]$$
$$= 4{,}320 \div \{[(100)\times(0.34)]\times(26.1)\} = 4.9\ \text{m}^2$$

c. For the stage of subcooling condensate:

$$\text{Removal of superheat}(q) = (m_{\text{con}})\times(C_{p,\text{con}})\times(\Delta T)$$
$$= (172)\times(0.13)\times(56.2-30) = 586\ \text{kJ/min}$$

$$\text{Temperature change of the coolant}(\Delta T) = q \div [(m_L)(C_{p,L})]$$
$$= 586 \div [(69.6)\times(4.182)] = 2.0\ \text{K}$$

$$\text{Temperature of the coolant entering this stage} = 22.0 - 2.0 = 20.0°\text{C}$$

$$\Delta T_{\text{logmin}} = \frac{(\Delta T_{\text{inlet}} - \Delta T_{\text{outlet}})}{\ln(\Delta T_{\text{inlet}}/\Delta T_{\text{outlet}})} = \frac{[(56.2-22.0)-(30-20]}{\ln[(56.2-22.0)/(30-20)]} = 19.7$$

$$\text{Required heat transfer area}(A_3) = q \div [(U)(\Delta T_{\text{log mean}})]$$
$$= 586 \div \{[(50)\times(0.34)]\times(19.7)\} = 1.7\ \text{m}^2$$

d. Overall temperature change of the coolant = 3.2 + 14.8 + 2.0 = 20 K (check)
e. The total surface area = 1.6 + 4.9 + 1.7 = 8.2 m^2

Discussion:

1. In this case, the area needed for the second stage (condensation of acetone) is much larger than that of the other two stages.
2. The coolant and the gas stream flow counter-currently. The temperature profile of the gas stream is 120°C (the influent) → 56.2°C (the end of the condensation stage after cooling of the superheated air and the beginning and the end of condensation) → 30°C (at the end of subcooling and as the effluent), while that of the coolant: 20°C (the influent) → 22.0°C (the end of the subcooling stage) → 36.8°C (after the stage of condensation) → 40°C (the effluent after cooling the superheated acone vapor to its boiling point).

7.7 BIOREACTORS

Bioreactors use microorganisms to degrade COCs in waste gas streams. Although bioreactors for air pollution control can be aerobic or anaerobic, aerobic is usually the mode. They are mainly for waste gas streams with relatively low COC concentrations, which are soluble in water and biodegradable. Biodegradation is especially applicable to most organic COCs and some inorganic COCs, such as hydrogen sulfide (H_2S) and nitrous oxide (N_2O). Bioreactors are also commonly used for odor control in wastewater treatment plants. Removal efficiencies range from 60% to >99% (USEPA, 2002, 2003b).

7.7.1 Principles of Bio-reaction

Aerobic biodegradation of hydrocarbons is an oxidation process in which microorganisms (M/O) use oxygen as an electron acceptor to degrade organics (the electron donors) for cell synthesis and energy. The final products will be CO_2 and H_2O, which are the same as those from thermal oxidation, except for a large heat release from thermal oxidation, while biomass will be generated in biodegradation instead. The other significant difference is that biodegradation typically occurs under ambient temperatures.

7.7.2 Bioreactors and Their Components

There are three main types of bioreactors for air pollution control: (i) biofilters, (ii) biotrickling filters, and (iii) bioscrubbers.

A biofilter can be as simple as a box that contains a plenum (an enclosed space used for airflow), a support rack above the plenum, and bed media on top of the support rack (Figure 7.19). Various materials are used for bed media, such as peat, compost, bark, and lava rock. The bed media are to allow M/O to form a biofilm where the biodegradation would take place. Typical beds are 3 to 5 ft (1 to 1.5 m) deep with a porosity of approximately 50% of water saturation. Water with nutrients can be irrigated on top of the biofilter bed, or the COC-laden gas will be humidified before being introduced.

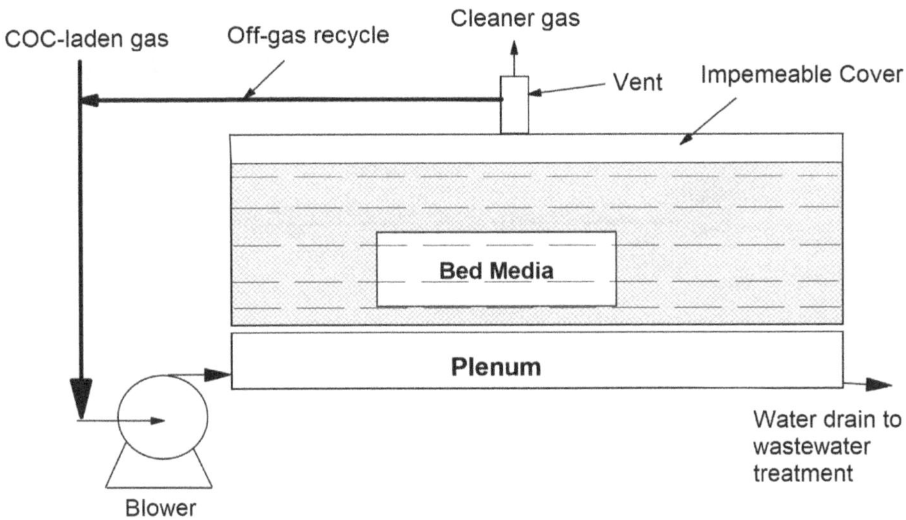

FIGURE 7.19 A biofilter system with off-gas recycle (Modified from USEPA, 2003b).

As the waste gas stream flows through the bed media, the COCs are absorbed by the moisture in the biofilm, come into contact with M/O, and get biodegraded. Typical operating temperatures of the bed are around 60 to 85°F (25–30°C), and the pH of the support media is typically 6 to 8. The residence time of the waste gas stream in the bed media ranges from 15 to 60 minutes (USEPA, 2002). COCs containing sulfur or nitrogen may generate acidic byproducts during biodegradation, requiring alkali to keep the pH in the desirable range. Sometimes, oyster shells are used as part of the bed media for neutralizing acid build-up, or soda ash (Na_2CO_3) is sprayed intermittently over the top of the media bed (USEPA, 2003b).

The biofilter in Figure 7.19 is equipped with a cover and a vent that allows the gas stream to be recycled to enhance removal efficiency. The biofilters can also be placed in the ground (Figure 7.20), which is common in wastewater treatment plants for odor control. They may appear as planters.

Figure 7.21 illustrates the process flow diagram of a biofilter, which includes pretreatment for particulate, temperature, and load control. The influent waste gas stream is humidified; additional water, nutrients, and pH buffers are intermittently added to achieve optimal conditions for M/O activities. Leachate collection may also be needed.

A biotrickling filter operates under the same principle as a trickling filter in wastewater treatment. A trickling filter in wastewater treatment contains aggregate, ceramic, or plastic media to support biofilm growth. A wastewater stream trickles down from the top, and its COCs get absorbed into the liquid in the biofilm. At the same time, clean air is induced through the bed and provides oxygen to the M/O in the biofilm for biodegradation. For a biotrickling filter in air pollution control, the COCs to be removed are in the waste gas stream, not in the water. When compared to biofiltration for air pollution control, water is recirculated, and nutrients and acids/bases can be added to maintain an environment conducive inside the biotrickling filter for

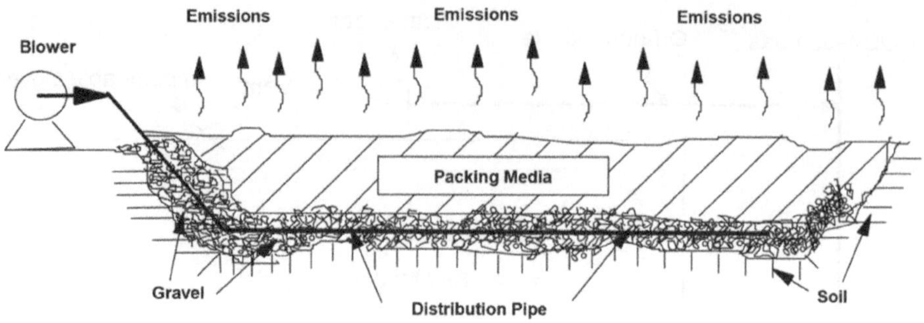

FIGURE 7.20 An in-ground biofilter system (Modified from USEPA, 2003b).

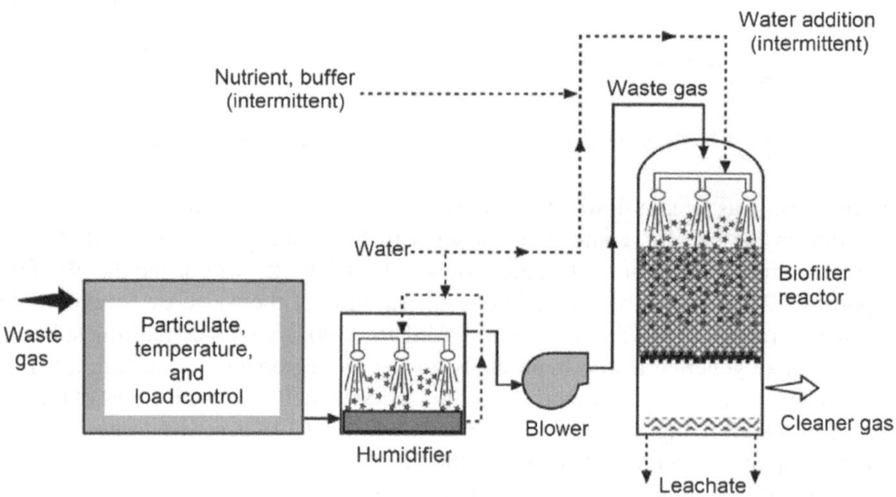

FIGURE 7.21 Process flow diagram of a biofilter system (Modified from USEPA, 2002).

optimal COC removal. It is less prone to clogging by biomass and to acid build-up, but a biotrickling filter would produce a wastewater stream that may require further treatment (Figure 7.22). Using ceramic or plastic media can achieve a porosity of up to 95%, which should significantly reduce the pressure drop of the gas flow through the media. For the same bed media depth, the biotrickling filter's pressure drop can be as low as one-fifth of a biofilter with natural packing materials (USEPA, 2003b).

As discussed, a biotricking filter is considered an enhancement of a biofilter, whereas a bioscrubber can be viewed as an enhancement of a biotrickling filter. Compared to biotrickling filters, the bioscrubbers improve the absorption of COCs in the liquid and lengthen the time for M/O to degrade the COCs. Absorption occurs in the scrubber. The effluent from the wet scrubber is then transferred to an aeration tank, where the COCs are degraded by suspended M/O. Water, nutrients, and buffer solutions can be added to provide optimal conditions for microbial activities. The

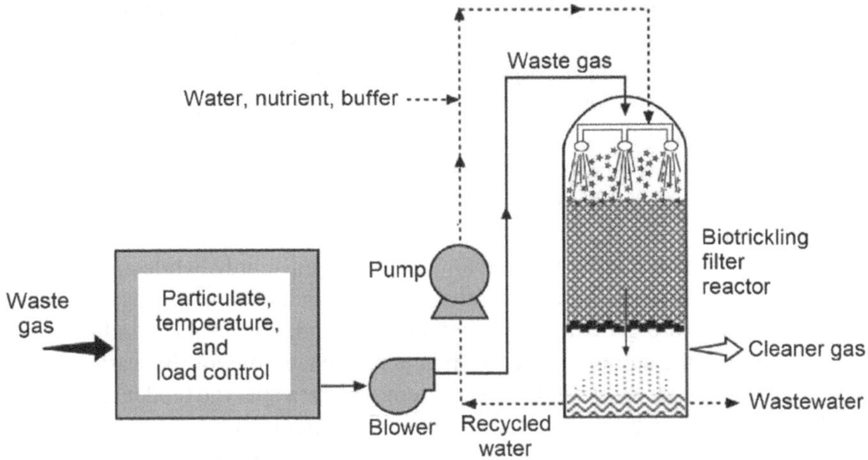

FIGURE 7.22 Process flow diagram of a biotrickling filter system (Modified from USEPA, 2002).

solution in the aeration tank can be recycled back to the scrubber (Figure 7.23). The potential for bio-clogging is minimal for bioscrubbers.

7.7.3 Factors Affecting the Performance of Bioreactors

Key parameters for the design of a bioreactor include concentrations and biodegradability of the COC, types of M/O present, media, residence time, presence of oxygen for aerobic biodegradation (or absence of oxygen for anaerobic biodegradation), pH, temperature, moisture, presence of available nutrients, and absence of toxins.

The temperature range can classify M/O for their optimal growth. A mesophile will grow best in moderate temperatures, typically between 20 and 45°C (68 and

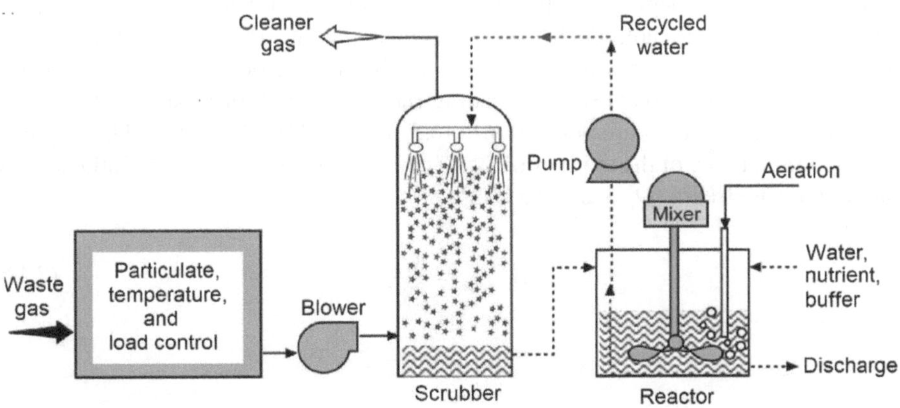

FIGURE 7.23 Process flow diagram of a bioscrubber system (Modified from USEPA, 2002).

113°F) [Note: the optimal temperature range for thermophilic M/O is 45 to 122°C (113 to 252°F), and that of psychrophilic M/O is <–15 to 10°C (<5 to 50°F)]. Most of the bioreactors count on mesophilic M/O to do the work. If the temperature of the waste gas stream is outside this range, it needs to be adjusted before entering the bioreactor. Humidifying the gas stream with water would lower its temperature and provide the required moisture. There are also optimal pHs for bacterial growth. The M/O active in most bioreactors prefers neutral or near-neutral pH. Chemicals to adjust or buffer the pH are often added.

Aerobic M/O cannot grow without an ample oxygen supply, while anaerobic bacteria live without oxygen. Since the rates of biodegradation of hydrocarbons under aerobic conditions are generally higher than those under anaerobic conditions, most bioreactors are in the aerobic mode. Sufficient residence time for the waste gas in the bed media is needed to achieve the desired removal for a specific biodegradation rate.

M/O needs moisture to survive, and moisture content is one of the main components of the biofilm of biofilters that absorb COCs from the gas stream. Although the higher moisture content of the bed media would facilitate more absorption, it would also reduce the passage of the gas stream. The media bed's degree of water saturation (the void occupied by water) should be kept between 40% and 60%. To achieve optimal efficiency, the waste gas stream should be 20 to 40°C and 95% relative humidity (USEPA, 1992).

Macronutrients (i.e., N, P, K, and others) are also needed for M/O growth. They may need to be added. Some compounds in the air stream, including COCs, in elevated concentrations, may become toxic to M/O. Their concentrations may need to be reduced before entering the bioreactors. In addition, some COCs may be more resistant to biodegradation, which should be considered.

7.7.4 Design Calculations for Bioreactors

Using biofilters is cost-effective for air streams with relatively low concentrations (<1,000 ppm as methane). The maximum influent organic COC concentrations are 3,000 to 5,000 mg/m^3 (1.9 to 3.0×10^{-4} lb/ft^3). The media bed should be maintained at 40 to 60% of the degree of saturation and a pH between 6 and 8. The typical gas loading rate (GLR) to biofilters has been designed to treat 1,000 to 150,000 m^3/h (600 to 88,000 cfm) waste gas, with the systems having 10 to 2,000 m^2 (110 to 21,500 ft^2) of filter media. The typical surface loading rate is 100 m^3/h of waste air stream per m^2 filter cross-sectional area or 100 m/h (5.5 ft/min) (USEPA, 1992). The required cross-sectional area of the biofilter ($A_{\text{biofilter}}$) can be determined from the influent flow rate of the gas stream (Q_{inf}) and the GLR:

$$A_{\text{biofilter}} = \frac{Q_{\text{inf}}}{\text{GLR}} \tag{7.42}$$

Example 7.25: Sizing Biofilters for Off-Gas Treatment

Referring to the remediation project described in Example 7.3, biofilters will treat an off-gas stream (Q=210 scfm=6.0 standard m³/min) containing 800 ppm of xylene. Determine the size of the biofilters needed for this project.

Solution:

a. The off-gas contains 800 ppm of xylenes, which is equivalent to 6,400 ppm of methane (each xylene molecule contains eight carbon atoms). This is beyond the typical range of <1,000 ppm as methane, and the maximum influent VOC concentrations of 3,000 to 5,000 mg/m³ have been reported in the literature. Since the xylene concentration (800 ppm of xylenes = 6,400 mg/m³) is high, dilution of the off-gas would be a safe approach. The optimal influent concentration should be determined from a pilot study. In this example, let us dilute the off-gas and let the influent flow rate to the biofilter become 12 standard m³/min (420 scfm).
b. Using the typical surface loading rate of 100 m³/h/m², the required cross-sectional area is:

$$A_{\text{biofilter}} = \frac{Q_{\text{inf}}}{\text{GLR}} = \frac{\left(12\dfrac{\text{m}^3}{\text{min}}\right)\left(60\dfrac{\text{min}}{\text{h}}\right)}{100\ \text{m}^3/\text{h}/\text{m}^2} = 7.2\ \text{m}^2 = \left(78\ \text{ft}^2\right)$$

c. Use 1.5 m (5 ft) as the thickness of the media bed as a conservative approach without conducting a pilot study.

Discussion: If the biofilter is constructed in a cylindrical shape, its diameter would be around 3.0 m (10 ft). Having more filters in parallel and in series would be favorable for operation and for regulatory compliance.

7.8 SUMMARY

1. Gaseous air pollutants can be organic or inorganic.
2. VOCs are often toxic and essential in photochemical smog formation.
3. Common processes to reduce VOC emissions from stationary sources include absorption, condensation, adsorption, biodegradation, thermal incineration, and catalytic incineration.
4. To effectively remove by absorption, the organic pollutants must be soluble in water or absorbent solution. Additives can be added to enhance absorption.
5. To have effective removal by condensation, the organic pollutants need to be condensed at relatively high temperatures to be economically feasible. If the contaminants are condensable and reusable, the process is a good alternative for pollutants with high concentrations.
6. For biodegradation, the organic pollutants need to be readily biodegradable. The process is also practically limited to low concentration levels. For high concentrations, dilution and larger footage would be required.

7. Adsorption and thermal processes generally apply to various organic pollutants. Thermal methods are preferable for high pollutant concentrations because adsorbents have limited capacities. On-site regeneration of adsorbents can be an alternative for high pollutant concentrations. The incineration of pollutants containing sulfur and chlorine would generate acidic gases.
8. Thermal and biological processes are destruction processes, while adsorption, absorption, and condensation are mass transfer processes. Wastes are usually generated as byproducts from the non-destructive processes, which may require additional treatment and/or handling.

EXERCISE QUESTIONS

1. The LEL and UEL of propane (C_3H_8) are 2.05% and 11.38% by volume, respectively. There is a propane–air mixture with an A/F ratio of 20.0 kg air/kg propane. Can it be ignited?
2. The stoichiometric mixture of methane (CH_4) and air contains 9.48% methane by volume. What is the *A*/*F* ratio for combustion of this mixture (in lb/lb or kg/kg)?
3. Assuming air contains 20% by volume O_2 and 80% N_2, for the combustion of 100 standard cubic meters of methane (CH_4) in air, determine:

$$CH_4 + 2O_2 \rightarrow CO_2 + 2H_2O$$

   a. the stoichiometric amount of O_2 needed for complete combustion,
   b. the stoichiometric amount of air needed for complete combustion,
   c. the stoichiometric mixture by volume (in % methane in the mixture), and
   d. the *A*/*F* ratio (in a mass/mass ratio).
4. To burn a pure methane (CH_4) stream of 100 standard cubic meters ($T_{standard}$ = 25°C) with <u>25% excess air</u> (20% O_2 and 80% N_2 by volume). Assuming the combustion is complete and the flue gas temperature is 300°C,
   a. the total influent flow rate to the oxidizer (in scfm) = ______________,
   b. the flow rate of the <u>flue gas</u> (in <u>acfm</u>) = ______________, and
   c. The CO_2 concentration in the flue gas (in %) = ______________.
5. Assuming air contains 20% by volume O_2 and 80% N_2, for combustion of 100 standard m³/min of ethane (C_2H_6) in air, determine:

$$C_2H_6 + 3.5O_2 \rightarrow 2CO_2 + 3H_2O$$

   a. the stoichiometric amount of O_2 needed for complete combustion,
   b. the stoichiometric amount of air needed for complete combustion,
   c. the stoichiometric mixture by volume (% ethane in the stoichiometric mixture), and
   d. the *A*/*F* ratio (in a mass/mass ratio).

6. Soil venting, coupled with GAC for off-gas treatment, is being considered for a site impacted by PCE (C_2Cl_4). The PCE removal efficiency by GAC is 95%, and the discharge limit to the atmosphere is 0.05 kg PCE/day. Two GAC units (2,500 kg GAC in each unit) are operated in series, and the actual adsorption capacity is 0.2 kg PCE/kg GAC.
   a. Determine the maximum allowable PCE loading rate to the GAC system (in kg/day).
   b. Estimate the breakthrough time if the system runs continuously (in days).
7. Soil venting with direct incineration for off-gas treatment is considered an alternative to bioremediation at a site. The off-gas (100 scfm) will be burned at 2,000°F (1,093°C) with an additional 20 scfm of supplementary fuel and dilution air. To achieve 99% destruction, the required residence is 1 second. The standard condition is $T=77$°F (25°C) and $P=1$ atm. What is the minimum required volume of the incineration chamber?
8. Off-gas from a process is to be treated by GAC adsorbers. The chlorobenzene concentration in the off-gas is 1,500 ppm. The gas flow rate is 20 m³/min @ $T=20$°C. Two 2,000-kg GAC adsorbers are to be used. Determine the maximum amount of chlorobenzene that can be held by one GAC adsorber before it is exhausted. The adsorption isotherm ($T=20$°C) for chlorobenzene concentrations ranging from 0.0001 to 0.05 psi is (q in kg/kg and P in psi):

$$q = 0.508P^{0.210}$$

9. The flow rate of the off-gas from a project is 100 m³/min @$T=77$°F (25°C) and $P=1$ atm. An incinerator or a catalytic incinerator should treat it before discharge. Determine:
   a. the minimum required volume of the incineration chamber if the design retention time is 0.3 seconds and the temperature in the incinerator is 900°C, and
   b. the minimum required volume of the catalyst if the design space velocity is 30,000/h.
10. A packed-column absorber is designed to remove 2-butanone from a gas stream ($Q=50$ m³/min). Determine:
    a. the minimum liquid-to-air ratio,
    b. the design liquid-to-air ratio, and
    c. the design liquid flow rate. Use the following information in calculations:
       - Henry's constant = 0.0274 atm/M;
       - absorption factor (AF) = 1.6;
       - $T=298$ K and $P=1$ atm.
11. For the absorber in Question #10, if the influent 2-butanone concentration is 0.3% and the required removal is 60%. The diameter of the column is 0.50 m, and $K_Ga=8.0$/s. Determine the packing height required.
12. Estimate the heat content of the gas stream in Question #10.
13. The acetone concentration in the influent gas stream is 600,000 ppm. How low does the operating temperature of a condenser need to be to achieve a

50% removal efficiency? The Antonine equation for acetone ($T = -12.9$ to 55.3°C) is shown below (USEPA, 2012a):

$$\log_{10}\left(P \text{ in mm-Hg}\right) = 7.11714 - \left[\frac{1210.595}{229.664 + T(\text{in } ^\circ\text{C})}\right]$$

14. For a gas stream containing 500 ppm of total petroleum hydrocarbon (TPH), what would be a good size of biofilters to treat this gas stream (flow rate = 10 standard m^3/min)?

BIBLIOGRAPHY

LADCO. (2009). *APTI 400: Introduction to Air Toxics - Student Manual*, Lake Michigan Air Directors Consortium (LADCO), Rosemount, IL.

USEPA. (1991). *Control Technologies for Hazardous Air Pollutants*, EPA/625/6-91/014, Office of Research and Development, US Environmental Protection Agency, Washington, DC.

USEPA. (1992). *Control of Air Emissions from Superfund Sites*, EPA/625/R-92/012, Office of Research and Development, United States Environmental Protection Agency, Washington, DC.

USEPA. (1994). *Control Techniques for Fugitive VOC Emissions from Chemical Process Facilities*, EPA/625/R-93/005, Office of Research and Development, United States Environmental Protection Agency, Washington, DC.

USEPA. (2000). *APTI 415: Control of Gaseous Emissions - Student Guide*, 3rd edition, Office of Air and Radiation, United States Environmental Protection Agency, Research Triangle Park, NC.

USEPA. (2002). *APTI 482: Sources and Control of Volatile Organic Air Pollutants - Student Manual*, 3rd edition, prepared by J.W. Crowder for Air Pollution Training Institute, United States Environmental Protection Agency, Research Triangle Park, NC.

USEPA. (2003a). *APTI 452: Principles and Practices of Air Pollution Control - Student Manual*, 3rd edition, Air Pollution Training Institute, United States Environmental Protection Agency, Research Triangle Park, NC.

USEPA. (2003b). *APTI 455: Using Bioreactors to Control Air Pollution*, EPA-456/R-03-003, The Clean Technology Center, United States Environmental Protection Agency, Research Triangle Park, NC.

USEPA. (2012a). *APTI 415: Control of Gaseous Emissions - Student Guide*, Office of Air and Radiation, United States Environmental Protection Agency, Research Triangle Park, NC.

USEPA. (2012b). *APTI 427: Combustion Source Evaluation - Student Manual*, Office of Air and Radiation, United States Environmental Protection Agency, Research Triangle Park, NC.

USEPA. (2025). *Air Pollutant Emissions Trend Data*, https://www.epa.gov/air-emissions-inventories/air-pollutant-emissions-trends-data, last updated on 04/28/2025.

8 Control of Nitrogen Oxides Emissions

8.1 INTRODUCTION

8.1.1 Nitrogen Oxides and NO_x

Nitrogen oxides represent a family of seven compounds: nitrous oxide (N_2O), nitric oxide (NO), nitrogen dioxide (NO_2), dinitrogen dioxide (N_2O_2), dinitrogen trioxide (N_2O_3), dinitrogen tetroxide (N_2O_4), and dinitrogen pentoxide (N_2O_5) (USEPA, 1999).

Nitric oxide (NO) is the main form of nitrogen oxides emitted from combustion. Once emitted into the atmosphere, NO will be rapidly oxidized to NO_2. Dinitrogen tetroxide (N_2O_4) is in constant equilibrium with NO_2, and it is considered part of NO_2 and not regulated separately. While the properties of NO_2 and N_2O_4 are very similar, those of NO are different. NO_2 is water soluble and reddish brown, while NO is relatively insoluble and colorless. In addition, NO does not absorb UV light, so it cannot initiate photochemical reactions. Nitrous oxide (N_2O), also called laughing gas, does not absorb UV light, and it is excluded from the family of NO_x in USEPA, probably because it is not a product of combustion, and it does not participate in photochemical reactions of ozone formation in the troposphere. Emissions of N_2O recently started being separately regulated because it is a potent greenhouse gas (GHG). It is one of the major GHGs and significantly impacts ozone layer depletion in the stratosphere (USEPA, 2014).

The USEPA regulates NO_2 as a surrogate for this family of nitrogen oxides because it is the most prevalent form in the ambient air. Most people and literature consider NO_x to be an acronym for the sum of nitric oxide and nitrogen dioxide; that is, "$NO_x = NO + NO_2$".

8.1.2 Sources and Emissions of NO_x

NO can be produced during thunderstorms from lightning that splits nitrogen molecules. Natural sources, such as fires and vegetation, can also emit NO_x into the atmosphere. Most NO_x in the atmosphere originates from anthropogenic sources, almost exclusively from combustion. Nearly all combustion sources emit NO_x. They are also released from nitric acid (HNO_3) plants and other industrial processes involving the use of nitric acid (USEPA, 2012a).

According to the USEPA's National Emissions Inventory (NEI), the total NO_x emission in the U.S. is 6.94 million tons in 2024 (USEPA, 2025). Combustion (from the transportation and stationary fuel combustion sectors) accounts for the majority, 80.5%, of the total emissions, of which approximately 60% is from transportation and the remaining is from stationary fuel combustion. Industrial and other

DOI: 10.1201/9781003532705-8

TABLE 8.1
Sources and Amounts of the U.S. NO_x Emissions in 2024

Source Category	(1,000 tons)	(%)
Stationary fuel combustion	**2,263**	**32.6**
Electric utility	751	10.8
Industrial	967	13.9
Others	545	7.9
Industrial and other processes	**1,150**	**16.6**
Chemical & allied product manufacturing	31	0.4
Metals processing	57	0.8
Petroleum & related industries	705	10.2
Other industrial processes	271	3.9
Solvent utilization	1	0.0
Storage & transport	3	0.0
Waste disposal & recycling	82	12
Transportation	**3,323**	**47.9**
Highway vehicles	1,745	25.1
Off-Highway	1,578	22.7
Miscellaneous	**204**	**2.9**
Wildfires	40	0.6
Others	164	2.4
Total	**6,940**	**100.0**

processes account for 16.6%, while wildfires and other miscellaneous sources contribute 2.9% of the total. Electric power plants produce about 1/3 for stationary combustion sources, and industrial boilers, incinerators, gas turbines, and engines in various industries contribute 2/3 of the NO_x emissions from this category. Regarding transportation, highway vehicles emitted a little more than those from off-highway vehicles (Table 8.1).

Figure 8.1 depicts trends of annual NO_x emissions in the U.S. since 1970. As shown, emissions have been decreasing, especially since 1990. Most of the reductions came from the transportation and the stationary fuel combustion sectors.

Example 8.1: Emission Rate of Nitrogen Oxides

What is the total NO_x emission rate (in lb/h) from a boiler emitting 360 ppm NO and 40 ppm NO_2 in a flue gas of 10,000 standard cubic feet per minute (scfm)?

Solution:

a. For NO_2 (MW = 46) at $P = 1$ atm and $T = 68°F$, using Eq. 4.18

$$1\text{ ppm} = \left(\frac{\text{MW}}{\text{MV (in ft}^3)}\right) \times 10^{-6}\,\frac{\text{lb}}{\text{ft}^3} = \left(\frac{46}{385}\right) \times 10^{-6}\,\frac{\text{lb}}{\text{ft}^3} = 1.20 \times 10^{-6}\,\frac{\text{lb}}{\text{ft}^3}$$

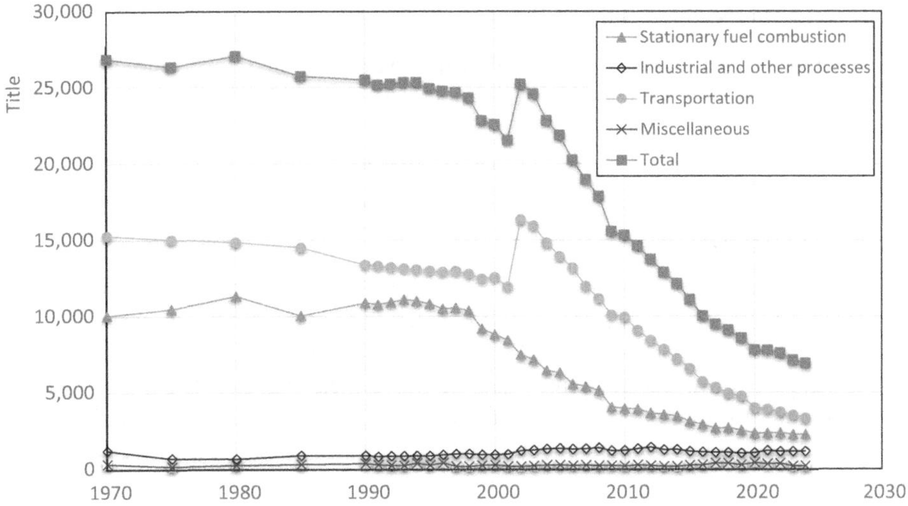

FIGURE 8.1 Annual trends of NO_x emissions from major source categories in the U.S. (1990–2024).

b. Mass emission rate of NO_x

$$= (360 + 40 \text{ ppm}) \infty (1.20 \infty 10^{-6} \text{lb/ft}^3/\text{ppm}) \infty \left(10{,}000 \text{ ft}^3/\text{min}\right)$$
$$= 4.8 \text{ lb/min} = \underline{288 \text{ lb } NO_x \text{ as } NO_2/\text{h}}.$$

Discussion:

1. The standard temperature for stack sampling is usually 68°F (20°C). The molar volume of an ideal gas is 385 ft^3 (or 24.05 L) at this temperature.
2. Although most NO_x in the flue gas would be in the form of NO (approximately 90 to 95%), instead of NO_2, the NO is readily oxidized to form NO_2 after being discharged to the atmosphere. Accordingly, the total mass emissions of nitrogen oxides from the system to the atmosphere are usually expressed in the form of NO_2 (USEPA, 2012b).

8.1.3 The Problems Associated with NO_x

NO_2 is a criteria pollutant; it is toxic and causes respiratory issues. NO_2 participates in *photochemical reactions* (i.e., chemical reactions that are initiated because of absorption of light) with volatile organic compounds (VOCs) and sunlight to produce another criteria pollutant, ozone (O_3), and other reaction products. *Photochemical smog,* a mixture of these pollutants, would form and appear as a brown haze, most often above cities and in summer. In addition, NO_2 and other NO_x would interact with water, oxygen, and other chemicals in the atmosphere to form nitric acid (HNO_3) as part of the acid rain (see Chapter 12 for more details).

8.2 STATIONARY COMBUSTION SYSTEMS

In Chapter 7, we have discussed thermal oxidation systems, emphasizing the destruction of organic compounds of concern (COCs). Since combustion is responsible for >80% of the anthropogenic NO_x emissions, let us be more familiar with combustion and combustion systems, especially regarding fuel combustion.

8.2.1 Diffusion and Premixed Flame

The combustion of fossil fuels to generate energy is a two-step process: (i) physical mixing/contact of the fuel and air, and (ii) chemical reactions between the fuel and oxygen in the air to convert the fuel to CO_2 and H_2O to release energy. For *diffusion flames*, liquid oil droplets, natural gas, or pulverized coal are directly introduced into the combustion zone (i.e., the fuel and the oxidizer (i.e., the air) are not premixed before combustion), but instead diffuse into each other in the flame zone. They include all systems that burn oil, solid fuel, and many natural gas systems. In other words, they include most furnaces and nearly all boilers (USEPA, 2014).

But for some burners, these steps take place in sequence. For example, gasoline vapor and air are mixed before entering an automobile engine and then ignited with a spark. If the air and gaseous fuel are completely premixed to form a gas mixture before being introduced to a combustion chamber, the created flame is called a *premixed flame*. All premixed flame burners can only use vaporized liquid or gaseous fuel (e.g., natural gas or propane). Premixed flame burners are more challenging to operate than diffusion flames because the air/fuel ratio needs to be kept within a relatively narrow range during operation. However, they provide the potential for less NO_x emissions.

8.2.2 Combustion Temperature

Excess air is the "excess" air beyond the stoichiometric (theoretical) amount needed for complete thermal oxidation of combustible compounds (also see Chapter 7). Figure 8.2 illustrates the combustion temperatures vs. the *air-to-fuel equivalence ratio* (λ). It is dimensionless that describes the ratio of the actual air-to-fuel ratio to the stoichiometric air-to-fuel ratio of a given fuel:

$$\lambda = (\text{Air/Fuel})_{\text{actual}} / (\text{Air/Fuel})_{\text{stoichiometric}} \tag{8.1}$$

Theoretically, the combustion temperature will be maximum when $\lambda = 1$ (i.e., no excess air; "*stoichiometric mixture*"). When $\lambda < 1$ (i.e., fuel-rich burning; "*rich mixture*"), the combustion cannot be complete. Thus, the combustion temperatures are lower. On the other hand, under fuel-lean conditions (i.e., $\lambda > 1$; "*lean mixture*"), the combustion may be complete. However, the energy generated from combustion needs to heat the excess air so that the combustion temperature will also be lower. The combustion temperature and the extent of excess air will significantly impact the formation of NO_x (and other COCs), which will be discussed in more detail later. It should be noted that the *fuel-to-air equivalence ratio* (ϕ), the inverse of the air-to-fuel equivalence ratio, is also used; and $\phi = 1/\lambda$.

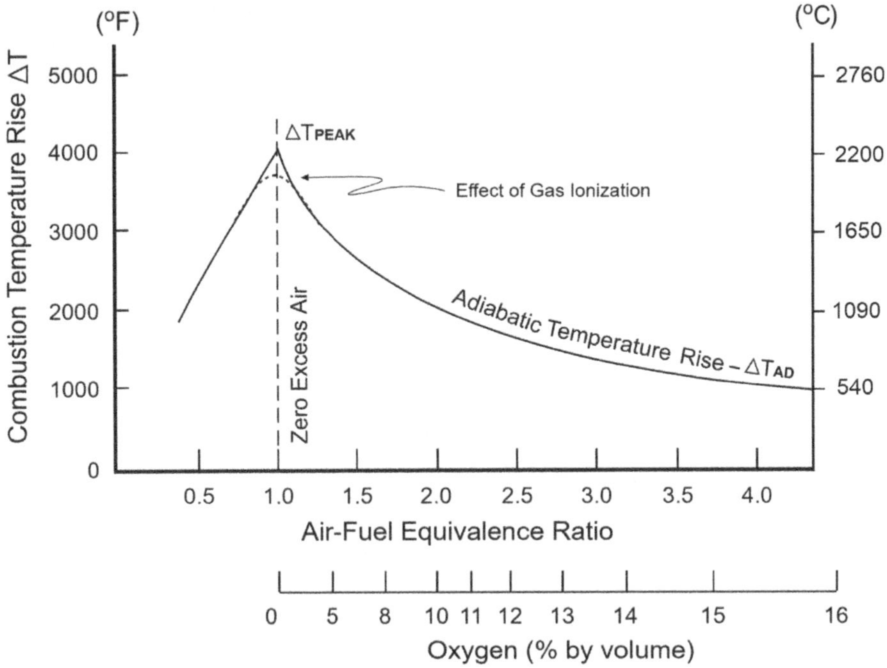

FIGURE 8.2 Combustion temperature vs. excess air (Modified from USEPA, 2014).

8.2.3 Stationary Combustion Systems

From the perspective of NO_x control, the regulated stationary combustion systems can be classified into three groups: (i) furnaces and boilers, (ii) reciprocating engines, and (iii) combustion turbines. A *boiler* is a device that heats water to produce steam or hot water for heating, power generation, or industrial processes. Boilers are of a wide range in size, from home furnaces to huge industrial/utility boilers (Figure 8.3).

An industrial/utility boiler consists of a *firebox* where the fuel is burned to produce heat, and the heat of the combustion gases is used to generate hot water or steam (Figure 8.4). Boilers can use virtually any fuel (e.g., natural gas, oil, or coal).

A *reciprocating engine* is an engine where pistons reciprocate (i.e., move back and forth) inside cylinders. Reciprocating engines are used for small- to medium-sized stationary power sources, typically <20 megawatts (MW). Most of them fire natural gas as the primary fuel, and some can use No. 2 fuel oil (i.e., a type of distillate fuel used primarily for heating and in some industrial applications) or diesel as an alternative. Due to high temperatures, NO_x emissions are roughly ten times higher than those from a boiler burning the same fuel. Reciprocating engines are factory-built, and their combustion systems and operating controls are not amenable to simple adjustment or alteration (USEPA, 2014). More details on reciprocating engines' operation and emission characteristics will be presented in Chapter 10.

A *combustion turbine* (or *gas turbine)* is an engine that compresses air, mixes it with fuel, burns it, and then uses the expanding hot gases to spin a turbine for power generation. Contrary to reciprocating engines, the combustion in turbines is in a steady-state mode. Air is drawn into a compressor to be pressurized and then enters the combustor to react with fuel to raise the gas stream temperature to about 1,500 to 2,000°F (815 to 1,090°C). The higher temperature increases the volume of the gas stream, which is then expanded through a turbine (Figure 8.5). The power generated by the turbine is used to run the compressor and for the output shaft power (e.g., for electricity generation). Their power levels can be as significant as 250 MW. Although combustion can burn liquid or gaseous fuel, it practically only burns clean fuel to avoid the accumulation of deposits formed from contaminants in the fuel on the turbine blades.

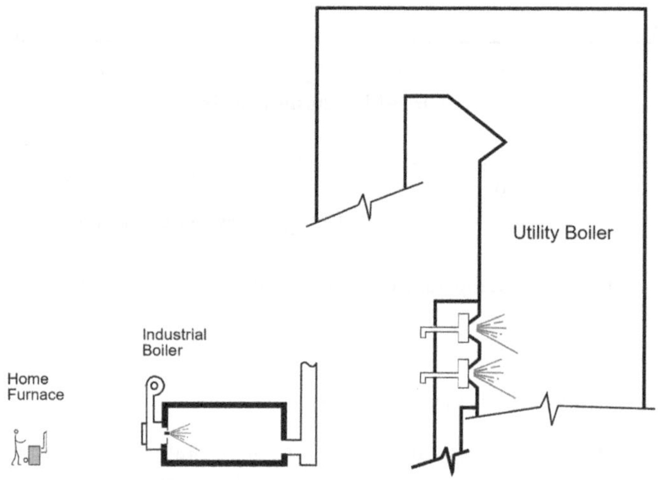

FIGURE 8.3 Comparative sizes of boilers (USEPA, 2012b).

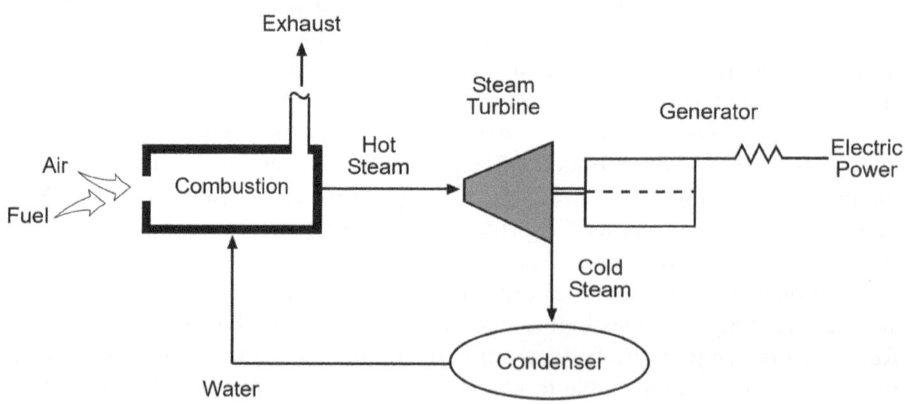

FIGURE 8.4 Boiler, turbine, and generator (USEPA, 2012b).

Thermal efficiency indicates how efficiently the heat from fuel combustion is converted to mechanical work. *A combined cycle* is a power generation system that combines two thermodynamic cycles to increase overall efficiency. As shown in Figure 8.6, the waste heat from a combustion turbine system is used to heat a boiler to drive a steam engine. This arrangement could increase the thermal efficiency from ~35% of the combustion turbine to ~50% of the combined cycle system (USEPA,

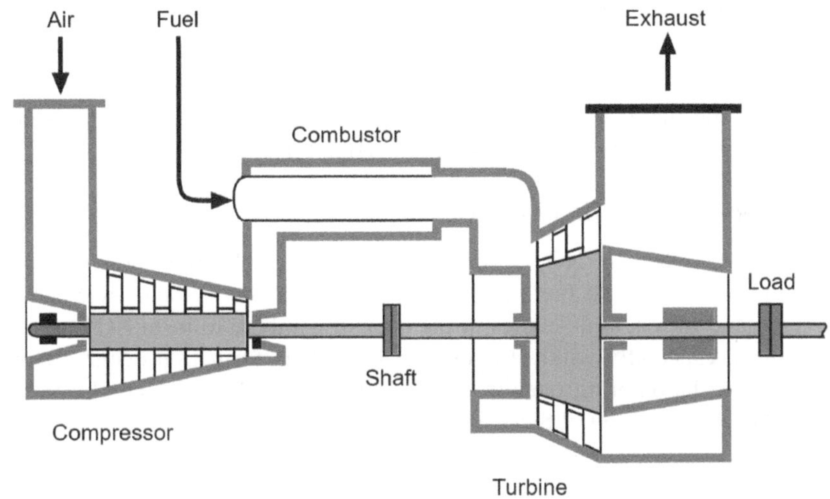

FIGURE 8.5 Schematic of a combustion turbine (USEPA, 2012b).

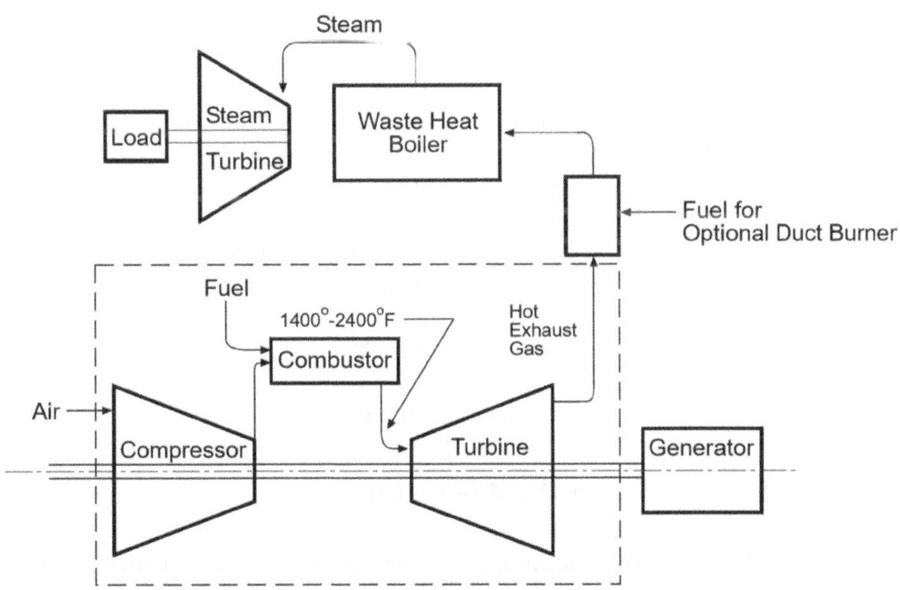

FIGURE 8.6 A combined cycle system (USEPA, 2012b).

2012b). *Cogeneration*, also known as *combined heat and power* (CHP), is a system that simultaneously generates electricity and useful heat from the same energy source; for example, the waste heat from an engine or a turbine is used directly for heating a building or a manufacturing process. Note: CHP/cogeneration focuses on generating both electricity and heat for direct use, while combined cycle plants focus solely on maximizing electricity generation efficiency.

8.3 FORMATION MECHANISMS OF NO_X

During combustion, NO_x emissions are formed by three complex chemical reactions: (i) thermal fixation of molecular nitrogen (thermal NO_x), (ii) oxidation of nitrogen contained in the fuel (fuel NO_x), and (iii) formation of NO_x due to the presence of partially oxidized organic species within the flame (prompt NO_x).

8.3.1 Formation of Thermal NO_x

Thermal NO_x refers to NO_x formed during the combustion process at high temperatures. When fossil fuel is burned with air in a combustion device, some of the oxygen and nitrogen molecules in the air will combine to form NO due to high temperatures, especially within the burner flame. The formation occurs primarily through the oxidation of nitrogen in the air (not in the fuel) during combustion. The simplified overall reaction of thermal NO_x formation is:

$$N_2 + O_2 \leftrightarrow 2NO \tag{8.2}$$

$$2NO + O_2 \leftrightarrow 2NO_2 \tag{8.3}$$

The formation of free radicals is essential in combustion. A *free radical* is an uncharged molecule having an unpaired valence electron, and it is typically highly reactive and short-lived. The free radicals most often involved in combustion reactions are •O, •H, •N, •OH, and hydrocarbons that have lost one or more hydrogen (e.g., methyl radical, $•CH_3$). Free radicals are also formed and participate in photochemical reactions in the atmosphere. Note: A dot is often added to the chemical formula of a free radical to represent the unpaired electron, such as •OH for hydroxyl radical and $^{•}CH_3$ for methyl radical. However, it is also often omitted in literature.

During combustion, for example, these free radicals can be formed as:

$$O_2 + M \rightarrow 2O + M \tag{8.4}$$

$$N_2 + M \rightarrow 2N + M \tag{8.5}$$

$$H_2O + M \rightarrow H + OH + M \tag{8.6}$$

An "M" is included on both sides of Eqs. 8.4–8.6. It is to show that another molecule is involved in allowing the reaction to occur, but is not chemically changed at the end of the reaction (similar to the role that hydrocarbon plays in photochemical reactions to form ozone as a catalyst).

In 1946, Y. B. Zeldovich proposed free-radical chain formation of thermal NO at high combustion temperatures. The mechanism assumes that O radicals attack N_2 to form N radicals (Eq. 8.7). The N radicals can form NO by reacting with O_2 (Eq. 8.8) or hydroxyl radicals (Eq. 8.9):

$$N_2 + O \overset{k_1}{\leftrightarrow} NO + N \tag{8.7}$$

$$N + O_2 \overset{k_2}{\leftrightarrow} NO + O \tag{8.8}$$

$$N + OH \overset{k_3}{\leftrightarrow} NO + H \tag{8.9}$$

The reactions shown in Eqs. 8.7 and 8.8 are mainly responsible for NO formation, while the reaction shown in Eq. 8.9 is only essential in near-stoichiometric and fuel-rich combustion zones. The reaction in Eq. 8.8 is considered the rate-limiting step because of its higher activation energy, and it will become significant only at temperatures >1,500°C (2,700°F) (USEPA, 2014).

Most thermal NO_x is formed in the highest temperature region of the combustion zone. The overall reaction (Eq. 8.2) proceeds rapidly in the forward direction (i.e., NO formation) at high temperatures. While it is a reversible process, the rate of the backward reaction (i.e., the decomposition of NO back to molecular nitrogen and oxygen) is extremely slow. When the gas begins to cool, the reaction should theoretically be reversed. However, it does not happen because of the absence of required free radicals and insufficient activation energy (USEPA, 2014). Instead, the NO reacts with more oxygen to form NO_2 (Eq. 8.3).

The main factors affecting the amount of thermal NO_x formed in combustion are the flame's temperature, residence time, and the amount of combustion gases in the peak temperature zone of the flame. Both NO and NO_2 are present in the effluent gas streams of most combustion sources. NO is the main form of NO_x emitted from traditional combustion devices, accounting for ≥95% of the total NO_x emissions. However, ≥50% of the NO_x from newer "low-NO_x" sources (e.g., turbines and engines with lean premix combustors) may be NO_2 (USEPA, 2014).

Although oxygen concentration and residence time also affect the formation of thermal NO_x, temperature is usually the dominating factor. Figure 8.7 illustrates the extent of thermal NO_x formation as a function of temperature. It shows that the formation of thermal NO_x peaks occurs between 1,900 and 2,000°C. It is the temperature range in which most stationary combustion sources operate (USEPA, 2014).

8.3.2 Formation of Fuel NO_x

Fuel NO_x is generated from the oxidation of organically bound nitrogen in the fuel itself during combustion. The extent of fuel NO_x formation depends on the amount of this type of nitrogen in the fuel. U.S. coals contain approximately 0.2%–3.5%

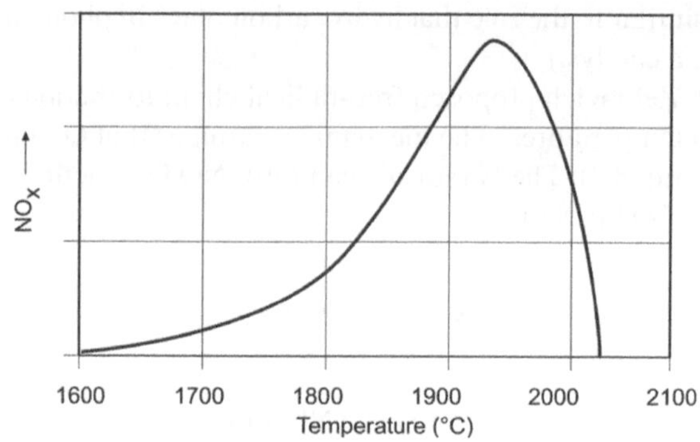

FIGURE 8.7 Thermal NO_x formation vs. temperature (USEPA, 2014).

nitrogen by weight; anthracite coals constitute the low end, while bituminous coals are the high end of this range. When distilling crude oil, nitrogen tends to accumulate in the heavy residuals, while distillates such as #2 fuel oil and diesel oil typically have little nitrogen. Natural gas contains no organic nitrogen.

Although the fuel NO_x formation would be less at lower flame temperatures, it would not be affected significantly. Reducing the oxygen level within the flame, not the average oxygen concentration in the combustion chamber, is critical in lowering fuel NO_x formation.

8.3.3 Formation of Prompt NO_x

Prompt NO_x is formed from intermediate nitrogen species (such as HCN and NH), which are produced from the reactions between the hydrocarbon radicals and the molecular nitrogen in the combustion air. The intermediate nitrogen species are then oxidized to NO. It occurs in the early phase of the flame front (or the fuel-rich flame zone) and is not sensitive to the peak gas temperature. While prompt NO_x is part of the total NO_x emissions from combustors, it is not usually considered when dealing with NO_x emissions control.

8.4 APPROACHES TO REDUCE NO_X EMISSIONS

Techniques used to minimize NO_x emissions from stationary combustion sources can be grouped into three categories: (i) fuel switching, (ii) combustion modifications, and (iii) add-on (back-end or end-of-the-pipe) control systems. Fuel switching is the use of fuels that would yield lower NO_x emissions because of the specific characteristics of their fuel nitrogen and combustion behaviors.

As mentioned, the flame temperature, oxygen concentration, and residence time affect thermal NO_x formation. The objective of combustion modifications is to create operating conditions that inhibit the NO_x formation reactions or cause the NO_x formed to be reduced back to molecular nitrogen (backward reaction in Eq. 8.2), while the gases are still within the combustor. The formation of thermal NO_x can be reduced in three ways:

i. reduction of the oxygen concentration in the maximum temperature zone of the combustor;
ii. reduction of the peak flame temperature; and/or
iii. reduction of the residence time of the combustion gases at peak temperature.

The *add-on NO_x control systems* inject chemical reagents (e.g., ammonia, urea) to reduce NO_x chemically to molecular nitrogen downstream of the combustors. In addition, oxidation of NO_x is also being used; it is to raise the oxidation state of nitrogen (e.g., N_2O_5) to become more water soluble. It can be accomplished by injecting hydrogen peroxide (H_2O_2) with the presence of a catalyst or ozone into the air flow. Non-thermal plasma, when used without a reducing agent, can also be used to oxidize NO_x. A scrubber must be added to absorb N_2O_5 emissions to the atmosphere (USEPA, 2003a).

8.5 FUEL SWITCHING

Fuel switching means changing the type of fuel used in a combustion system to reduce emissions, lower costs, or improve combustion efficiency. Fuel NO_x formation in boilers can be reduced by switching to a different fuel. Three potential approaches are (i) conversion of fuel supply with a lower nitrogen content or the one with fuel nitrogen in a chemical form less likely to form NO_x; (ii) firing the boiler with natural gas (which has no organically bound nitrogen) during the periods when the ambient ozone levels are high (e.g., hot summer days), and (iii) *gas-cofiring*, that is simultaneous combustion of coal and gas (USEPA, 2003b).

8.6 COMBUSTION MODIFICATION

There are five main categories of combustion modification techniques, especially for industrial boilers; they are:

i. low excess air combustion,
ii. off-stoichiometric combustion,
iii. flue gas recirculation (FGR),
iv. low-NO_x burners, and
v. gas reburning.

These approaches are to reduce peak temperatures and oxygen concentrations as well as the residence time under these peak conditions so that the formation of thermal NO_x (and fuel NO_x to a lesser extent) would be minimized.

8.6.1 Low Excess Air Combustion

A certain amount of excess air is required for combustion systems to ensure complete fuel combustion. Excess air levels practiced in industrial and utility boilers depend on the types of fossil fuel and the boiler operating load. The typical excess levels are 3 to 5%, 5 to 20%, and 20 to 50% for gas-fired (near full load), oil-fired, and pulverized coal-fired boilers, respectively (USEPA, 2012a). When the excess air is too low, emissions of CO, smoke, and *products of incomplete combustion* (PICs) in the boiler exhausts would increase (Figure 8.8). However, higher levels of excess air can lead to higher flame temperatures, which promote NO_x formation.

There is an optimum excess air level where the formation of NO_x is reduced, and concentrations of CO and PICs are not excessive (Figure 8.9). In addition, the optimal air-to-fuel ratio may also lead to reduced fuel consumption, as less heat is lost in the form of hot exhaust gas. To maintain this optimal level, the control system may need to be upgraded (e.g., installing an oxygen analyzer) to precisely and continuously control the airflow (USEPA, 2012b). NO_x reductions between 16 and 20% can be achieved on gas- and oil-fired boilers, and 20% (on average) for coal-fired utility boilers (USEPA, 2003b).

8.6.2 Off-Stoichiometric Combustion

Off-stoichiometric combustion (or *staged combustion*) refers to a combustion process where the air-to-fuel ratio is intentionally adjusted to be either above or below the stoichiometric ratio. In the fuel-rich zone (i.e., the equivalence ratio (λ) < 1), fuel is fired with less than the stoichiometric amount of air. The combustion temperature will be lower because the combustion will not be complete. The other zone is air-rich (λ >1), where the balance of the combustion air is introduced to combust the fuel (Figure 8.10). Both thermal and fuel NO_x formation can be reduced because of oxygen deficiency in the fuel-rich zone, and temperatures in both the fuel-rich and air-rich zones are lower than those in single-stage combustion. As shown, primary

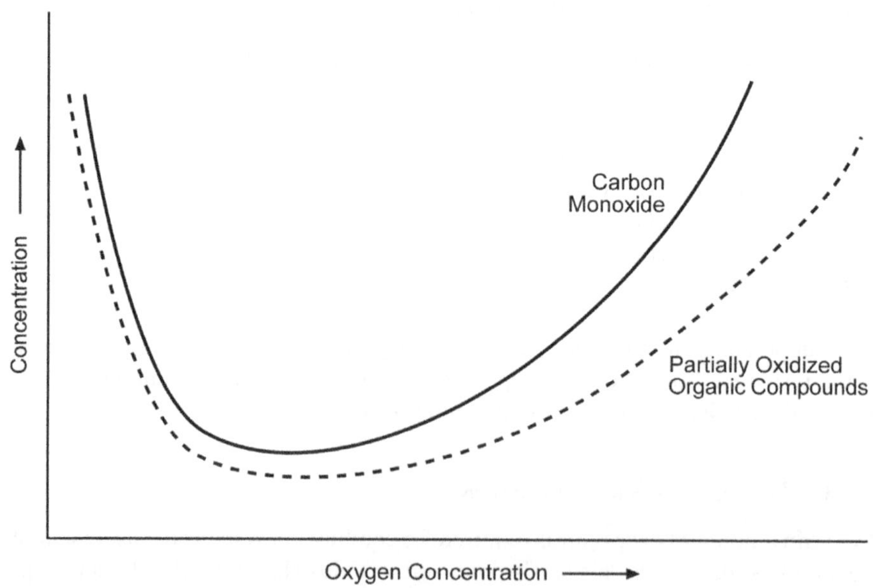

FIGURE 8.8 Concentrations of CO and PICs vs. oxygen concentrations (USEPA, 2003b).

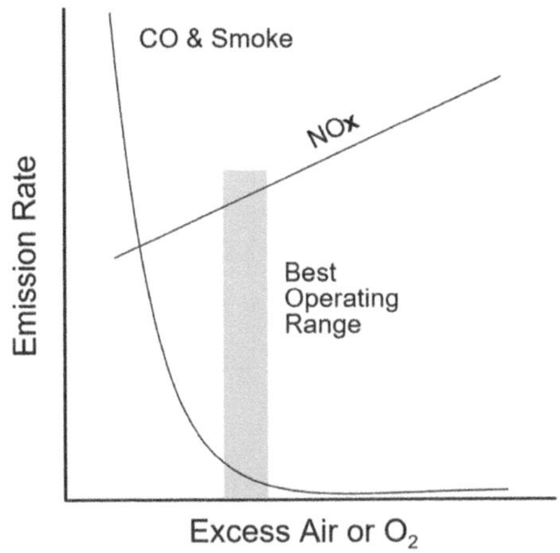

FIGURE 8.9 NO_x, CO, and PIC emissions vs. excess air (USEPA, 2014).

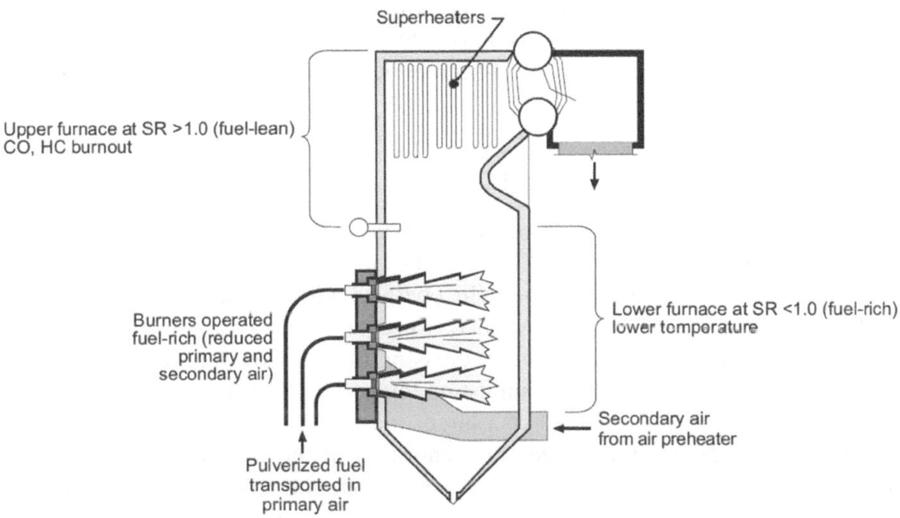

FIGURE 8.10 Off-stoichiometric combustion (USEPA, 2003b).

and secondary air are added to the combustion system. *Primary air* is the air supplied to the burners with the fuel to control the rate of combustion and the amount of fuel that can be burned. *Secondary air* is the air added to control the combustion efficiency and/or completeness of fuel burned.

Off-stoichiometric combustion can be accomplished by employing *over-fire air* (OFA) ports. They are air injection nozzles located above the burners. The burners are operated fuel-rich, while the OFAs, working as air-rich, maintain the remainder of the combustion. Figure 8.11a illustrates that in some boilers, the burners are operated in a staggered operation, called *biased firing*, with some burners in the fuel-rich

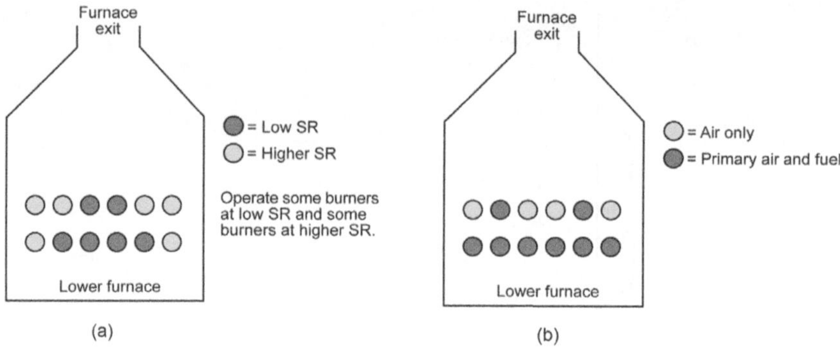

FIGURE 8.11 Burner arrangement: (a) biased firing and (b) out-of-service (USEPA, 2003b).

mode employing a low stoichiometric ratio (SR) and others in the air-rich mode. When some burners are operated on air only (the air-rich mode), this modification is called *burners-out-of-service* (BOOS), as shown in Figure 8.11b.

The NO_x emission reductions, using off-stoichiometric combustion, are approximately 30 to 40% for the gas- and oil-fired boilers and 30 to 50% for the coal-fired boilers (USEPA, 2012b).

8.6.3 Flue Gas Recirculation (FGR)

Flue gas recirculation (FGR), also called *exhaust gas recirculation* (EGR), works by recycling the portion (10 to 30% typical) of the flue gas back to the main combustion chamber (Figure 8.12). The FGR operates without changing the flow rates of the inlet air, the fuel, and the exhaust. However, the flow rate through the combustor is increased due to recycling. Increasing the gas flow (and the mass) in the combustion zone decreases the combustion temperature because the same amount of supplied energy is distributed to a larger mass. The lower oxygen in the recycle stream also dilutes the oxygen concentration in the combustion zone, which reduces the temperature and oxygen concentration, and, consequently, lowers the NO_x formation.

NO_x emission reduction of 40 to 50% is achievable with recirculation of 20 to 30% of the exhaust gas in gas- and oil-fired boilers. If the recirculation rate is too high, the flame may become unstable, so emissions of CO and PICs may increase (USEPA, 2003b). The extent of NO_x emission reduction of the coal-fired boilers employing FGR is generally less.

8.6.4 Low-NO_x Burner

The approach of using BOOS for off-stoichiometric combustion will not be doable for small boilers with one or two burners, and many smaller boilers may not have a proper location for the installation of OFA ports. Many manufacturers have developed designs for burners themselves to reduce NO_x formation. Some low-NO_x burners

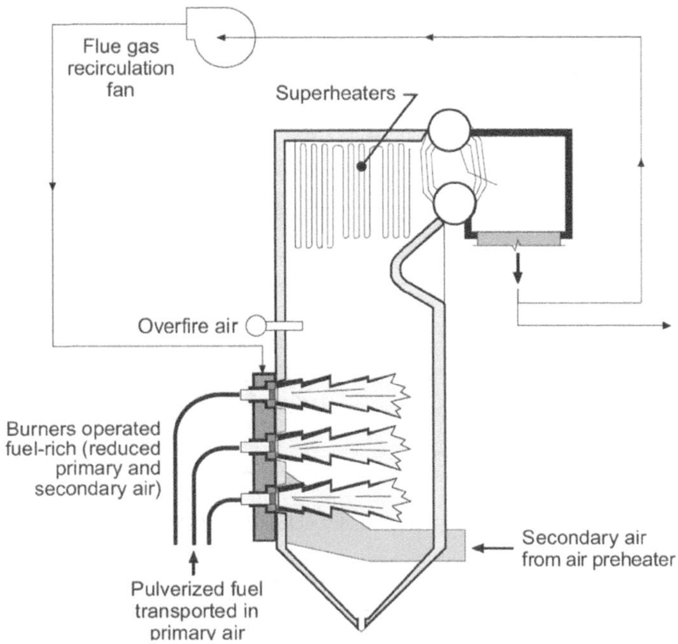

FIGURE 8.12 Flue gas recirculation (USEPA, 2003b).

are designed to have fuel-rich and air-rich regions (i.e., similar to the approach of off-stoichiometric combustion). In contrast, others are designed to control the flame shape to minimize the reaction between nitrogen and oxygen at peak flame temperatures (USEPA, 2012a).

The schematic of a dual-register low-NO_x burner is illustrated in Figure 8.13 as an example. The term "*dual-register*" refers to the burner with two distinct air registers/zones for better control of air distribution and mixing for combustion. For this burner, the pulverized coal and primary air mixture is controlled to delay the fuel combustion slightly (i.e., to create a fuel-rich mixture). The remaining combustion air is introduced as the secondary air through two concentric air zones that surround the nozzle (i.e., air-rich). The flame produced by this low-NO_x nozzle is elongated compared to the flame produced by a conventional burner. The formation of thermal and fuel NO_x is reduced due to the lower peak flame temperature and less oxygen availability in the flame. Most low-NO_x burners have demonstrated a NO_x reduction of 25 to 40%, while some reportedly > 60% (USEPA, 2003b).

8.6.5 Gas Reburning

Gas reburning refers to introducing natural gas or another fuel into the boiler or furnace in stages. Figure 8.14 illustrates a gas reburning system [Note: a *superheater* in a boiler is a device that heats saturated steam to superheated steam]. In addition to the coal-fired burners located at the bottom (the "*burning*" *zone*), separate

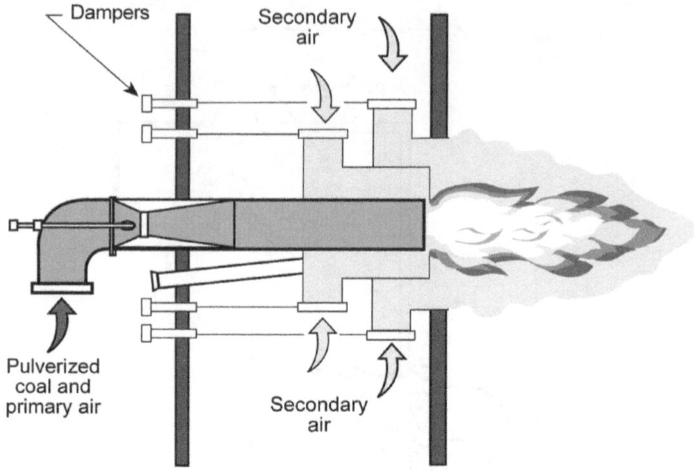

FIGURE 8.13 A dual-register low-NO_x burner (USEPA, 2003b).

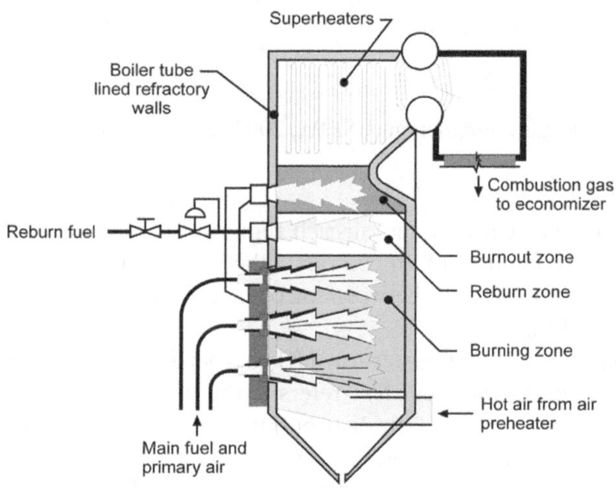

FIGURE 8.14 A gas reburning system (USEPA, 2003b).

gas-fired burners are added in the upper portion of the combustion chamber (the *"reburn" zone*). In the reburn zone, more fuel (i.e., natural gas) is injected without accompanying air to create a fuel-rich zone. No additional NO_x will be formed within this zone due to a lack of oxygen and lower temperatures. The temperatures are below the threshold for NO_x formation but sufficient to ignite the supplemental fuel. The hungry fuel species will then take oxygen from previously formed NO_x by reducing it to nitrogen molecules. OFA can then be added to the *"burnout" zone* to combust the PICs generated in the reburn zone. Not much NO_x will be formed in this zone because temperatures will not be high (the heat is being lost to the wall of the furnace). Gas reburning is essentially a practice of off-stoichiometric combustion (USEPA, 2014).

All existing coal- and oil-fired units can employ gas reburning, provided they have enough room above their original burners to install gas burners and the OVA ports. Fifty to seventy percent of NO_x reductions at the full boiler load have been demonstrated (USEPA, 2003b).

8.7 FLUE GAS TREATMENT

In addition to combustion modification to reduce NO_x formation, the formed NO_x can be removed from the exhaust (flue gas) stream by back-end control. Two techniques commonly used in stationary boilers are to reduce NO_x in the flue gas to nitrogen molecules before discharge to the atmosphere. They are selective non-catalytic reduction (SNCR) and selective catalytic reduction (SCR) processes. In addition to these two commercially available approaches, some emerging technologies employ principles of reduction, absorption, or oxidation (USEPA, 2014).

8.7.1 Selective Non-Catalytic Reduction (SNCR)

Ammonia (NH_3) is in a reduced state (valence state of nitrogen = −3) when compared to NO_2 (valence state of nitrogen = +4). Under ambient conditions, pure ammonia is a colorless, pungent-smelling, and caustic gas. It is highly soluble in water to form ammonium hydroxide (NH_4OH) as:

$$NH_{3(g)} + H_2O_{(aq)} \rightarrow NH_4OH_{(aq)} \rightarrow NH^+_{4(aq)} + OH^-_{(aq)} \quad (8.10)$$

There are several commercial versions of *selective non-catalytic reduction* (SNCR). In general, a more reduced nitrogen compound (mainly ammonia (NH_3) or urea ($CO(NH_2)_2$) is injected into the post-combustion zone of a boiler to react with the hot flue gas (Figure 8.15) [Note: an *economizer* in a boiler is a heat exchanger that capture waste heat from the flue gas and used it to preheat the feedwater to the boiler]. The injected ammonia or urea will react with NO and NO_2 in a complex set of high-temperature gaseous phase reactions to reduce NO_x to nitrogen molecules. The process is called "*selective reduction*" because it uses a reagent to react with NO_x. The overall reactions for NO reduction are:

$$4NO + 4NH_3 + O_2 \rightarrow 4N_2 + 6H_2O \quad (8.11a)$$

$$2NO_2 + 4NH_3 + O_2 \rightarrow 3N_2 + 6H_2O \quad (8.101b)$$

$$4NO + 2CO(NH_2)_2 + O_2 \rightarrow 4N_2 + 4H_2O + 2CO_2 \quad (8.12a)$$

$$2NO_2 + 2CO(NH_2)_2 + O_2 \rightarrow 3N_2 + 4H_2O + 2CO_2 \quad (8.12b)$$

The reactions shown in Eqs. 8.11 and 8.12 are very temperature sensitive. As shown in Figure 8.16, the best conversion occurs in a temperature range of approximately 1,600 to 1,900°F (870 to 1,037°C). Above this temperature range, ammonia or urea

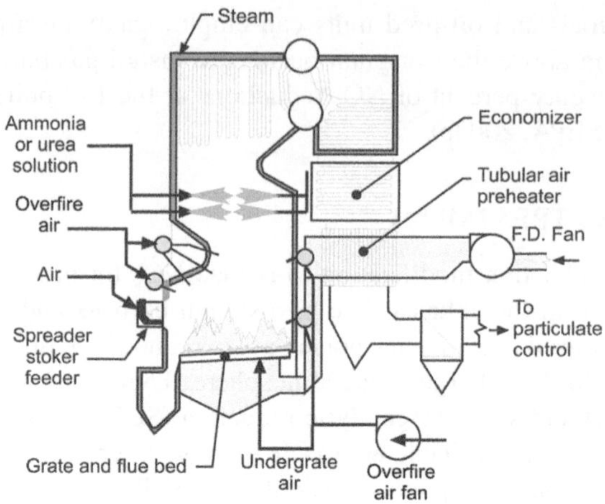

FIGURE 8.15 Schematic of an SNCR system (USEPA, 2003b).

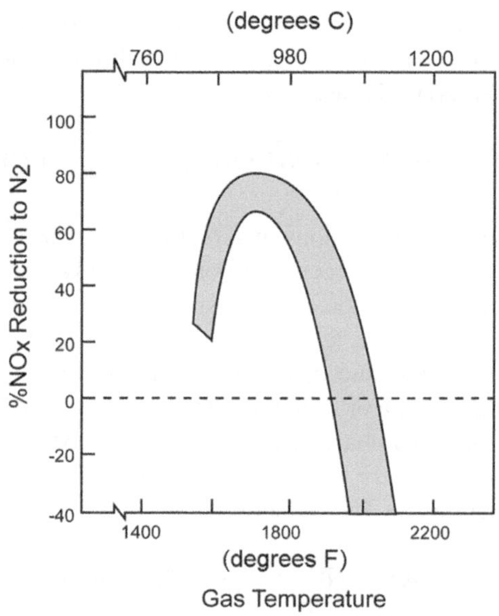

FIGURE 8.16 Temperature sensitivity of SNCR reactions (Modified from USEPA, 2003b).

will be increasingly oxidized to form NO_x. Below this temperature range, more ammonia will pass through the system unreacted to generate ammonia emissions called *ammonia slip*. An ammonia slip with an ammonia concentration >10 ppm is undesirable because ammonia itself is an air pollutant, and it can react with chlorine or sulfur compounds to form a light-scattering particulate plume.

The injected amount of the chemical reagent is critical. The stoichiometric molar ratio of NH_3:NO is 1:1, while that of $CO(NH_2)_2$:NO is 0.5:1. Although injection at

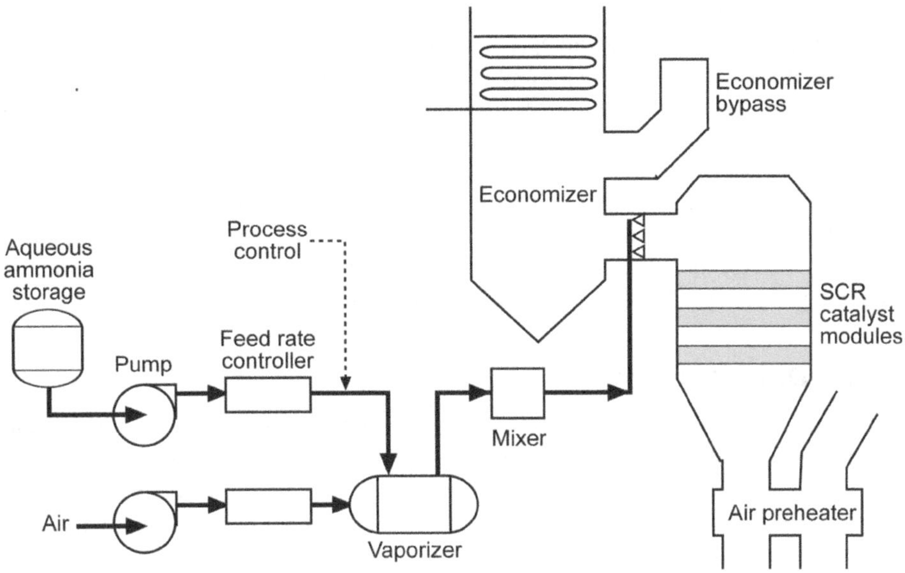

FIGURE 8.17 Schematic of an SCR system (USEPA, 2003b).

higher ratios would increase NO_x reduction, the risk of ammonia slip increases. The reagent is generally injected at about 90% of its stoichiometric amount to avoid excessive ammonia slip. Other design and operating considerations include the injection points' location and the reagent's proper distribution – typical NO_x reduction efficiencies of SNCR range from 20 to 60% (USEPA, 2003c).

8.7.2 Selective Catalytic Reduction (SCR)

Similar to SNCR in principle, a catalyst module is added in *selective catalytic reduction* (SCR) (Figure 8.17). The primary function of the catalyst is to reduce the temperature required for chemical reduction reactions (Eqs. 8.11 and 8.12). Typical catalysts include vanadium pentoxide (V_2O_5), titanium dioxide (TiO_2), tungsten trioxide (WO_3), and noble metals. The temperature needed for efficient reduction would depend on the type of catalyst used, typically in the range of 550 to 750°F (290 to 400°C). Temperature variation of as little as ±50°F (±30°C) can impact its NO_x reduction performance. SCR systems could deliver NO_x reduction of 60 to 90% (USEPA, 2003d). Ammonia oxidation at higher temperatures and ammonia slips at lower temperatures, as in SNCR, are also of concern.

8.8 PROCESS SELECTION AND DESIGN CONSIDERATIONS

8.8.1 Process Selection

As discussed, there are three different NO_x formation mechanisms: fuel, thermal, and prompt. The NO_x emission reductions can be achieved mainly by switching fuels, treating flue gas, and modifying combustion. They are not exclusive to others. A combined approach can be applied if it is found effective. In addition to economic considerations, all these approaches have advantages and limitations. Fuel switching mainly reduces fuel NO_x, not thermal NO_x and prompt NO_x. The primary considerations should be the cost, availability of fuel with low nitrogen content, and effectiveness in NO_x emission reduction. For combustion modifications, the primary considerations are the feasibility of upgrading the existing combustor, the cost of the upgrade or purchasing new devices, and their effectiveness. For the back-end control, the capability of maintaining optimal operating conditions to minimize/avoid ammonia slip and the formation of NO_x must be considered. The bottom line is that the selected NO_x emission reduction technology needs to meet the stringent NO_x emission requirement.

8.8.2 Design Calculation

Characterizing fuel used in combustion is essential to learn its energy content and estimate its potential pollutant emissions. In a *proximate analysis*, moisture content, ash content, and volatile matter are analyzed, and fixed carbon is the balance. In an *ultimate/elemental analysis*, C, H, S, and N are determined from chemical analyses. The oxygen content is the balance:

$$\%O = 100 - (\%C + \%H + \%S + \%N + \%Ash) \tag{8.13}$$

The fuel's energy content can be estimated using Dulong's formula (Eqs. 2.23 and 2.24). The stoichiometric amount of oxygen required for complete combustion can be calculated by using the following equations:

$$C + O_2 \rightarrow CO_2 \tag{8.14}$$

$$H + \frac{1}{4}O_2 \rightarrow \frac{1}{2}H_2O \tag{8.15}$$

$$S + O_2 \rightarrow SO_2 \tag{8.16}$$

$$N + O_2 \rightarrow NO_2 \tag{8.17}$$

For example, Eq. 8.14 indicates that one mole of O_2 will be needed to combust one mole of C; the reaction will generate one mole of CO_2. Therefore, the stoichiometric oxygen amount for complete combustion will be (on a molar basis):

$$\text{moles of } O_2 = \text{moles of } \left\{C + \frac{1}{4}H + S + N - \frac{\text{fuel O}}{2}\right\} \tag{8.18}$$

Assuming oxygen concentration in the air is 21%, the stoichiometric air amount required for complete combustion can be found as (on a molar basis):

$$\text{moles of air} = \frac{\text{moles of oxygen}}{0.21} \tag{8.19}$$

Example 8.2: Stoichiometric Air for Coal Combustion

For a coal firing rate of 100 metric tons/h, what is the minimum air required (in standard m^3/h) to complete the combustion? Results of coal ultimate analysis (by wt.) are shown below:

- C: 65.0%
- H: 8.0%
- O: 7.0%
- S: 1.0%
- N: 1.5%
- Moisture: 4.0%
- Ash content: 13.5%

Solution:

a. The mass loading rates of C, H, O, S, N, H_2O, and ash are 65, 8, 7, 1, 1.5, 4, and 13.5 tons/h, respectively.
b. The molar loading rates of C, H, O, S, N, and H_2O are 5,420 (= 65,000/12), 8,000 (= 8,000/1), 438 (= 7,000/16), 31 (= 1,000/32), 107 (= 1,500/14), and 222 (= 4,000/18) kg-mole/h, respectively (by dividing the mass loading rate by the corresponding atomic mass or MW).
c. Use Eq. 8.18 to find the stoichiometric oxygen requirement:

$$O_2 = \left\{5,420 + \frac{8,000}{4} + 31 + 107 - \frac{438}{2}\right\} = 7,340 \text{ kg-mole/hr}$$

d. Use Eq. 8.19 to find the stoichiometric air requirement ($T=25°C$):

$$\text{stoichiometric air} = \frac{7,340}{0.21} = 35,000\ \frac{\text{kg-mole}}{\text{hr}} = (35,000)(24.46)$$
$$= 8.6 \times 10^5 \text{ standard m}^3/\text{hr}$$

Discussion:

1. In this example, standard temperature = 25°C (77°F), typically for combustion. The molar volume of an ideal gas = 24.46 m^3/kg-mole.
2. If 40% of the excess air is used, the airflow rate will be 140% of the value derived in part (d).

Example 8.3: Flue Gas Flow Rate for Coal Combustion

For the combustion described in Example 8.2, what would the flow rate of the flue gas be (in kg-mole/h and standard m^3/h)?

Solution:

a. Each mole of C, H, S, and N in the fuel will produce 1, ½, 1, and 1 mole of CO_2, H_2O, SO_2, and NO_2, respectively.
b. The moles of nitrogen in the influent gas = (79%) of the influent air =

$$= (0.79) \times (35{,}000) = 27{,}600 \text{ kg-mole/h}$$

c. Total moles of the flue gas = moles of {CO_2, H_2O, SO_2, NO_2, N_2, and O_2}

$$= 5{,}420 + (222 + 8{,}000/2) + 31 + 107 + 27{,}600 + 0 = 37{,}400 \text{ kg-mole/h}$$

d. Flue gas flow rate = (37,400) × (24.46) = 9.15×10^5 standard m^3/h.

Discussion:

1. The moisture in the flue gas came from two sources: moisture in the coal and water produced from the combustion of H in the coal. Moles of water in the flue gas = 222 (the moisture content of the fuel) + 8,000/2 (from the combustion of H) = 4,222 kg-mole/h.
2. Nitrogen in the supply air is assumed to be inert (i.e., does not produce NO_x).
3. It is assumed that the combustion is complete. There's no oxygen in the flue gas because the oxygen in the influent is at a stoichiometric amount.
4. If 25% excess air is added, similar calculations can be made, but oxygen will be present in the flue gas (35,000 × 21% × 25%), and nitrogen will also be in a larger amount, [(0.79) × (35,000)] × 125% for using 25% excess air.
5. Using a spreadsheet is a good approach to solving questions of this nature.

Example 8.4: Pollutant Concentrations in the Flue Gas of Coal Combustion

For the combustion described in Example 8.3, what would be the concentrations of CO_2 and NO_2 in the flue gas for stoichiometric and complete combustion? What would the mass emission rate of CO_2 be (in tons/h)?

Solution:

a. From Example 8.3, the molar flow rates of CO_2, NO_2, and flue gas are 5,420, 107, and 37,400 kg-mole/h, respectively.

$$[CO_2] = (5{,}420) \div (37{,}400) = 14.5\%$$

$$[NO_2] = (107) \div (37{,}400) = 2.86 \times 10^{-3} = 2{,}860 \text{ ppm}$$

b. Mass emission rate of CO_2

$$= (5{,}420 \text{ kg-mole/h})(44 \text{ kg/kg-mole}) = 238{,}500 \text{ kg/h} = 238.5 \text{ ton/h.}$$

Discussion:

1. From these calculations, the CO_2 emission rate of burning 100 tons/h of coal is 239 tons/h. It implies that burning one ton of coal will emit ~2.4 tons of CO_2.
2. The calculated NO_2 concentration in the flue gas is 2,860 ppm, which is relatively high. However, the only NO_2 source in this calculation is from the oxidation of all fuel nitrogen (i.e., only fuel NO_x).

Monitoring of flue gas from a combustion device is needed for (i) regulation compliance (e.g., the effluent concentration of NO_x) and (ii) optimizing the performance of the device (e.g., the oxygen concentration). Flue gas is monitored through continuous emission monitoring (CEM) for real-time immediate indication of emissions and routine source testing (Kuo, 2015; Kuo & Dow, 2017). *Orsat analysis* is a method often used in combustion studies to determine the composition of flue gas resulting from a combustion process. It primarily measures the concentrations of oxygen (O_2), carbon monoxide (CO), and carbon dioxide (CO_2).

Since water vapor content can vary significantly and affect volumetric gas composition, the effluent concentrations of flue gas are often reported on *a dry-air basis*. It provides a consistent basis for comparing emissions as well as combustion efficiency. The conversion between the concentration of gases/particulates on a wet basis, which includes water vapor (C_{wet}), and the concentration on a dry basis (C_{dry}) can be done by incorporating the fraction of moisture in the flue gas, $B_{moisture}$ (in %), as:

$$C_{\text{wet}} = \frac{C_{\text{dry}}}{\left(1 - B_{\text{moisute}}\right)} \tag{8.20}$$

Note: the units of C_{wet} and C_{dry} should be the same.

Example 8.5: Flue Gas Concentrations on a Dry Basis and Combustion Efficiency

A wet flue gas sample contains 5%, 1%, 10%, and 8% by volume of O_2, CO, CO_2, and H_2O, respectively.

a. What are the concentrations of CO, CO_2, and O_2 on a dry basis?
b. What would be the combustion efficiency?

Solution:

a. The dry-basis concentration of CO can be found by using Eq. 8.19:

$$1\% = \frac{C_{\text{dry}}}{(1-8\%)} \rightarrow C_{\text{dry,CO}} = \underline{0.92\%}$$

b. Similarly, the dry-basis concentration of $CO_2 = (10\%) \times (1-8\%) = \underline{9.2\%}$.
   The dry-basis concentration of $O_2 = (5\%) \times (1-8\%) = \underline{4.6\%}$.
c. Using Eq. 7.24 to find the combustion efficiency (CE):

$$\text{CE (\%)} = \frac{[CO_2]}{[CO_2]+[CO]} \times 100 = \frac{9.2\%}{9.2\% + 0.92\%} \times 100 = 90.9\%$$

Discussion:

1. Combustion efficiency is independent of oxygen concentration and moisture in the flue gas.
2. Using concentrations of CO and CO_2 on a dry or wet basis will yield the same value of CE.
3. Orsat analysis yields values on a dry basis, since the sample is dried before measurement.

The readings of the effluent concentrations will be reduced due to any in-leakage (i.e., unintended infiltration of air or other gases into a system or enclosed space) throughout the process or the increased excess air level. The discharge standards generally require the concentration readings to be corrected using oxygen, carbon dioxide, or excess air to report emissions at a specified dilution level (e.g., 7% O_2, 12% CO_2, and 50% excess air). The corrected concentrations for these conditions are:

$$C_{@7\%O_2} = C_{\text{dry}} \times \left[\frac{20.9 - 7}{(20.9 - \%O_2)}\right] \tag{8.21a}$$

$$C_{@12\%CO_2} = C_{\text{dry}} \times \left[\frac{12}{\%CO_2}\right] \tag{8.22b}$$

$$C_{@50\%EA} = C_{\text{dry}} \times \left[\frac{100 + \text{EA}(\text{in }\%)}{150}\right] \tag{8.23c}$$

where C_{dry} is the concentration of pollutant in the flue gas on a dry basis [Note: a value of 21 is used in Eq. 8.21, instead of 20.9, in PE/Environmental Reference Book] (NCEES, 2025).

Example 8.6: Corrected Flue Gas Concentrations

The NO_x and PM_{10} concentrations of a flue gas sample from an industrial boiler with ESP and SCR are 100 ppm and 10 mg/Nm^3, respectively, and the flue gas contains 3% O_2 and 11% CO_2 by volume. What are their concentrations corrected to 7% O_2 and 12% CO_2?

Solution:

a. From Eqs. 8.21a and 8.21b,

$$C_{\text{corrected}} = C_{\text{dry}} \times \left[\frac{20.9 - 7}{(20.9 - \%O_2)}\right] \times \left[\frac{12}{\%CO_2}\right] = C_{\text{dry}} \times \left[\frac{20.9 - 7}{(20.9 - 3)}\right] \times \left[\frac{12}{11}\right] = C_{\text{dry}} \times 0.847$$

b. $NO_{x,\ corrected} = 100\ ppm \times 0.847 = 84.7\ ppm$
c. $PM_{10,\ corrected} = 10 \times 0.847 = 8.47\ mg/Nm^3$

Discussion:

1. Nm^3 = normal cubic meter = cubic meter at standardized conditions (i.e., P=1 atm and T=0°C here).
2. The correction factor (0.847 here) is independent of the types of pollutants.

For SNCR of NO with ammonia, the stoichiometric molar ratio of NH_3:NO = 1:1. With a design molar ratio and the mass loading rate of NO, the ammonia addition rate can be readily estimated. Ammonia is generally injected at about 90% of its stoichiometric amount to avoid excessive ammonia slip. Other design and operating considerations include the locations of the injection points and the reagent's proper distribution (USEPA, 2003b).

Example 8.7: Ammonia Addition of an SNCR System

A boiler employs an SNCR system to reduce its NO_x emissions. The NO concentration at the boiler outlet is 200 ppm (T=25°C). A pilot study found that dosing ammonia at 90% of the stoichiometric NH_3: NO ratio would reduce 40% of NO_x emissions. Calculate the ammonia addition rate if the flue gas flow rate is 3,000 standard m^3/min (T=25°C).

Solution:

a. MW of NO = 14 + 16 = 30. At 25°C,

$$200\ \text{ppm} = (200)\left[\frac{30}{24.42}\right] = 246\ \text{mg/m}^3$$

b. The mass loading rate of NO

$$= \left(246\ \text{mg/m}^3\right) \times \left(3{,}000\ \text{m}^3/\text{min}\right) = 0.737\ \text{kg/min} = 737\ \text{g/min}$$

c. The molar loading rate of NO

$$= \left(737\ \text{g/min}\right) \div \left(30\ \text{g/g-mole}\right) = 24.6\ \text{g-mole/min}$$

d. At a molar ratio of 0.9, the molar loading rate of ammonia

$$= (0.9) \times (24.6) = 22.1\ \text{g-mole/min}$$

e. The mass feeding rate of NH_3

$$= \left(22.1\ \text{g-mole/min}\right) \times \left(17\ \text{g/g-mole}\right) = \underline{376\ \text{g/min}}$$

8.9 SUMMARY

1. There are three different NO_x formation mechanisms from fuel combustion: thermal, fuel, and prompt NO_x.
2. Most NO_x in the flue gas is NO, but NO will be oxidized to NO_2 readily after being emitted.
3. Formation of thermal and prompt NO_x can be reduced, but not avoided. Reduction in formation can be achieved by lowering the combustion temperature, oxygen concentration, and residence time in the flame.
4. Emission reduction of fuel NO_x can be achieved by fuel switching.
5. Combustion modification can reduce thermal NO_x formation with low excess air combustion, off-stoichiometric combustion, flue gas recirculation, low-NO_x burners, and gas reburning.
6. Back-end control can reduce NO_x to molecular nitrogen. Two common approaches are SNCR and SCR. Ammonia and urea are commonly used as reducing agents.

EXERCISE QUESTIONS

1. What is the total NO_x emission rate (in lb/h) from a boiler emitting 500 ppm NO and 60 ppm NO_2 in a flue gas of 50,000 standard cubic feet per minute (scfm)?
2. For a coal firing rate of 120 metric tons/h, what would be the minimum amount of air required (in standard m^3/h) to complete the combustion? Results of coal ultimate analysis (by wt.) are shown below:
   - C: 60.0%
   - H: 8.0%
   - O: 9.0%
   - S: 2.0%
   - N: 1.0%
   - Moisture: 5.0%
   - Ash content: 16.0%
3. For the coal combustion described in Question #2, what would the flue gas flow rate be (in kg-mole/h and standard m^3/h)?
4. For the coal combustion described in Question #2, what would the concentrations of CO_2 and NO_2 be in the flue gas? What would be the mass emission rate of CO_2 (in ton/h)?
5. For the coal combustion described in Question #2, if 50% excess air is applied, what would the flow rate of the flue gas be (in kg-mole/h and standard m^3/h)?
6. For the coal combustion described in Question #2, if 50% excess air is applied, what would the concentrations of CO_2 and NO_2 be in the flue gas? What would the mass emission rate of CO_2 be (in ton/h)?
7. A wet flue gas sample contains 7%, 1%, 12%, and 6% by volume of O_2, CO, CO_2, and H_2O, respectively.
   a. What are the concentrations of CO, CO_2, and O_2 on a dry basis?
   b. What would be the combustion efficiency?

8. The NO_x and PM_{10} concentrations of a flue gas sample from an industrial boiler with ESP and SCR are 100 ppm and 10 mg/Nm^3, respectively, and the flue gas contains 5% O_2 and 14% CO_2 by volume. What are their concentrations corrected to 7% O_2 and 12% CO_2?
9. A boiler employs an SNCR system to reduce its NO_x emissions. The NO_x concentration at the boiler outlet is 300 ppm. A pilot study found that dosing ammonia at 95% of the stoichiometric NH_3: NO ratio would reduce 45% of NO_x emissions. Calculate the ammonia addition rate if the flue gas flow rate is 2,500 standard m^3/min.

BIBLIOGRAPHY

Kuo, J. (2015). *Air Quality Issues Related to Using Biogas from Anaerobic Digestion of Food Waste*, CEC-500-2015-037, California Energy Commission, March 2015.

Kuo, J.; Dow, J. (2017). "*Biogas Production from Anaerobic Digestion of Food Waste and Relevant Air Quality Implications*," *J. Air Waste Manage. Assoc.* 67(9), 1000–1011. doi: 10.1080/10962247.2017.1316326.

NCEES. (2025). *PE/Environmental Reference Handbook (Version 1.6)*, National Council of Examiners for Engineering and Surveying (NCEES), Greenville, SC.

USEPA. (1999). *Nitrogen Oxides (NO_x), Why and How They Are Controlled*, EPA 456/F-99-006R, Office of Air Quality, United States Environmental Protection Agency, Research Triangle Park, NC.

USEPA. (2003a). *APTI 452: Principles and Practices of Air Pollution Control - Student Manual*, 3rd edition, Air Pollution Training Institute, United States Environmental Protection Agency, Research Triangle Park, NC.

USEPA. (2003b). *APTI 455: Inspection of Gas Control Devices and Selected Industries - Student Manual*, 2nd revision, Air Pollution Training Institute (APTI), Environmental Research Center, United States Environmental Protection Agency, Research Triangle Park, NC.

USEPA. (2003c). *Selective Non-Catalytic Reduction (SNCR)*, Air Pollution Control Technology Fact Sheet, EPA-452/F-03-031, United States Environmental Protection Agency.

USEPA. (2003d). *Selective Catalytic Reduction (SCR)*, Air Pollution Control Technology Fact Sheet, EPA-452/F-03-032, United States Environmental Protection Agency.

USEPA. (2012a). *APTI 415: Control of Gaseous Emissions - Student Guide*, Office of Air and Radiation, United States Environmental Protection Agency, Research Triangle Park, NC.

USEPA. (2012b). *APTI 427: Combustion Source Evaluation - Student Manual*, 3rd edition, prepared by ICES Ltd. for Air Pollution Training Institute, United States Environmental Protection Agency, Research Triangle Park, NC.

USEPA. (2014). *APTI 418: NO_x Emission Control from Stationary Sources - Student Manual*, prepared by NESCAUM in 2009 (revised and reorganized by Doyle B.W. & Slot S. in 2014) for the United States Environmental Protection Agency.

USEPA. (2025). *Air Pollutant Emissions Trend Data*, https://www.epa.gov/air-emissions-inventories/air-pollutant-emissions-trends-data, last updated on 04/28/2025.

9 Control of Sulfur Oxides Emissions

9.1 INTRODUCTION

9.1.1 The Problems

Sulfur dioxide (SO_2) is used as the indicator of a larger group of gaseous sulfur oxides (SO_x) because of its much higher concentrations in the ambient air. Most people consider SO_x to be the sum of SO_2 and sulfur trioxide (SO_3), but some also include sulfuric acid (H_2SO_4) in this group.

SO_x can exert adverse impacts on human health and the environment. Short exposure to SO_x can hurt the human respiratory system and make breathing difficult, especially for sensitive receptors. They can harm vegetation at high concentrations by damaging foliage and decreasing growth. They can also react with other compounds in the atmosphere to form fine particles that contribute to particulate matter (PM) pollution, causing additional health problems, damaging the façades of buildings, statues, and monuments, and impairing visibility. They are also significant contributors to acid rain, which can harm sensitive ecosystems.

9.1.2 Sources of the Problems

All the fossil fuels contain sulfur compounds to some extent. Table 9.1 shows the sources and emissions of SO_2 in the U.S. in 2024. The total SO_2 emission rate was about 1.7 million tons in 2024. Burning fossil fuel by power plants (38.7%) and industrial boilers (13.1%) represents more than half of anthropogenic SO_2 emissions into the atmosphere. The other significant source category is industrial and other processes (37.1%), including petroleum and related industries (20.0%), chemical and allied product manufacturing (4.3%), and metals processing (3.2%). Transportation accounts for only 1.7% of the total emissions, while the miscellaneous sector, including wildfires, accounts for 7.6% (USEPA, 2025).

Figure 9.1 depicts trends of annual SO_2 emissions in the U.S. since 1970. As shown, the emissions have been decreasing. Most of the reductions came from the source categories of stationary fuel combustion and the industrial and other processes.

9.1.3 Overview of Alternatives to the Problems

Typical SO_x concentrations in flue gases from coal- or oil-burning boilers are 0.1% or less, which makes recovery of SO_2 for beneficial reuses uneconomical. *Flue gas desulfurization* (FGD), essentially an absorption process, is the most common approach for controlling SO_x emissions from combustion. Due to the acidic nature

DOI: 10.1201/9781003532705-9

TABLE 9.1
SO_2 Emission Rates and Source Categories in the U.S. in 2024

Source Category	(1,000 tons)	(%)
Stationary fuel combustion	**897**	**53.6**
Electric utility	649	38.7
Industrial	219	13.1
Others	29	1.7
Industrial and other processes	**621**	**37.1**
Chemical & allied product manufacturing	72	4.3
Metals processing	53	3.2
Petroleum & related industries	335	20.0
Other industrial processes	126	7.5
Solvent utilization	0	0.0
Storage & transport	1	0.1
Waste disposal & recycling	34	2.0
Transportation	**29**	**1.7**
Highway vehicles	10	0.6
Off-Highway	19	1.1
Miscellaneous	**127**	**7.6**
Wildfires	36	2.1
Others	91	5.4
Total	**1,674**	**100.0**

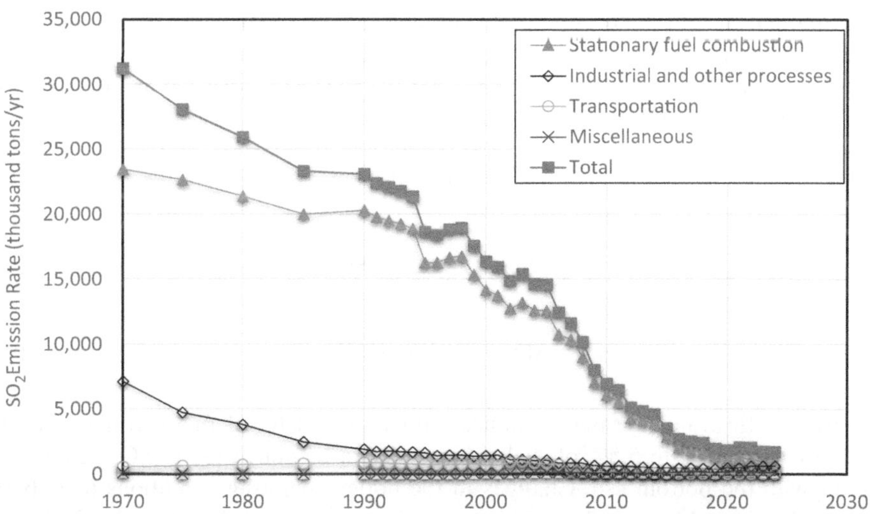

FIGURE 9.1 Annual trends of SO_2 emissions from major source categories in the U.S. (1990–2024).

of SO_x, alkaline reagents, such as hydrated lime ($Ca(OH)_2$), limestone ($CaCO_3$), and sodium carbonate (Na_2CO_3), are used in FGD to absorb/scrub SO_2 out of gas streams. These reagents can be added as wet slurry or dry powders. In addition to FGD, combustion processes can be modified by injecting alkaline reagents directly into the combustion chamber or by fluidizing the chamber (USEPA, 2000).

Since SO_2 emissions from fossil fuel combustion originate from the fossil fuel, alternatives such as using lower sulfur-content fuel and removing sulfur compounds from fuel are also used. For flue gas with higher SO_2 concentrations (e.g., from metal ore extraction), recovery of SO_2 for beneficial reuse should be considered. The subsequent sections will present more discussion on FGD and these other alternatives for reducing SO_x emissions.

9.2 CHEMISTRY OF SULFUR AND SULFUR OXIDES

In its elemental state (i.e., S), sulfur is relatively inert and harmless to humans. Human and all living organisms need some sulfur in their bodies. Many sulfur oxides (e.g., SO_2, SO_3, H_2SO_4) are air pollutants. On the other hand, reduced sulfur compounds (RSC), such as hydrogen sulfide (H_2S), methanethiol (CH_3SH), dimethyl sulfide (C_2H_6S), and dimethyl disulfide ($C_2H_6S_2$), pose odor threats. H_2S is also very toxic.

Sulfur compounds in coal include pyrite (FeS_2), sulfates, and organic sulfur compounds. In burner flames, approximately 94 to 95% of the total sulfur compounds are converted to SO_2 (Eq. 9.1), while another 0.5 to 2% are converted to SO_3 (Eq. 9.2). It should be noted that both Eqs. 9.1 and 9.2 indicate only the overall stoichiometry of the reactions; many simultaneous reactions are occurring in a combustion chamber. As indicated by Eq. 9.1, burning 1 kg of S will generate 2 kg of SO_2 because the stoichiometric molar ratio of $S{:}SO_2 = 1{:}1$ and the molecular weight of SO_2 (64) is twice that of S (32).

Although SO_2 (a polar compound) is more soluble than SO_3 (a non-polar compound) in water under ambient conditions, SO_3 will preferentially combine with a water molecule to form sulfuric acid (H_2SO_4) when the flue gas is cooled below approximately 600°F (316°C) (Eq. 9.3):

$$S + O_2 \rightarrow SO_2 \tag{9.1}$$

$$SO_2 + \frac{1}{2}O_2 \rightarrow SO_3 \tag{9.2}$$

$$SO_3 + H_2O \rightarrow H_2SO_4 \tag{9.3}$$

Figure 9.2 illustrates the mass balance, fate, and transfer of fuel sulfur in a boiler. As shown, about 3 to 5.5% of fuel sulfur will not be converted to SO_x, leaving the boiler with the bottom ash. Changes in the boiler operating conditions (e.g., boiler oxygen levels and burner characteristics) will not significantly affect the fraction of fuel sulfur leaving with the ash. Still, they will influence the ratio of SO_2 to SO_3 in the flue gas (USEPA, 2003a).

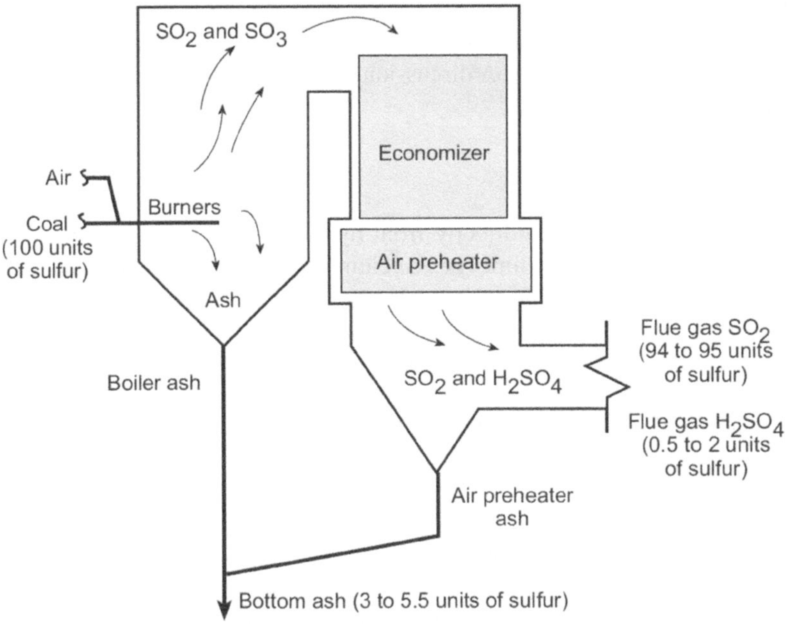

FIGURE 9.2 Fate and transfer of fuel sulfur in a boiler (USEPA, 2003a).

9.3 WET FLUE GAS DESULFURIZATION

FGD systems capture SO_2 before the gas is released into the atmosphere, and many different FGD processes have been developed. FGD processes can be wet or dry. *Wet FGD* processes remove >90% of SO_2 emissions using a caustic liquid absorbent. A *dry FGD* process is a two-step process that uses a wet or dry spray of alkaline reagents to react with SO_2 and form dry particles, which are then collected in a fabric filter, Electrostatic Precipitator (ESP), or cyclone. Removal efficiencies of dry FGD systems range from ~50% to >90%, depending on the reagent used and the system configuration/operation. Wet FGD systems are more common, representing approximately 70% of boilers' uses (USEPA, 2000, 2012). Wet scrubbing with caustic chemical reagents is also effective for other acidic gases such as hydrogen sulfide (H_2S) and hydrochloric acid (HCl) from air streams.

The FGD systems can be further grouped into non-regenerative and regenerative. The *non-regenerative* systems produce sludge/waste that needs to be appropriately disposed of, sometimes called *throw-away* FGD processes. On the other hand, the *regenerative* FGD processes reuse the absorbent after cleaning or generate salable end-products such as concentrated SO_2, H_2SO_4, elemental S, and gypsum (USEPA, 2003a, 2003b).

This section describes seven common wet FGD processes: four non-regenerative systems (i.e., lime, limestone, sodium carbonate, and dual alkali) and three regenerative (i.e., Wellman-Lord, magnesium oxide, and citrate). Both lime and limestone scrubbing processes can be regenerative or non-regenerative. Regenerative systems are becoming less popular because of higher capital costs and insufficient market demand for recovered products. A discussion of dry FGD processes will be presented in the next section.

9.3.1 Lime Scrubbing

An alkaline slurry, prepared directly from hydrated lime (i.e., calcium hydroxide, $Ca(OH)_2$) or by slaking quicklime (i.e., calcium oxide, CaO), is used as the scrubbing reagent in *lime-scrubbing FGD* [Note: *slaking* is to add water to powder quicklime, put it in an oven, and then pulverize it with water to form hydrated lime (Eq. 9.4)]. It has about 20% of the U.S. market. The slurry is sprayed into the absorber at a stoichiometric molar ratio of ~1:1 to absorb SO_2 into the slurry droplets (Eq. 9.5). Calcium sulfite ($CaSO_3$) and calcium sulfate ($CaSO_4$, gypsum) will form as precipitates (Eqs. 9.5 and 9.6) and be removed as solids (after thickening and vacuum filtration). Before disposal, the solids are often stabilized by adding fly ash (Figure 9.3).

$$CaO + H_2O \rightarrow Ca(OH)_2 \tag{9.4}$$

$$SO_2 + Ca(OH)_2 \rightarrow CaSO_3 + H_2O \tag{9.5}$$

$$CaSO_3 + \frac{1}{2}O_2 \rightarrow CaSO_4 \tag{9.6}$$

It should be mentioned that a mixture of significant quantities of calcium sulfite and sulfate is usually not desirable for final disposal. Therefore, the oxidation of calcium sulfite to calcium sulfate (i.e., Eq. 9.6) is either enhanced or suppressed.

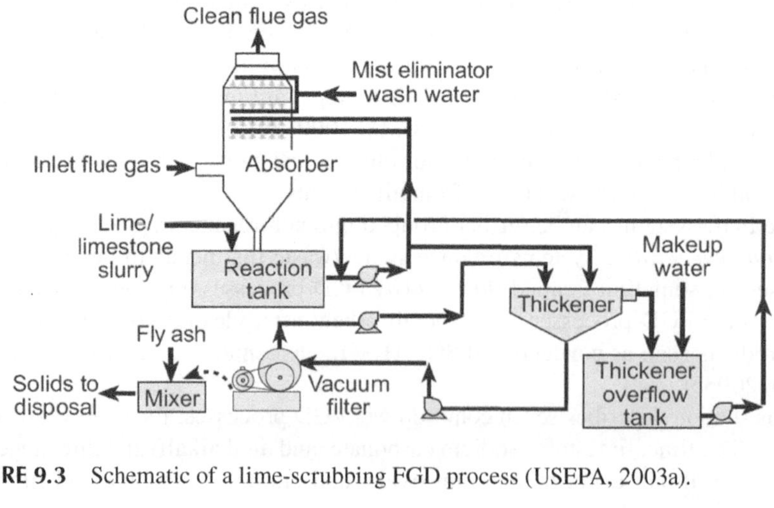

FIGURE 9.3 Schematic of a lime-scrubbing FGD process (USEPA, 2003a).

9.3.2 Limestone Scrubbing

The *limestone scrubbing FGD* process is essentially the same as lime scrubbing, except that limestone ($CaCO_3$) is used. The limestone is first crushed and ball-milled to sizes finer than 200 mesh (<0.074 mm) before being made into a slurry with water. SO_2 will first react with water to form sulfurous acid (H_2SO_3), neutralized by $CaCO_3$, and oxidized by O_2 to form gypsum ($CaSO_4$). Limestone is abundant and relatively less expensive; it represents ~45% of the U.S. market. The primary reactions in sequence are:

$$SO_2 + H_2O \rightarrow H_2SO_3 \tag{9.7}$$

$$H_2SO_3 + CaCO_3 \rightarrow CaSO_3 + CO_2 + H_2O \tag{9.8}$$

$$CaSO_3 + \frac{1}{2}O_2 \rightarrow CaSO_4 \tag{9.9}$$

9.3.3 Sodium Carbonate Scrubbing

Similar to both lime and limestone FGD systems, soluble sodium compounds (i.e., sodium hydroxide (NaOH) or sodium carbonate (soda ash, Na_2CO_3)) are used to remove SO_2 from the flue gas to form sodium bisulfite ($NaHSO_3$) or sodium sulfite (Na_2SO_3) (see Eqs. 9.10a and 9.10b); they can be further oxidized into sodium bisulfate ($NaHSO_4$) and sodium sulfate (Na_2SO_4) (see Eqs. 9.11a and 9.11b). The chemical reagent (e.g., soda ash) is added to a recirculation tank at a rate controlled by the solution pH and the dissolved solids content. A portion of the recirculation stream is continuously removed for disposal or treatment to reduce its ash content and sulfate level. The primary reactions are:

$$SO_2 + Na_2CO_3 + \rightarrow NaHSO_3 + CO_2 \tag{9.10a}$$

$$SO_2 + Na_2CO_3 \rightarrow Na_2SO_3 + CO_2 \tag{9.10b}$$

$$NaHSO_3 + \frac{1}{2}O_2 \rightarrow NaHSO_4 \tag{9.11a}$$

$$Na_2SO_3 + \frac{1}{2}O_2 \rightarrow Na_2SO_4 \tag{9.11b}$$

One of the advantages of this process is that the reaction product, sodium sulfite (Na_2SO_3) or sodium sulfate (Na_2SO_4), is very water-soluble. Hence, the scaling and plugging of the system are minimal compared to lime or limestone scrubbing. In addition, the smaller footprint and ease of operation make them accepted by ~4% of the U.S. market despite the cost of the sodium-based reagents being more expensive than that of the calcium-based ones (USEPA, 2012).

9.3.4 Dual Alkali Scrubbing

In a *dual alkali scrubbing* (or *double alkali scrubbing*) *FGD* system, two types of alkalis are added to two separate liquid recirculation loops (Figure 9.4). One loop is for the absorption of SO_2 and the other is for the regeneration of the scrubbing liquid. First, soluble alkali (such as Na_2CO_3 or NaOH) is used to capture the SO_2 from the flue gas to form $NaHSO_3$ or Na_2SO_3 (see Eqs. 9.10a and 9.10b) – the main scrubbing loop.

A portion of the scrubbing fluid is withdrawn to a regeneration loop. In this regeneration loop, calcium hydroxide ($Ca(OH)_2$) is added to convert the reaction products of the SO_2 absorption loop (i.e., sodium bisulfite or sodium sulfate) to calcium sulfite ($CaSO_3$) and sodium hydroxide (NaOH); and NaOH returns to the scrubbing loop as the effective reagent (see Eqs. 9.12a and 9.12b). Make-up sodium carbonate (Na_2CO_3) is added to the recirculation loop. Calcium sulfite can be further oxidized to form calcium sulfate (Eq. 9.13). The byproducts formed in the regeneration loop, $CaSO_3$ and $CaSO_4$, are removed as solids for disposal or beneficial reuse. The primary reactions in the regeneration loop are:

$$NaHSO_3 + Ca(OH)_2 \rightarrow CaSO_3 + NaOH + H_2O \tag{9.12a}$$

$$Na_2SO_3 + Ca(OH)_2 \rightarrow CaSO_3 + 2NaOH \tag{9.12b}$$

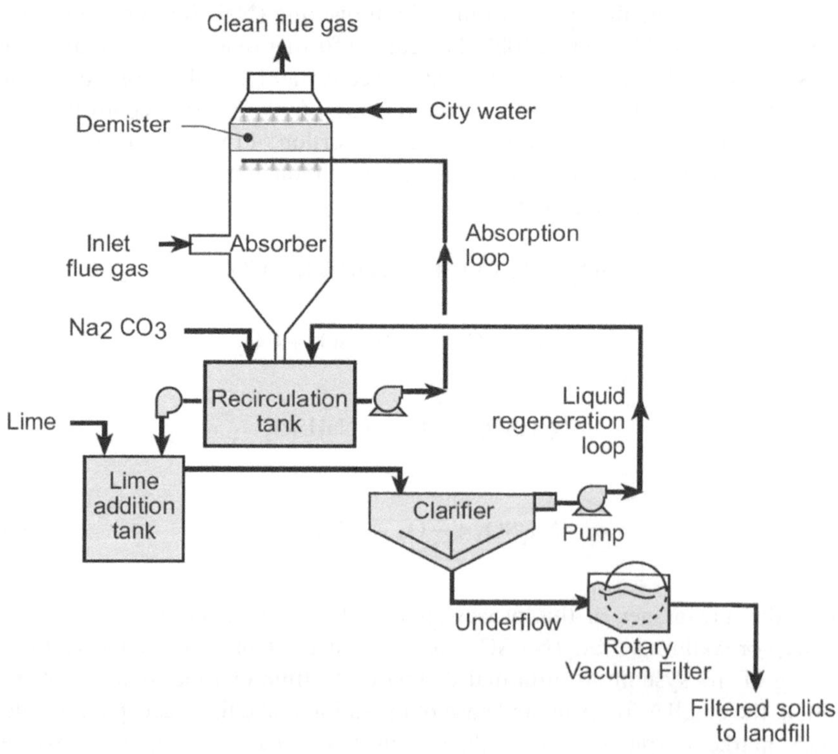

FIGURE 9.4 Schematic of a dual alkali-scrubbing FGD process (USEPA, 2003a).

$$CaSO_3 + \frac{1}{2}O_2 \rightarrow CaSO_4 \tag{9.13}$$

The system is more complex and comes with a higher capital cost. This type of FGD system has ~3% of the U.S. market, mainly used in power plants and industrial facilities that require high-efficiency SO_2 removal with heavy scaling potential (USEPA, 2012).

9.3.5 Magnesium Oxide Scrubbing

Magnesium oxide scrubbing is a wet regenerative FGD process where magnesium oxide (MgO) is used to form magnesium hydroxide ($Mg(OH)_2$) to remove SO_2 from the flue gas. The primary reactions in the sequence are:

$$MgO + H_2O \rightarrow Mg(OH)_2 \tag{9.14}$$

$$SO_2 + Mg(OH)_2 \rightarrow MgSO_3 + H_2O \tag{9.15}$$

The magnesium sulfite can be oxidized to magnesium sulfate as:

$$MgSO_3 + \frac{1}{2}O_2 \rightarrow MgSO_4 \tag{9.16}$$

The spent scrubbing liquid is thickened and centrifuged to obtain the magnesium sulfite solids, which are dried and calcined to generate magnesium oxide (MgO) and concentrated SO_2 (bottom of Figure 9.5) [Note: it is the backward reactions shown

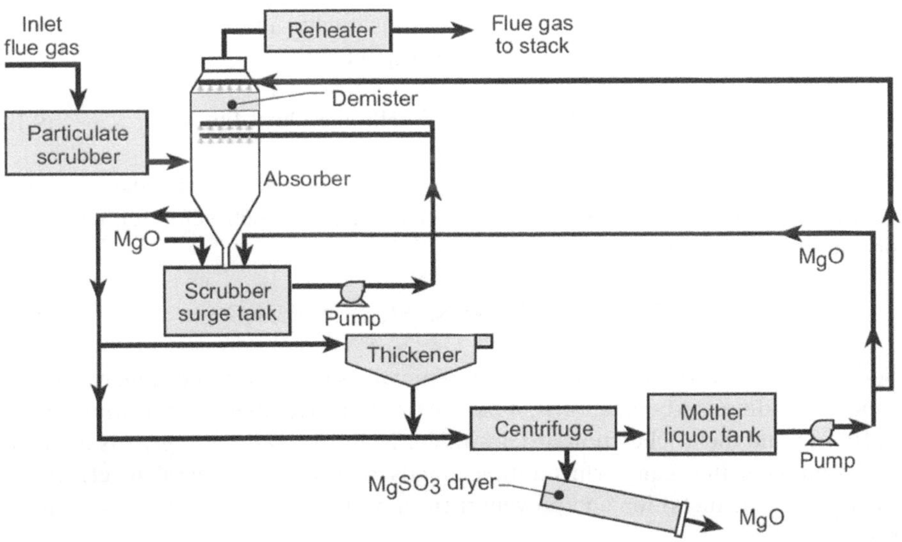

FIGURE 9.5 Schematic of an MgO-scrubbing FGD process (USEPA, 2003a).

in Eqs. 9.14 and 9.15]. The produced MgO can be slaked with water to produce $Mg(OH)_2$ (Eq. 9.14), which is returned to the scrubber. The concentrated SO_2 can be liquefied, sold as an intermediate chemical, or used to produce H_2SO_4 or elemental S (USEPA, 2003a).

9.3.6 Wellman–Lord Scrubbing

Wellman–Lord scrubbing is a regenerative FGD process that removes SO_2 using a sodium-based alkali solution and recovers pure SO_2 instead of producing waste. A sodium sulfite solution is used to absorb SO_2 to form sodium bisulfite ($NaHSO_3$) in the absorption process as follows:

$$SO_2 + Na_2SO_3 + H_2O \rightarrow 2NaHSO_3 \tag{9.17}$$

The spent scrubbing solution is sent to an evaporator to convert sodium bisulfite to sodium sulfite (for recycling) and concentrated SO_2 gas for beneficial reuses (e.g., making sulfuric acid or converted to elemental sulfur) (USEPA, 2003a):

$$2NaHSO_3 \xrightarrow{\text{heat}} Na_2SO_3 + H_2O + SO_2 \tag{9.18}$$

It regenerates and reuses scrubbing chemicals. It is more suitable for high SO_2 concentrations and when SO_2 recovery is economically justified.

9.3.7 Citrate FGD Process

The *citrate FGD process* is a regenerative, aqueous-based method. A solution of citrate ions $\left(C_6H_5O_7^{-3}\right)$ and citric acid ($C_6H_8O_7$) is used to attain a higher solubility of SO_2 in the solution by forming bisulfite ion (Eq. 9.19).

$$SO_2 + Na_3C_6H_5O_7 + H_2O \rightarrow NaHSO_3 + \text{Other Products} \tag{9.19}$$

The solution can then be reacted with H_2S gas to produce elemental S and to regenerate the citrate solution (Eq. 9.20). H_2S gas can come from petroleum refineries. The primary reactions are:

$$NaHSO_3 + 2H_2S \rightarrow Na_2SO_3 + 3S + 3H_2O \tag{9.20}$$

Sodium citrate is not as effective in SO_2 removal as other alkaline reagents such as lime and sodium carbonate. Still, it can be used in specialized applications where buffering and metal chelation are desirable. Note: A *chelating agent* is a chemical compound that can form multiple bonds with a single metal to effectively "sequester" the metal ion and prevent it from participating in unwanted chemical reactions.

9.4 DRY FLUE GAS DESULFURIZATION

Dry FGD is a two-step process. The SO_2-containing flue gas from a boiler is brought into contact with a dry or slightly wet alkali reagent, usually calcium hydroxide. The reaction products (i.e., $CaSO_3$ or $CaSO_4$) are dry and subsequently collected in a particulate control device such as a bag house or an ESP. Consequently, dry waste is produced with handling properties similar to fly ash. The reagent can be delivered into the flue gas in a slurry form (spray-dryer absorption) or as a dry powder (dry-injection absorption).

9.4.1 Spray-Dryer Absorption

Spray-dryer absorption is a semi-dry FGD where hot flue gas meets a mist of finely atomized lime slurry, prepared in a ball-mill slaker, then into a spray dryer (Figure 9.6) [Note: a *spray dryer* is a device to produce dry powder from liquid or slurry materials by rapidly drying them with a hot gas]. Rotary atomizers or two-fluid nozzles are often employed to disperse the lime slurry into the flue gas. Typically, the spray dryer is operated at a lime stoichiometry of 0.9 (for low S-content coal) to 1.5 (for high S-content coal). The amount of water fed into the dryer is carefully controlled to avoid complete saturation of the flue gas. Simultaneous heat and mass transfer between SO_2 in the flue gas and lime in the slurry results in absorption reactions (Eq. 9.5) and gas drying. Some solids will settle to the bottom of the dryer, while the remaining solids will travel to the downstream particulate removal device. Typical residence time in a spray dryer is 8–12 seconds (USEPA, 2000, 2012).

9.4.2 Dry-Injection Absorption

Dry-injection absorption uses superfine alkali powders, <325 mesh (<44μ), with a large specific surface area for effective absorption. The main disadvantage is that the

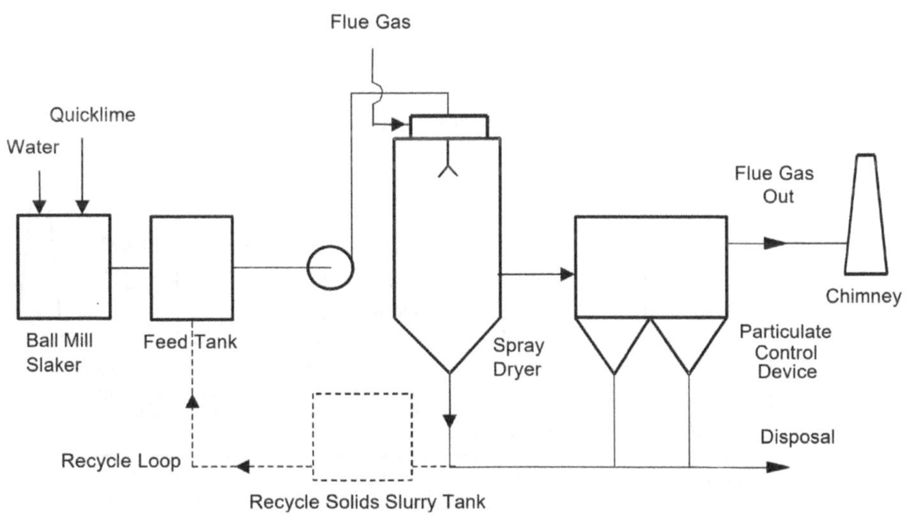

FIGURE 9.6 Schematic of spray-dryer absorption (USEPA, 2000).

typical calcium hydroxide feed rate is three to four times the stoichiometric amount, which is much higher than the spray dryer's. Its main advantage is a minor capital cost (e.g., a dedicated absorber, such as a sprayer dryer, may not be needed).

Figure 9.7 shows a *duct-sorbent-injection* process in which the alkali reagent, usually hydrated lime, is injected directly into the duct downstream of the boiler's preheater. Water may be injected separately upstream or downstream of the reagent injection point to humidify the flue gas and facilitate absorption. The distance between these two injection points is critical for adequate contact between the water droplets and the reagent powders. Reagents in a slurry form can also be used for injection instead (USEPA, 2000).

The dry reagent can also be injected directly into the furnace in the optimum temperature region above the flame (Figure 9.8). Due to the high temperature (~1,000°C), calcium hydroxide will decompose to yield a large specific surface area for absorption. The primary reactions are (USEPA, 2000):

$$Ca(OH)_2 \xrightarrow{\text{heat}} CaO + H_2O \tag{9.21}$$

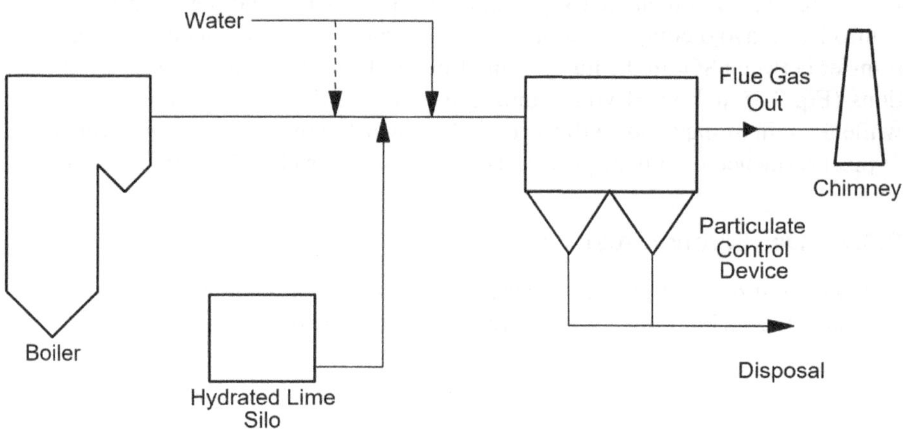

FIGURE 9.7 Schematic of a duct-sorbent-injection process (USEPA, 2000).

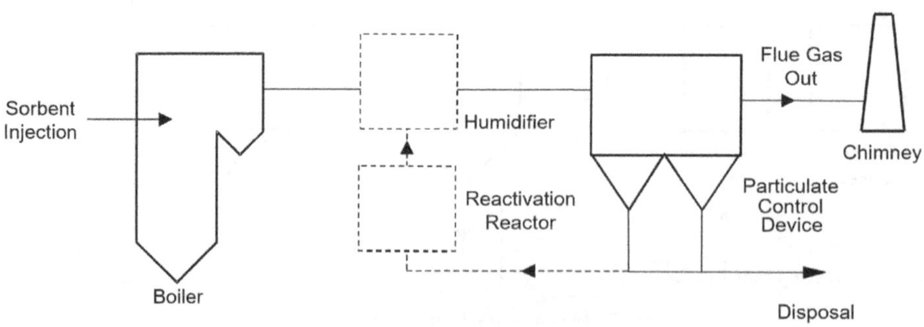

FIGURE 9.8 Schematic of a furnace sorbent injection process (USEPA, 2000).

$$CaO + SO_2 + \frac{1}{2}O_2 \rightarrow CaSO_4 \tag{9.22}$$

The reaction product ($CaSO_4$) and any remaining unreacted reagent will leave the furnace with the flue gas and enter the particulate removal device.

Figure 9.9 illustrates a *circulating-fluidized-bed* process in which a dry reagent, usually calcium hydroxide, meets a humidified flue gas in the reactor. The reagent is fluidized because of the up-flow movement of the flue gas. The reactor provides ample contact between the reagent and the flue gas for absorption. Continuous abrasion of the reagent particles results in the exposure of fresh, unreacted reagents to SO_2 (USEPA, 2000).

9.5 ALTERNATIVES TO END-OF-THE-PIPE CONTROL

In addition to the end-of-pipe control to remove SO_2 from the flue gas, several other viable alternatives exist. They include (i) removal of reduced sulfur compounds, (ii) coal liquefaction and gasification, (iii) coal cleaning, (iv) use of low-sulfur coal, and (v) sulfur recovery for beneficial uses.

9.5.1 Removal of Reduced Sulfur Compounds in Natural Gas and Petroleum

Reduced sulfur compounds (e.g., H_2S) are present (i) in many natural gas deposits, (ii) as byproduct gases produced in oil refineries, or (iii) in the fuel gas produced by coal gasification. H_2S can be readily absorbed into a liquid solution such as

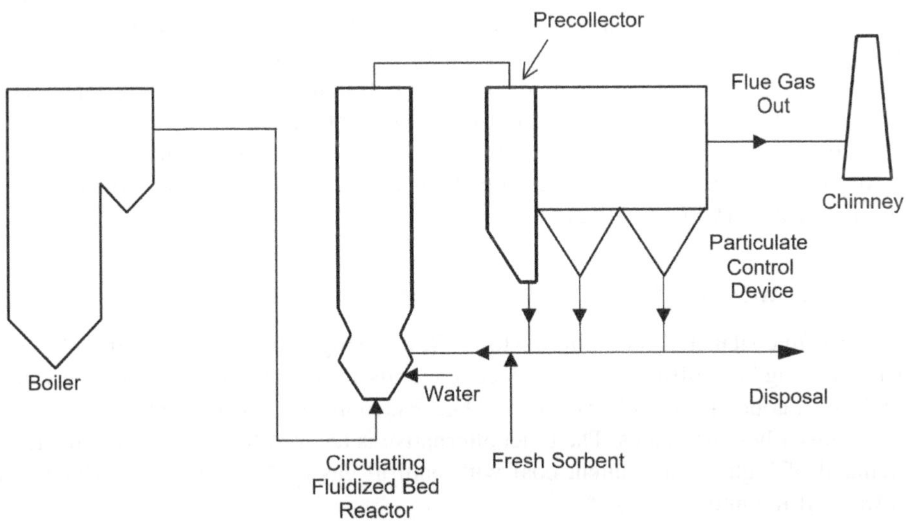

FIGURE 9.9 Schematic of a circulating-fluidized-bed process (USEPA, 2000).

monoethanolamine, $HO(CH_2)_2NH_2$. The absorbed H_2S can be catalytically oxidized under controlled conditions to elemental S by the Claus process:

$$2H_2S + O_2 \rightarrow 2S + 2H_2O \quad (9.23)$$

9.5.2 Coal Liquefaction and Gasification

Coal gasification produces combustible gas from coal through three basic steps: (i) pretreatment, (ii) gasification, and (iii) gas cleaning. The pretreatment step involves coal pulverizing and washing. Some sulfur compounds, such as pyrite, can be removed. The pulverized coal is gasified in a pyrolytic reactor. During gasification, the sulfur content in the coal will be converted to H_2S, which is then converted to elemental S by the Claus process (Eq. 9.23). The synthetic gas produced is sulfur-free and can be burned without SO_2 emissions.

Coal liquefaction is the process of changing solid coal into synthetic oil. One approach is to gasify the coal (similar to coal gasification), followed by condensation to convert the gas into oil. H_2S can be removed from the synthetic gas. The other approach is to use a solvent and hydrogen gas to liquefy pulverized coal and then process the liquid into heavy fuel oil (USEPA, 2003a). Both coal liquefaction and gasification turn coal into a product that can be transmitted through pipelines and produce cleaner-burning fuel with post-treatment.

9.5.3 Coal Cleaning

Physical coal cleaning has been used for decades to reduce coal's impurities, such as sulfur, ash content, and other non-combustible materials, to increase the heating value, reduce SO_2 and CO_2 emissions, and improve combustion efficiency. In coal cleaning processes, the coal is crushed into a specific size to facilitate the separation of impurities, washed, and separated by cyclones, air classifiers, or magnetic separators. It utilizes the density differences between the coal and the impurities. Approximately 40 to 90% of the pyrite sulfur content can be removed (USEPA, 2003a). Chemical coal cleaning is also used to remove sulfur compounds in coal. Several approaches are being developed, including microwave and hydrothermal desulfurization (USEPA, 2012).

9.5.4 Low-Sulfur Fuel

Low-sulfur coal usually contains 0.5 to 1% S, while high-sulfur coal contains 3%–5% sulfur. Using low-sulfur coal can reduce SO_2 emissions. However, both heating value and sulfur content should be considered because many low-sulfur-content coals have much lower heating values. The other alternative is to substitute a portion of the heat demand of high-sulfur-content coal with natural gas or No. 2 fuel oil, which has lower sulfur content.

Hydrodesulfurization is a refinery process in which sulfur compounds in crude oil are converted to H_2S (which is subsequently removed) using hydrogen and catalysts.

Common types of low-sulfur fuels include ultra-low-sulfur diesel, low-sulfur fuel oil, and low-sulfur gasoline.

9.5.5 Recovery for Beneficial Reuses

The SO_2 concentrations in the off-gases from metal smelting generally range from 2 to 40%. Using FGD to remove SO_2 for such high concentrations would be costly. Such gases can be economically converted to valuable H_2SO_4 through the following processes (de Nevers, 2000):

$$SO_2 + \frac{1}{2}O_2 \xrightarrow{\text{vanadium catalyst}} SO_3 \quad (9.24)$$

$$SO_3 + H_2O \rightarrow H_2SO_4 \quad (9.25)$$

9.6 PROCESS SELECTION AND DESIGN CONSIDERATIONS

9.6.1 Process Selection

The fuel itself is the only sulfur source for SO_x formation from fossil fuel combustion. Therefore, using fuel of low sulfur content is a direct way to reduce SO_x emissions. However, some fuels with lower S-content may have a less heating value. Consequently, emission reductions by using low S-content fuel may not be significant. The first consideration in process selection is whether the SO_x concentrations are high enough to warrant beneficial uses of the reaction products. This may only apply to waste gas streams of very high SO_x concentrations from metal ore smelting. The fundamental principle of end-of-the-pipe control is adding alkali agents to react with acidic sulfur oxides. Since the processes use chemicals, the trade-off between the capital and O&M costs should be considered. There is a variety of commercially available wet and dry FGD technologies. The scrubbing agent's availability and cost should always be considered. For a specific application, pilot-scale testing is recommended to evaluate the process feasibility and gather data for site-specific design.

9.6.2 Design Calculations

The amount of SO_x formation from fuel combustion is related to the sulfur content of the fuel used. The maximum mass emission and concentration can be estimated from the fuel sulfur content, as in the following examples:

Example 9.1: SO_x Emissions from Coal Combustion

Refer to the coal-firing boiler in Example 8.2, determine the mass emission rate and the concentration of SO_2 in the flue gas. The coal-firing rate is 100 metric tons/h using the stoichiometric amount of air. Results of coal ultimate analysis (by wt.) are shown below:

- C: 65.0%
- H: 8.0%
- O: 7.0%
- S: 1.0%
- N: 1.5%
- Moisture: 4.0%
- Ash content: 13.5%

Solution:

a Mass loading rate of S to the boiler = (100)(1%) = 1 ton/h = 1,000 kg/h
b Molar flow rate of S to the boiler = (1,000/32) = 31 kg-mole S/h
= the molar flow rate of SO_2 in the flue gas
= 31 kg-mole SO_2/h = (31)(32 + 16 × 2) = 2,000 kg SO_2/h = 2 ton/h
c From Example 8.2, total moles of the flue gas = 37,510 kg-mole/h
d SO_2 concentration in the flue gas = (31)/(37,520)
= 8.26×10^{-4} = 826 ppm

Discussion:

1. This is a worst-case scenario, assuming that all the sulfur in the fuel is converted into SO_2.
2. Every kg of S in the fuel can form 2 kg of SO_2.

Example 9.2: Switching to Low-Sulfur-Content Coal

A power generation plant uses high-sulfur-content (1.5% by wt.) coal, with a heating value of 12,000 Btu/lb (5,160 kJ/kg). There is a lower sulfur-content (1.0% by wt.) coal available, but it has a lower heating value of 9,000 Btu/lb (3,870 kJ/kg). How much SO_2 emission (in percentage) will there be if the plant switches from the existing coal to the new source?

Solution:

Basis: For one metric ton of the current coal:

a. Mass of S in one ton coal = (1,000) × (1.5%) = 15 kg
b. Heating value of one ton coal = (5,160) × (1,000) = 5.16×10^6 kJ
c. Mass of coal from the new source to have the same heating value

$$= (5.16 \times 10^6 \text{kJ}) \div 3,870 \text{ kJ/kg} = 1,333 \text{ kg} = 1.333 \text{ ton}$$

d. Mass of S in this new batch of coal = (1,333) × (1%) = 13.33 kg
e. Percentage reduction in SO_2 emission = (15 − 13.33) ÷ 15 = 11.1%

Discussion:

1. The mass of S was used in the calculations, and the result will be the same if the mass of SO_2 is used.

2. The reduction is only 11.1%, even though the sulfur content of the new coal is one-third less.

9.7 SUMMARY

1. Stationary fuel combustion and industrial and other processes are the primary anthropogenic sources of SO_x emissions.
2. The sulfur in SO_x generated from fossil fuel combustion is from the fuel itself.
3. Using fuel with low sulfur content directly reduces SO_x emissions from fuel combustion.
4. SO_x concentrations in flue gases may not warrant resource recovery unless the concentrations are relatively high.
5. FGD is a scrubbing/absorption process in which various caustic reagents react with acidic sulfur oxides to remove them.

EXERCISE QUESTIONS

1. A facility operates several boilers that burn fuel oil. The fuel oil burned at the facility is 150,000 gallons/yr. The fuel oil (specific gravity = 0.90) contains 0.5% of S (by weight). Assuming that all the S included in the fuel oil would be converted to SO_2 and that the emission control devices remove 95% of SO_2 generated during combustion, determine the amount of SO_2 emitted to the atmosphere from these industrial boilers annually (in tons/yr).
2. The city has one coal-fired power plant with a 90% efficient SO_2 scrubber and one oil-fired power plant without a scrubber for SO_2 emission control. The coal has a heat content of 13,000 BTU/lb and contains 3.5% (by wt.) sulfur, while the oil has a heat content of 6,000,000 BTU/barrel (one barrel = 42 gallons) and contains 0.9% (by wt.) sulfur. The oil has a specific gravity of 7.5 lb/gal. Assuming that all sulfur is oxidized to SO_2 in the combustion process, determine the SO_2 emission rate per unit amount of energy input from each plant (in lb SO_2/BTU).
3. For a coal-firing rate of 120 metric tons/h, what is the minimum amount of air required (in standard m^3/h) to complete combustion? The results of coal ultimate analysis (by wt.) are shown below:
   - C: 60.0%
   - H: 7.0%
   - O: 8.0%
   - S: 2.5%
   - N: 1.0%
   - Moisture: 5.5%
   - Ash content: 16.0%
4. For the coal combustion described in Question #3, what would the flue gas flow rate be (in kg-mole/h and standard m^3/h)?

5. For the coal combustion described in Question #3, what would the concentrations of CO_2 and SO_2 be in the flue gas? What would the mass emission rates of CO_2 and SO_2 be (in tons/h)?
6. For the coal combustion described in Question #3, if 40% excess air is applied, what would be the flow rate of the flue gas (in kg-mole/h and standard m^3/h)?
7. For the coal combustion described in Question #3, if 40% excess air is applied, what would the concentrations of CO_2 and SO_2 be in the flue gas? What would the mass emission rates of CO_2 and SO_2 be (in tons/h)?

BIBLIOGRAPHY

de Nevers, N. (2000). *Air Pollution Control Engineering*, 2nd edition, McGraw-Hill Companies, Inc.

USEPA. (2000). *Controlling SO_2 Emissions: A Review of Technologies*, EPA/600/R-00/093, prepared by Srivastava, R.K. for Office of Research and Development, Washington, DC.

USEPA. (2003a). *APTI 455: Inspection of Gas Control Devices and Selected Industries - Student Manual*, 2nd revision, Air Pollution Training Institute (APTI), Environmental Research Center, United States Environmental Protection Agency, Research Triangle Park, NC.

USEPA. (2003b). *Flue Gas Desulfurization (FGD) – Wet, Spray Dry, and Dry Scrubbers*, Air Pollution Control Technology Fact Sheet, EPA-452/F-03-034, United States Environmental Protection Agency.

USEPA. (2012). *APTI 415: Control of Gaseous Emissions - Student Guide*, Office of Air and Radiation, United States Environmental Protection Agency, Research Triangle Park, NC.

USEPA. (2025). *Air Pollutant Emissions Trend Data*, https://www.epa.gov/air-emissions-inventories/air-pollutant-emissions-trends-data, last updated on 04/28/2025.

10 Control of Mobile Source Pollution

10.1 OVERVIEW OF MOBILE SOURCE AIR POLLUTION

10.1.1 Mobile Sources of Air Pollution

Mobile source air pollution refers to air pollution from moving vehicles/engines – anything that moves and burns fuel. These sources are divided into two categories by USEPA (2025a):

- On-road vehicles:
    - Passenger cars and trucks
    - Commercial trucks and buses
    - Motorcycles
- Non-road vehicles and engines:
    - Aircraft
    - Locomotives
    - Marine vessels
    - Recreation vehicles (snowmobiles, all-terrain vehicles, etc.)
    - Heavy equipment
    - Small engines and tools (such as lawnmowers)

10.1.2 Pollutants Emitted by Mobile Sources

The leading cause of air pollutant emissions from mobile sources is fuel combustion. Consequently, the main compounds of concern (COCs) are CO, CO_2, hydrocarbons (HC), NO_x, and SO_2. Emissions of particulate matter (PM) are also of concern, especially from diesel engines and some off-road vehicles and equipment using diesel fuel. Table 10.1 shows the emissions (in thousand tons) of CO, NO_x, VOCs, and SO_2 in the U.S. in 2024. The contributions of the transportation sector are highlighted, and they are responsible for 54%, 48%, 15%, and 1.7% of the total emissions of CO, NO_x, VOCs, and SO_2, respectively. As shown, transportation (highway and off-highway) is the primary source of CO emissions and contributes a significant portion of NO_x emissions (USEPA, 2025b).

The emission of SO_2 from the transportation sector is expected to be small because the sulfur content in most of the fuels used for mobile sources is low. Reductions of transportation emissions would help reduce the corresponding emissions of CO, NO_x, and VOCs. Interestingly, the "off-highway" emissions are equally important as the "highway" emissions. The SO_2 emission from the "off-highway transportation" is even larger than that from the "highway transportation", which may be because fuels

DOI: 10.1201/9781003532705-10

used in off-highway devices have larger sulfur contents. The USEPA has established a long list of compounds emitted by mobile sources (USEPA, 2006).

The percentages of emissions from mobile sources are even more significant in urban areas because of the more frequent use of automobiles. The emissions of NO_x and VOCs contribute to the formation of photochemical smog, which is a prevailing problem in many metropolitan areas.

In previous chapters, we have discussed the trends and sectors' contributions to emissions of VOCs, NO_x, and SO_2. This is an excellent place to discuss CO emissions (CO_2 emissions will be addressed in Chapter 12).

Table 10.2 provides a breakdown of CO emissions from different sectors in the U.S. in 2024. The total CO emission was about 45 million tons in 2024. As shown, the percentage emissions are 54.1%, 28.3%, 10.5%, and 7.1% from the sectors of transportation, miscellaneous, stationary fuel combustion, and industrial and other processes, respectively. Wildfire contributed approximately one-third of the CO emissions from the miscellaneous sector. For the transportation sector, highway vehicles contributed about 50%. USEPA (2025c) presents general information on the impacts of transportation on air pollution and climate change.

Figure 10.1 illustrates trends of annual CO emissions in the U.S. from 1970 to 2024. The annual CO emission has decreased significantly from >204 million tons in 1970 to about 45 million tons in 2024, a reduction of about 75%. The transportation sector is the primary contributor to this reduction. Most of the decrease in transportation can be attributed to the use of effective technology in areas such as engine control, the use

TABLE 10.1
Emissions (in 1,000 tons/yr) of CO, NO_x, VOCs, and SO_2 in the U.S. in 2024

	CO	NO_x	VOCs	SO_2
Stationary fuel combustion	4,730	2.263	629	897
Industrial and other process	3,168	1,150	6,950	621
Transportation – Highway	*12,322*	*1,745*	*896*	10
Transportation – Off-Highway	*11,983*	*1,578*	*990*	*19*
Miscellaneous	12,734	204	3,317	127
Total	44,937	6,940	12,782	1,674
Transportation/total	**54%**	**48%**	**15%**	**1.7%**

TABLE 10.2
CO Emissions from Different Source Categories in the U.S. in 2024

Source Category	**(1,000 tons)**	**(%)**
Stationary fuel combustion	**4,730**	**10.5**
Electric utility	409	0.9
Industrial	875	1.9
Others	3,446	7.7

(*Continued*)

TABLE 10.2 (*Continued*)
CO Emissions from Different Source Categories in the U.S. in 2024

Source Category	(1,000 tons)	(%)
Industrial and other processes	**3,168**	**7.1**
Chemical & allied product manufacturing	128	0.3
Metals processing	345	0.8
Petroleum & related industries	750	1.7
Other industrial processes	461	1.0
Solvent utilization	1	0.0
Storage & transport	6	0.0
Waste disposal & recycling	1,477	3.3
Transportation	**24,305**	**54.1**
Highway vehicles	12,322	27.4
Off-Highway	11,983	26.7
Miscellaneous	**12,734**	**28.3**
Wildfires	3,692	8.2
Others	9,042	20.1
Total	**44,937**	**100.0**

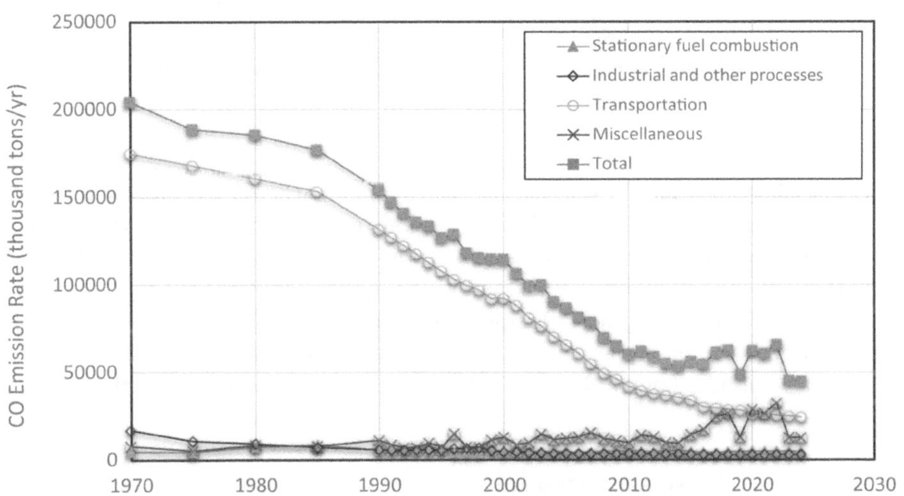

FIGURE 10.1 Trends of annual CO emissions in the U.S. (1970–2024).

of catalysts to reduce emissions from tailpipes, the use of improved or alternative fuels, and changes in communication patterns. However, transportation has become an indispensable part of our lifestyle. A balance between the reduction of emissions from the transportation sector and the quality of life may need to be achieved.

10.2 EMISSIONS FROM MOBILE SOURCES

10.2.1 Sources of Emissions from Automobiles

Motor vehicles are by far the most significant mobile sources of air pollution. Pollutants from motor vehicles are released into the atmosphere in three main areas of an automobile: (i) evaporative emissions, (ii) refueling losses, and (iii) exhaust (Figure 10.2). *Evaporative losses* from automobiles happen when fuel (i.e., gasoline, diesel) evaporates/escapes into the atmosphere in three ways. *Diurnal emissions* are due to increases in fuel evaporation as the temperature rises during the day, heating the fuel tank and venting its vapor. *Running losses* occur while the engine runs and everything (e.g., fuel lines, injectors, and tanks) gets hot, causing more petroleum vapor to form and leak. The fuel evaporation continues, even when the car is parked, until the engine reaches ambient temperatures. The *hot soak emissions* refer to the fuel evaporation from the engine and fuel system during this cooling-down period. Since fuel vapors are always present in fuel gas, *refueling losses* occur when these vapors are forced out when the tank is filled with liquid fuel (USEPA, 2003). The emissions from both evaporative losses and refueling losses should be the VOCs contained in the fuel.

The most significant emission from an automobile is the fuel combustion in its engine, which is vented through the exhaust pipe. This exhaust accounts for approximately 90% of the total emissions by an automobile. In addition to unburned HC, the exhaust also contains CO, CO_2, NO_x, and some additional hazardous compounds (e.g., acetaldehyde, formaldehyde, and 1,3-butadiene) that are not present in the fuel (USEPA, 2003). The extent of pollutant emissions depends on the design and operations of the engines.

10.2.2 Reciprocating Engines

An internal combustion (IC) engine is a type of engine where fuel combustion occurs within the engine itself, especially in a combustion chamber, converting chemical energy into mechanical energy. A *reciprocating engine* is a type of engine in which pistons reciprocate (i.e., move back and forth) inside cylinders, and it is a common configuration for IC engines. Reciprocating engines are used in cars or trucks. Combustion in reciprocating engines is not under steady state. A vehicle can have a 4-, 6-, or 8-cylinder engine. Figure 10.3 shows the typical sequence of the events of

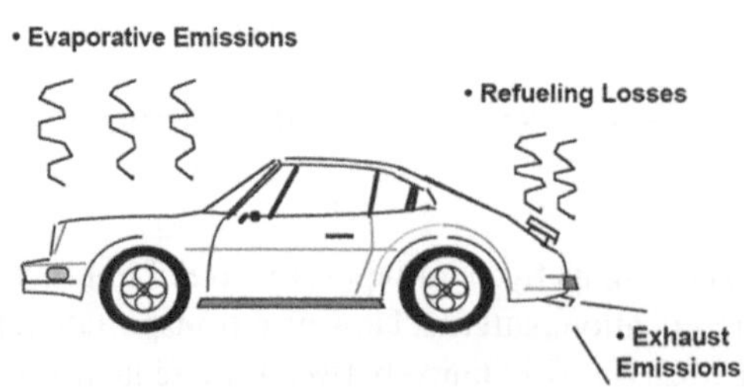

FIGURE 10.2 Emission sources of an automobile (USEPA, 1994).

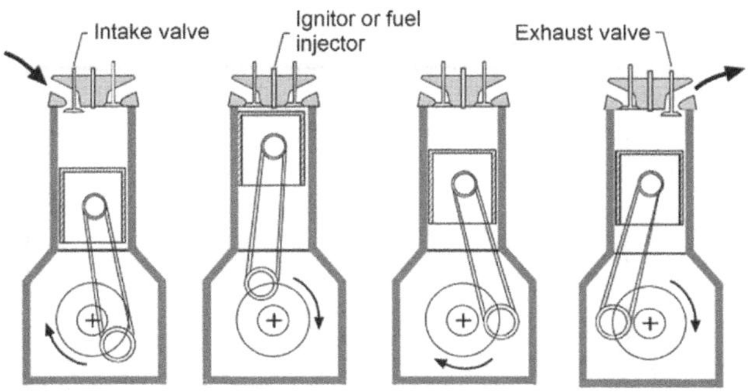

FIGURE 10.3 Operating sequence of a reciprocating engine (USEPA, 2012).

the classic four-stroke cycle (intake → compression → power → exhaust). First, the intake valve opens; the piston draws in an air–fuel mixture (the *intake stroke*). Then, the intake valve closes, and the mixture is compressed on the up-stroke (the *compression stroke*). Ignition occurs when the piston is near the top. Third, the *power stroke* follows as combustion drives the piston down. Lastly, the exhaust gases are pushed out on the final up-stroke (the *exhaust stroke*).

Fuel can get into the cylinder of a reciprocating engine in two ways: spark ignition (SI) and compression ignition. In SI, the air and fuel are premixed before being drawn into the cylinder on the down-stroke and ignited with an electric spark. It is a premixed flame. Diesel engines employ *compression ignition*, where air is drawn into the cylinder and the fuel oil is injected as the piston approaches the top. The fuel auto-ignites when the air gets super-hot due to the high compression. It is a diffusion flame.

10.2.3 Factors Affecting Pollutant Formation and Emissions

Figure 10.4 illustrates the hypothetical relationship between the air–fuel (A/F) ratio and emissions from an engine. The stoichiometric A/F ratio in the figure is around 16 (mass/mass). The rich burn refers to insufficient air to burn up all the fuel present. On the other hand, the lean burn refers to the air present exceeding the stoichiometric amount. An IC engine can operate in one of three modes: rich-burn, stoichiometric, and lean-burn, at different times. As shown, when an IC engine operates in the lean-burn mode, CO, NO_x, and *non-methane hydrocarbon* (*NMHC*) emissions are lower. The fuel economy is also better. However, the power will be relatively weaker. When more significant power is required (e.g., starting, passing, and idling), the IC engine will operate in the rich-burn mode. The fuel economy worsens in this mode, and the air pollutant emissions are more significant.

Carbon monoxide. CO production from engine combustion increases sharply with a decreasing A/F ratio because of the absence of sufficient oxygen to ultimately convert the HC fuel into CO_2. The extent of CO formation depends mainly on the A/F ratio. Using a catalyst control system, CO

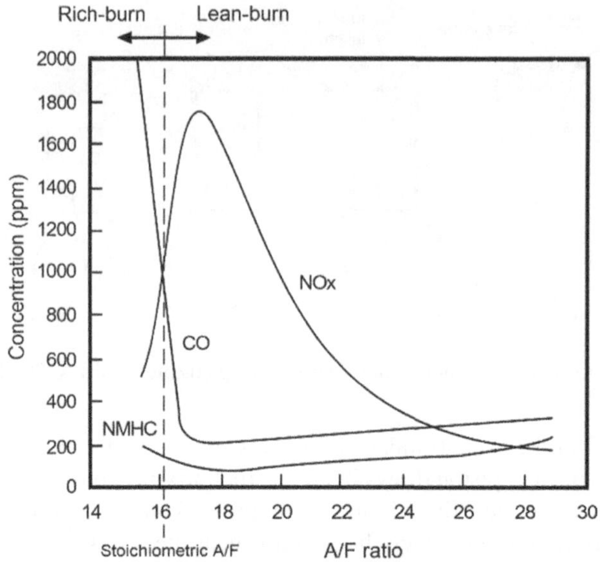

FIGURE 10.4 Emissions vs. A/F ratio (USEPA, 2012).

emissions in new SI engines are controlled by maintaining the combustion closer to the stoichiometric A/F ratio (USEPA, 2003).

Nitrogen oxides. As mentioned in Chapter 8, the formation of NO_x depends on the peak combustion temperature, the residence time of the air–fuel mixture in the flame, and the level of oxygen present in the combustion zone. SI engines can reach temperatures conducive to NO_x production. However, NO_x formation can be reduced by (i) operating in a lean-burn mode, (ii) retarding or delaying the spark, or (iii) lowering the engine compression ratio. Operating in the lean-burn mode would reduce the maximum combustion temperature to lower NO_x production. Delaying the spark is to let ignition occur late in the compression stroke, reducing the time available for producing higher temperatures. In the compression stroke, the piston moves upward, and the air–fuel mixture is compressed. The engine *compression ratio* is the ratio of the total volume of the combustion chamber when the piston is at the bottom of its stroke to the volume when the piston is at the top of its stroke. Lowering this ratio will produce lower combustion chamber temperatures (USEPA, 2003).

Hydrocarbons. Unburned HC are those fuel components that remain unoxidized after the combustion process. Factors contributing to incomplete combustion include (i) poor air–fuel mixing, (ii) inadequate combustion temperatures, and (iii) rapid quenching of flames. Quenching of flame can occur near cold cylinder walls, crevices, and gaps in the combustion chamber. Unburned HC emissions can result from the absorption of fuel components on carbon deposits and lubricating oil, as well as from liquid fuel (particularly during the cold start). The extent of HC emission depends mainly on the A/F ratio, cylinder compression ratio, engine speed, and spark timing. Approximately 1.5 to 2% of the gasoline fuel leaves an SI engine unburned (USEPA, 2003).

Particulate matter. Diesel engines have more potential to generate PM, and most PM is unburned carbon particles. The *soluble organic fraction* (*SOF*) refers to the portion of PM, particularly in diesel engine exhaust, that is composed of organic compounds soluble in organic solvents. The SOF consists of unburned HC that has condensed into liquid droplets or onto unburned carbon particles. While primarily associated with diesel engines, SOF can also be present in emissions from other combustion sources such as gasoline engines and industrial processes. In addition, burning fuel containing sulfur may generate sulfate particles. Reductions in PM emissions are being achieved with improvements in combustion and fuel systems, the introduction of electronic fuel systems, and increasing fuel injection pressures (USEPA, 2003).

10.3 CONTROL OF EMISSIONS FROM MOBILE SOURCES

10.3.1 Control of Evaporative and Fueling Losses

Figure 10.5 illustrates a *canister system*, filled with activated carbon, that is typically connected to the fuel tank and other parts of the fuel system to reduce evaporative emissions from a vehicle. When a car is parked, gas vapors from the engine and fuel tank flow into the canister. When the vehicle moves, the stored vapor in the canister is purged back into the engine intake manifold to be combusted.

Stage II vapor recovery systems in a gas station collect gasoline vapors forced out of the gas tank during fueling with a vapor recovery nozzle. This system complements *Stage I vapor recovery*, which captures vapors displaced from underground storage tanks (USTs) during fuel delivery. Stage II specifically targets the vapors emitted from the vehicle's fuel tank during refueling. There are flexible bellows around the nozzle spout or a port outside it. A vacuum is introduced to collect vapors entering the front end of the nozzle and transport the vapors back to a UST (Figure 10.6).

10.3.2 Control of Tailpipe Emissions

Modern automobiles have advanced electronic control units that manage ignition timing, fuel injection, and more to ensure optimal combustion and less pollution.

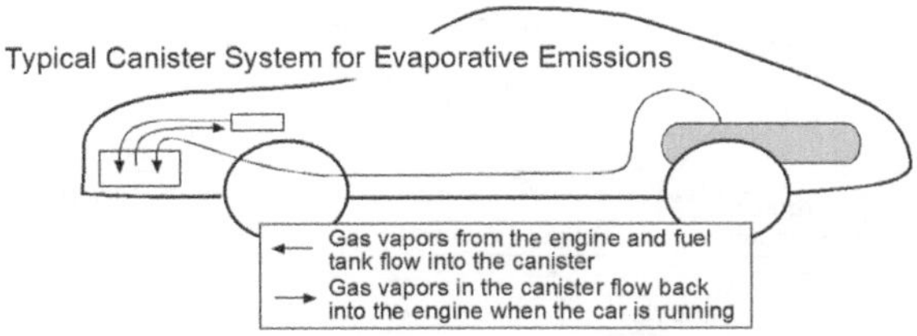

FIGURE 10.5 Canister to reduce evaporative losses from a vehicle (USEPA, 1994).

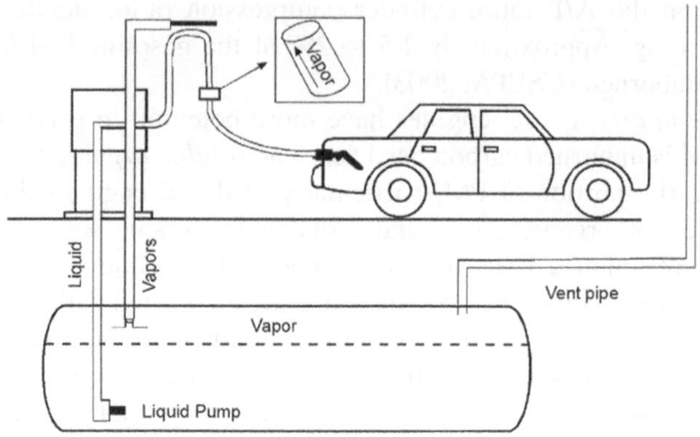

FIGURE 10.6 Stage II gasoline vapor recovery system (USEPA, 2003).

Figure 10.7 illustrates a control system to reduce tailpipe emissions. It generally includes a catalytic converter that hosts a "three-way" catalyst, a computer, and an oxygen sensor. The *three-way catalyst* serves three functions: (i) convert CO to CO_2, (ii) convert HC to CO_2 and H_2O, and (iii) reduce NO_x to N_2 and O_2. The oxygen sensor monitors the oxygen level in exhaust gases, and the readings will be fed back to the engine control system.

Diesel particulate filters (DPFs) trap and remove the PM from diesel exhaust. These filters trap soot and other particles, which are periodically burnt off through regeneration processes. Diesel oxidation catalysts reduce CO and HC emissions in diesel engines by oxidizing these pollutants into CO_2 and H_2O. In addition, selective catalytic reduction (SCR) is achieved by injecting urea-based fluid into the diesel engine exhaust to convert NO_x to N_2 (see Chapter 9 for SCR).

Optimizing engine parameters, such as ignition timing, air–fuel ratio, and boost pressure, can improve combustion efficiency and reduce emissions. Regular maintenance of engines and lubrication systems can help reduce fuel consumption and the formation of organic compounds during combustion.

10.3.3 Reformulated Gasoline

Reformulated gasoline (*RFG*) is blended to burn more cleanly than conventional gasoline. It is designed to produce fewer smog-forming and toxic pollutants in the

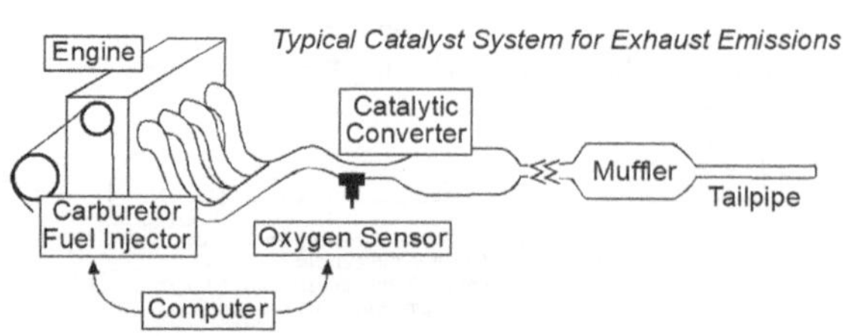

FIGURE 10.7 Exhaust control for an automobile (USEPA, 2003).

ambient air. RFG contains lower levels of VOCs and sulfur than traditional gasoline, and it often includes oxygenates to improve combustion efficiency and reduce emissions. RFG is required in cities with high smog levels and is optional elsewhere. The 1990 Clean Air Act Amendment mandated the RFG program.

Reid vapor pressure (RVP) is a standard measure of the volatility of gasoline; it measures the pressure exerted by a gasoline (or other volatile petroleum product) at 100°F (37.8°C) in a closed container. A higher RVP indicates greater volatility. The USEPA regulates the RVP of gasoline sold at retail stations during the summer ozone season (June 1 to September 15) to reduce evaporative emissions from gasoline.

Oxygenates are fuel additives that contain oxygen, usually in the form of alcohol or ether. They can enhance fuel combustion and thereby reduce exhaust emissions, especially CO and VOCs, and some also boost gasoline octane numbers. The CAA requires using oxygenated gasoline in areas where their winter-time CO levels exceed NAAQS. Note: the *octane number* is determined using standardized testing methods, and the tests simulate different engine conditions to evaluate fuel performance. Octane numbers typically range from 87 to over 100. Fuels with higher octane numbers allow engines to operate at higher compression ratios and improve efficiency and power.

In addition, the Mobile Source Air Toxics (MSAT) rules reduce emissions of HAPs, such as benzene, formaldehyde, 1,3-butadiene, and diesel particulate matter, from on-road and non-road vehicles. The Tier 2 Gasoline Sulfur program, finalized in 2000, is to reduce the sulfur content of gasoline by up to 90%. Lower sulfur levels would improve the efficiency of catalytic converters and other emission control systems, in addition to lowering SO_x emissions.

10.3.4 Alternative Fuels

Alternative fuels are energy sources for transportation that substitute traditional fossil fuels such as gasoline and diesel. According to USEPA (2025d), alternative fuels include gaseous fuels such as hydrogen; natural gas (i.e., compressed natural gas (CNG) and liquified natural gas (LNG)) and propane; alcohols such as ethanol, methanol, and butanol; vegetable- and waste-derived oils; and electricity. They may be used in a dedicated system that burns a single fuel or mixed with other fuels, including traditional gasoline or diesel, such as in hybrid-electric or flexible-fuel vehicles. *Flexible-fuel vehicles* are designed to run on more than one type of fuel, usually a blend of gasoline and ethanol. Many of these alternative fuels are renewable energy sources derived from natural processes that are continuously replenished, such as sunlight, wind, water, and organic matter. For example, ethanol is produced from the fermentation of sugars found in crops such as corn, sugarcane, and cellulosic biomass, and renewable natural gas is made by capturing methane from landfills, agricultural operations, wastewater treatment plants, and food-processing facilities. Using alternative fuels may or may not reduce emissions of air pollutants.

10.3.5 Other Measures

For a given automobile, the emissions are proportional to its distance traveled. Reducing vehicle usage should minimize pollutant emissions. It can be achieved by carpooling, using more public transportation, or telecommuting. However, having these alternatives available may often be challenging.

10.4 SUMMARY

1. The transportation sector accounts for 54%, 48%, 15%, and 1.7% of the total emissions of CO, NO_x, VOCs, and SO_2, respectively, in the U.S. in 2024.
2. Photochemical smog problems in metropolitan areas are often attributed to automobile pollutant emissions.
3. Many technical measures have been developed to reduce the formation and emission of pollutants from mobile sources.
4. There are USEPA programs on reformulated gasoline, RVP, winter oxygenates, gasoline sulfur, and mobile source air toxics to help reduce emissions from mobile sources.
5. Non-technical measures, such as carpooling and telecommuting, are also viable alternatives to cut mobile source emissions.

EXERCISE QUESTIONS

1. The USEPA estimated that the CO emission of average passenger cars was 20.9 g/mile (13.06 g/km) in 2000. In 2000, the USEPA started to implement more stringent emission standards for light-duty vehicles, and all the new cars and light trucks were required to meet the CO emission standard of 3.4 g/mile (2.1 g/km) by the end of 2007.
   a. Compute CO emission in g/mile (g/km) for an automobile traveling at 55 mi/h (88.5 km/h) on a highway. The exhaust flow rate is 300 ft^3/min (8.5 m^3/min), and the exhaust contains 0.08% of CO by volume after the catalytic converter (all based on 680°F (360°C) and 1 atmosphere).
   b. Will this vehicle meet the new CO emission standard?
2. Consider an urban area with a population of 150,000 and 50,000 vehicles, trafficking within a 100 km^2 (10 km × 10 km) area, with an average driving distance of 12 km during the 2 hours from 8 to 10 a.m. daily. Assuming that each vehicle emits 2.0 g/km of CO and the prevailing wind (4 m/s) blows perpendicular to this square area carrying a negligible concentration of CO, estimate the ambient CO concentration in that area during that period (the mixing height during that time of the day = 400 m) by using the area-source box model.

BIBLIOGRAPHY

USEPA. (1994). *Automobile Emissions: An Overview*, Fact Sheet OMS-5, EPA 400-F-92-007, Office of Mobile Sources, U.S. Environmental Protection Agency.

USEPA. (2003). *APTI 452: Principles and Practices of Air Pollution Control - Student Manual*, 3rd edition, Air Pollution Training Institute, United States Environmental Protection Agency, Research Triangle Park, NC.

USEPA. (2006). *Expanding and Updating the Master List of Compounds Emitted by Mobile Sources - Phase III*, EPA 420-R-06-005, Office of Transportation and Air Quality, U.S. Environmental Protection Agency.

USEPA. (2012). *APTI 427: Combustion Source Evaluation - Student Manual*, 3rd edition, prepared by ICES Ltd. for Air Pollution Training Institute, United States Environmental Protection Agency, Research Triangle Park, NC.

USEPA. (2025a). *Learn about How Mobile Source Pollution Affects Your Health.* https://www.epa.gov/mobile-source-pollution/learn-about-how-mobile-source-pollution-affects-your-health#mobile-sources, last updated 02/25/2025.

USEPA. (2025b). *Air Pollutant Emissions Trend Data*, https://www.epa.gov/air-emissions-inventories/air-pollutant-emissions-trends-data, last updated on 04/28/2025.

USEPA. (2025c). *Transportation, Air Pollution, and Climate Change*, https://www.epa.gov/transportation-air-pollution-and-climate-change/overview-air-pollution-transportation, last updated 04/23/2025.

USEPA. (2025d). *Alternative Fuels*, https://www.epa.gov/renewable-fuel-standard/alternative-fuels, last updated 06/24/2025.

11 Fugitive Emissions and Control

11.1 INTRODUCTION

11.1.1 Definition

The USEPA defines *fugitive emissions* as emissions of "those that could not reasonably pass through a stack, chimney, vent, or another functionally equivalent opening". The decision to consider an emission source as fugitive or non-fugitive is a factual determination made by the permitting authority.

Fugitive emissions come in many forms, including fine particles, aerosols, volatile organic compounds (VOCs), and greenhouse gases such as refrigerants and methane. Regarding air pollution control, fugitive emissions control focuses on VOCs and volatile hazardous air pollutants (VHAPs). Fugitive emissions of greenhouse gases (GHGs) are also gaining increasing attention (AGA, 2008; USEPA, 2025). According to Climate Chance (2018), the extraction and, to a lesser extent, the processing and transportation of fossil fuels are the primary sources of fugitive emissions in the energy sector.

For methane, VOCs, and VHAPs, the sources of fugitive emissions include equipment leaks, evaporation, and process vents. The USEPA has determined that leaking equipment is the largest source of emissions from petroleum refineries and chemical manufacturing facilities. Approximately 70,400 and 9,400 tons/year of VOCs and VHAPs have been emitted from equipment leaks. They exceed emissions from storage vessels, transfer operations, or process vents (USEPA, 2007). This chapter discusses the significant sources of equipment leaks, the detection of the leaks, an estimate of fugitive emissions, and alternatives to reduce the fugitive emissions from equipment leaks.

11.1.2 Regulations Affecting Fugitive Emissions

Regulations relevant to fugitive emissions from equipment leaks can be found in various USEPA regulations, mainly in NSPS, NESHAPs, and hazardous organic NESHAPs (HON). Under these regulations, multiple industries need to control their fugitive VOC emissions. For example, fugitive emissions from the Synthetic Organic Chemical Manufacturing Industry (SOCMI) accounted for >35% of its total VOC emissions; the HON regulations were implemented to control them (USEPA, 2000).

Based on vapor pressure, the USEPA has classified VOCs into four categories (see Table 11.1). Since the vapor pressure is a function of temperature, the temperature of a process stream will affect the category under which a VOC will belong. *Equipment in VOC service* refers to equipment that contains/contacts a process fluid with at least 10% VOC by weight. If the fluid is in the gaseous or vapor state at operating conditions, the equipment is *in gas/vapor service* (USEPA, 2000).

DOI: 10.1201/9781003532705-11

TABLE 11.1
Categories of VOCs

Category	Vapor Pressure
Very volatile	>380 mm-Hg
Volatile	0.1–380 mm-Hg
Semi-volatile	0.1–10^{-7} mm-Hg
Non-volatile	<10^{-7} mm-Hg

Source: USEPA (2000)

11.1.3 Sources of Fugitive Emissions

A typical refinery or chemical plant can emit 600 to 700 tons of VOCs annually from leaking equipment, mainly because the equipment is pressurized. The following components are the primary sources of emissions subject to leak regulations (USEPA, 2007):

- *Pumps.* They are used to transfer fluid. Leaks typically occur at the seal.
- *Valves.* They are used to control the flow of fluid. Leaks usually occur at the stem or gland area of the valve, and are caused by a failure of the valve packing or O-ring.
- *Connectors.* They are used to join piping and process equipment. Common ones include flanges and fittings, and gaskets/blinds are usually installed between the flanges to provide the seal. Leaks are commonly caused by gasket failure and improper installation of the flange.
- *Sampling components.* They are used to obtain samples within a process. Leaks commonly occur when the sampling line is purged for sampling.
- *Compressors.* They increase fluid pressure and provide motive force. Leaks most often come from the seals.
- *Pressure relief devices* protect equipment from exceeding the maximum allowable working pressure. Typical examples are rupture disks and pressure relief valves (PRVs). Common causes for leaks from PRVs include improper valve seating, operating too close to the set point, or worn/damaged seals, while that for rupture disks is improper installation of disk gaskets.
- *Open-ended lines (OELs).* OELs are pipes or hoses open to the atmosphere. The leaks occur at the point where the line opens to the atmosphere.

Previous USEPA studies found that valves and connectors accounted for >90% of the emissions from leaking equipment. New information suggests that OELs and sampling connections may account for 5 to 10% of the total emissions (USEPA, 2007).

11.1.4 Control of Emissions from Equipment Leaks

Since there are different causes for equipment leaks, no single emission control technique applies to all equipment leaks. It requires a combination of the best engineering practices, material selection, operation and maintenance (O&M), and monitoring.

The general approaches can be categorized into: (i) equipment practices and (ii) work practices.

Equipment practices involve the use of equipment to eliminate fugitive emissions. Examples include the uses of (i) leak-less (or low-emission) technology for pumps and valves; (ii) plugs and blinds for OELs; (iii) soft seats for PRVs; (iv) closed-loop sampling; and (v) enclosure of the seal area/vent to a combustion control device (USEPA, 1994). Others include uses of (i) specially designed valves with a low-emission packing system, (ii) gaskets with low permeability and high chemical resistance, and (iii) double mechanical seals with barrier fluids for pumps and compressors. In addition, the selected equipment should be adequately rated for the operating conditions, and corrosion-resistant materials should be used in a corrosive environment.

Proper installation and assembly of equipment will also minimize future potential for fugitive emissions. Consider using welding instead of flanges to reduce potential leak points. Align the pipes, flanges, and fittings properly to avoid unnecessary mechanical stress, which could lead to leaks. Applying the correct bolt torque and tightening sequence on flanges during the application would have a more uniform gasket load to prevent leaks.

Work practices refer to plans and procedures taken to reduce fugitive emissions. The primary work practice is *leak detection and repair* (*LDAR*) of potential emission sources. Many facilities are required by NSPS, NESHAPs, SIPs, the Resource Conservation and Recovery Act (RCRA), and other state or local requirements to implement LDAR programs. In addition, preventive maintenance and proper operator training are also essential, in addition to implementing an LDAR program. More discussions on LDAR are provided in the next section.

11.2 LDAR PROGRAM

11.2.1 Objectives of LDAR Programs

A well-designed and implemented site LDAR program would have the following benefits (USEPA, 2007):

- Reducing losses of product/material
- Increasing safety for facility personnel
- Decreasing exposure for the neighboring community
- Potentially reducing emission fees
- Avoiding enforcement actions

11.2.2 Elements of an LDAR Program

The following are the five elements of an LDAR program (USEPA, 2007):

- Identifying components
- Leak definition
- Monitoring components
- Repairing leaking components
- Record keeping

Each regulated component should be assigned a unique identification (ID) number, recorded in a log, and located on the piping and instrumentation diagrams (P&ID) or process flow diagrams. Monitoring intervals vary according to the applicable regulation and the component type. The leak detection can be made by using (i) optical gas imaging, such as infrared (IR) cameras, (ii) sniffers or gas analyzers such as photoionization detectors (PID), flame ionization detectors (FID), and volatile organic analyzers (VOA), and (iii) soap bubble test or ultrasonic leak detectors. They can be as frequent as weekly. The primary method for monitoring and detecting leaking equipment components is Federal Reference Method 21 (FRM 21), found in Appendix A of 40 CFR 60.

A *leak* is detected whenever the measured concentration by an approved portable analyzer (see section below) exceeds the threshold standard (i.e., *leak definition*) for the applicable regulations. Leak definitions vary by regulation, component type, service type (e.g., light liquid, heavy liquid, gas/vapor), and monitoring interval. Most NSPS have a liquid definition of 10,000 ppm, while many NESHAPs use 500 or 1,000 ppm as the leak definition (USEPA, 2007, 2008). A leak can also be defined based on visual inspections or observations.

Leaking components should be repaired as soon as practicable, but not later than a specified number of calendar days after the leak is detected. Detailed records for LDAR should be maintained and updated.

11.2.3 Federal Reference Method 21

Federal Reference Method (FRM) 21 specifies a hand-held instrument to determine VOC leaks from process equipment. It defines analyzer performance specifications but does not recommend specific analyzers or manufacturers. The specifications include monitor response, measurement range, scale resolution, response time, safety, probe dimensions, response factor, and accuracy (USEPA, 2017).

11.3 PROTOCOL FOR EQUIPMENT LEAK EMISSION ESTIMATES

There are many equipment components at a specific facility. The quantity of fugitive emissions from each leaking component is impractical, if not impossible. Using emission factors (EFs) and activity factors to estimate fugitive emissions from a specific facility or an industry sector is a commonly adopted approach. An *EF* expresses the amount of pollutant emitted per unit of *activity factor*, which is how many devices (e.g., number of valves, pumps, PRVs) or how much or how often the equipment operates, e.g., types of furnaces, rate and types of fuels used, and hours of operations (USEPA, 1986).

$$\text{Emissions} = \text{Activity Factor} \times \text{Emission Factor} \tag{11.1}$$

The author of this book conducted a study in 2012 to develop EFs for estimating fugitive emissions from California natural gas industry equipment. The project was funded by the California Energy Commission (CEC). Approaches and findings of the study were extracted and used here to illustrate types, derivation, and uses of EFs for fugitive emissions from equipment leaks. Details can be found in CEC (2012) and Kuo et al. (2015).

11.3.1 Types of Emission Factors

Several EFs and correlation equations have been generated to estimate fugitive emissions from equipment leaks (USEPA, 1995). These EFs can be categorized into (i) facility, (ii) equipment, and (iii) component. Using a *facility-level EF* to estimate fugitive emissions is the most straightforward approach, requiring only information on the type and capacity of a specific facility. The *equipment-level EF* approach is one level up and requires an accurate count of all significant pieces of equipment. The *component-level* EFs can be further categorized into four groups, in the order of increasing data requirements and accuracy for emission estimates: (i) average EF, (ii) screening ranges, (iii) correlation equation, and (iv) unit-specific correlation (USEPA, 1995; API, 2009).

The component-level average EFs for emission estimates need only the population data of components at a given facility. In contrast, the other three component-level EF approaches require screening data from an LDAR program. Screening data are collected using a portable monitoring instrument to sample air from potential leak interfaces on individual equipment components. A *screening value* (SV) measures the concentration of leaking compounds in the ambient air, typically in parts per million by volume (ppm), by a portable instrument. The instrument generally has a lower limit (the detection limit) and a higher limit (the *pegged SV*). The screening-ranges approach divides the SVs into two or more ranges, assigning an EF to each range. In older regulations, 10,000 ppm was often used as the leak definition (USEPA, 1996). The approach is also known as the "leak/no leak" approach, with the "leakers" having SVs ≥10,000 ppm and "non-leakers" having SVs <10,000 ppm. The correlation-equation approach uses the SV of a leaking component and established correlation equations between SVs and leak rates for emission estimates. Each correlation equation is specific to the component type. Using this approach, the emission rate for each leaking component is calculated individually, and the facility needs to keep track of the findings from the LDAR activities. The unit-specific correlation approach is similar, except that the correlations are unit-specific and site-specific. This approach can be expensive and would seldom justify developing unit-specific correlations to support estimates for GHG emissions (API, 2009).

11.3.2 Derivations of EFs

The portable detection devices used in the 2012 CEC study had a maximum detection limit of 50,000 ppm. The data sets were divided into four ranges: 0 to 999, 1,000 to 9,999, 10,000 to 49,999, and ≥50,000 ppm. The first group, "0 to 999 ppm", was designated as such because all the measured leak rates in this group were below the detection limit of 0.01 ft^3/min (Note: 1 ft^3/min of CH_4 = 59.98 lb/d = 1.316 kg/h = 9.9 06 ton/yr). The second group, "1,000 to 9,999 ppm", reflects the typical leak definition, 10,000 ppm, in earlier regulations. The third group (10,000 to 49,999 ppm) and the last group (≥50,000 ppm) were chosen to take advantage of the maximum detection value (50,000 ppm) of the portable detection devices used in the study.

The screening-range EFs are summarized in Table 11.2. As shown, the geometric means of the leak rates increase with SVs.

Table 11.3 shows the ranges of component leak rates along with the pegged EFs for ≥10,000 and ≥50,000 ppm. The pegged emission rate is the mass emission rate with an SV that is above the maximum detection limit of a portable screening device. The pegged EFs can be similar or different based on the detection limits chosen.

The component-level correlation-equation approach is often used to estimate the emission rate from a leaking component using a corresponding SV, especially for facilities with LDAR programs.

The correlation equations are commonly expressed as:

$$LR_{\text{methane}} = a \times (SV)^b \tag{11.2}$$

where LR_{methane} is the methane leak rate (e.g., in ft^3/min), a and b are constants developed from the correlation fitting, and SV is the screening value (e.g., in ppm).

Table 11.4 shows the correlation equations obtained for leak rate in ft^3/min for different components tested from the 2012 CEC study. The calculated leak rates for the same SV differ for different component types for 100 ppm ≤ SV ≤ 49,999 ppm.

TABLE 11.2
Leak Rates (in ft^3/min) vs. SVs

SV (ppm)	100 to 999	1,000 to 9,999	10,000 to 49,999	≥50,000
Flanges	0.005	0.0067	0.0293	0.0747
Manual valves	0.005	0.005	0.0136	0.0197
OELs	—	0.005	0.005	0.2874
Seals	0.005	0.0527	0.1217	0.3907
Threaded connections	0.005	0.0068	0.0143	0.0168

Source: Kuo et al. (2015)

TABLE 11.3
Pegged EFs in lb CH_4/d

	≥10,000 ppm	≥ 50,000 ppm
Flanges	2.81	4.48
Manual valves	1.01	1.18
OELs	7.14	17.24
Others	6.01	6.94
Seals	21.79	23.43
Threaded connections	0.93	1.01

Source: Kuo et al. (2015).

A *wellhead* is the uppermost part of a natural gas well located at production or storage facilities. For each wellhead surveyed, all components were screened from the ground up. The number of wellheads surveyed was 128 and 44 in the production and storage sectors, respectively. Table 11.5 shows the component-level average EFs of wellheads, which were calculated by using the leaking percentages and the mean leak rates from the field survey.

To derive a fugitive EF for a wellhead as a whole, the EF of each component was first multiplied by the average number of that component per wellhead. The multiplication products of all components were then summed to become the average EF of a wellhead. The wellhead EFs derived from the 2012 CEC study are shown in Table 11.6. As shown, the EFs of the same type of equipment in different sectors may differ.

TABLE 11.4
Correlation Equations for Leak Rate in ft³/min

	Correlation Equation	R^2
Flanges	$LR_{methane} = 0.00004 \times (SV)^{0.6549}$	0.4277
Manual valves	$LR_{methane} = 0.0006 \times (SV)^{0.2811}$	0.1204
Seals	$LR_{methane} = 0.0001 \times (SV)^{0.6816}$	0.4504
Threaded connections	$LR_{methane} = 0.0006 \times (SV)^{0.3031}$	0.1150

Source: Kuo et al. (2015).

TABLE 11.5
Component-Level EFs (in Tonne CH_4/yr/Component) of Wellheads

	Production	Storage
Flanges	2.10×10^{-3}	—
Manual valves	5.85×10^{-4}	9.09×10^{-3}
OELs	—	1.04×10^{-2}
Seals	5.87×10^{-3}	—
Threaded connections	6.20×10^{-4}	1.27×10^{-3}

Source: Kuo et al. (2015).

TABLE 11.6
Equipment-Level EFs (in Tonne CH_4/yr) for Wellheads

Production	Storage
3.83×10^{-2}	3.49×10^{-1}

Source: Kuo et al. (2015).

11.3.3 Uses of EFs

Example 11.1 below illustrates the uses of EFs to estimate fugitive emissions from equipment.

Example 11.1: Use of Emission Factors

A gas company has ten identical natural gas production wells. Each wellhead has five flanges, eight manual valves, one seal, and 30 thread connections. Use the EFs in Tables 11.5 and 11.6 to estimate the annual fugitive emissions from these ten wellheads.

Solution:

a. Use the component-level EF in Table 11.5 to estimate the fugitive emission from each wellhead

$$= (2.10\times10^{-3})(5) + (5.85\times10^{-4})(8) + (5.87\times10^{-3})(1) + (6.20\times10^{-4})(30)$$

$$= 3.97\times10^{-2}\,\text{tonne } CH_4/\text{yr/wellhead}$$

$$\text{Fugitive emissions from 10 wellheads} = (3.97\times10^{-2})(10)$$
$$= 0.397 \text{ tonne } CH_4/\text{yr/wellhead}$$

b. Use the equipment-level EF in Table 11.6 to estimate the fugitive emissions from 10 wellheads

$$= (3.83\times10^{-2})(10)$$
$$= 0.383 \text{ tonne } CH_4/\text{yr/wellhead}$$

Discussion: Estimates from the two approaches are similar because the component counts are very close to the average counts used in deriving the wellhead EF.

11.4 SUMMARY

1. Fugitive VOC emissions from petroleum refineries and chemical manufacturing facilities can be significant.
2. The sources of these fugitive emissions are mainly leaks from equipment and components of pressurized systems.
3. Control of fugitive emissions can be achieved through equipment practices and work practices.
4. A LDAR program is a good work practice to manage and control fugitive emissions from equipment leaks.
5. EFs are commonly used to estimate fugitive emissions from components, equipment, facilities, or sectors for the emission inventory.
6. Fugitive emissions of greenhouse gases are gaining increasing attention.

EXERCISE QUESTIONS

1. EFs for particulate matter (PM) from multiple hearth sewage sludge incinerators are 10 and 4 lb/ton sludge without and with a cyclone for particulate removal (USEPA AP-42). A site burns 50 tons of sludge daily and employs a cyclone system to remove PM from the flue gas before discharge to the atmosphere.
   a. What is the daily PM emission rate using the USEPA's EF value (in lb/day)?
   b. Based on these two EF values, what is the assumed PM removal by cyclone (in %)?

BIBLIOGRAPHY

American Gas Association. (2008). *Greenhouse Gas Emissions Estimation Methodologies, Procedures, and Guidelines for the Natural Gas Distribution Sector*, a report prepared by Innovative Environmental Solutions, Inc. for the American Gas Association.

American Petroleum Institute. (2009). *Compendium of Greenhouse Gas Emissions Methodologies for the Oil and Gas Industry*, American Petroleum Institute.

CEC. (2012). *Estimation of Methane Emissions from the California Natural Gas System*, CEC-500-2014-072, a report prepared by Kuo, J. for California Energy Commission, Sacramento, CA.

Climate Chance. (2018). *Fugitive Emissions: A Blind Spot in the Fight against Climate Change*, Climate Chance, Paris, France.

Kuo, J.; Hicks, T.C.; Drake, B.; Chan, T.F. (2015). "*Estimation of Methane Emission for California Natural Gas Industry*," *J. Air Waste Manag. Assoc.* 65(7), 844–55.

USEPA. (1986). *Emission Factors for Equipment Leaks of VOC and HAP*, EPA/450/3-86/002, United States Environmental Protection Agency, Research Triangle Park, NC.

USEPA. (1994). *Control Techniques for Fugitive VOC Emissions from Chemical Process Facilities*, EPA/625/R-93/005, Office of Research and Development, United States Environmental Protection Agency, Washington, DC.

USEPA. (1995). *Protocol for Equipment Leak Emission Estimates*, EPA/453/R-95-017, United States Environmental Protection Agency, Research Triangle Park, NC.

USEPA. (1996). *Methane Emissions from the Natural Gas Industry*, prepared by Gas Research Institute (GRI), EPA/600/R-96/080, United States Environmental Protection Agency.

USEPA. (2000). *APTI SI 380: Introduction to Fugitive Emissions Monitoring*, prepared by North Carolina State University for the United States Environmental Protection Agency.

USEPA. (2007). *Leak Detection and Repair - A Best Practices Guide*, Office of Compliance, United States Environmental Protection Agency, Washington, DC.

USEPA. (2008). *APTI 435: Atmospheric Sampling Course – Student Manual*, 5th edition, prepared by Tidewater Operations Center of C^2 Technologies for Air Pollution Training Institute, United States Environmental Protection Agency, Research Triangle Park, NC.

USEPA. (2017). *Method 21- Volatile Organic Compound Leaks*, United States Environmental Protection Agency, https://www.epa.gov/emc/method-21-volatile-organic-compound-leaks

USEPA. (2025). *Fugitive Emissions - New Source Review Permitting*, https://www.epa.gov/nsr/fugitive-emissions, last updated 04/02/2025.

12 Ozone, Acid Rain, and Greenhouse Gases

12.1 PHOTOCHEMICAL SMOG

Photochemical reactions start with the absorption of a photon of energy ($h\nu$) by an atom, molecule, ion, or free radical. The equation below illustrates the first step of a photochemical reaction of a species (A):

$$A + h\nu \rightarrow A^{*} \tag{12.1}$$

where A^* is the excited state of species A. This excited species may reduce its energy level through dissociation, fluorescence, reaction with another species, or collisional deactivation by transferring energy to another energy-absorbing species (USEPA, 2002). [Note: *Fluorescence* is a physical phenomenon where a substance absorbs electromagnetic radiation (e.g., light) at one wavelength of higher energy (e.g., UV light) and emits light at a longer wavelength (e.g., visible light)].

The energy for atmospheric photochemical reactions comes from the sun. Light intensity is a critical parameter for photochemical reactions. The wavelength of visible light is from ~390 to 700 nm. Plank's equation relates the energy of a photon (E) to its wavelength (λ), or frequency (ν), as follows:

$$E = h\nu = h\left(\frac{C}{\lambda}\right) \tag{12.2}$$

where h is the Planck's constant (6.626×10^{-34} J·s) and C is the speed of light (3.00×10^{8} m/s). As shown, a photon with a shorter wavelength will have a larger frequency and a larger energy.

For the photochemical reactions in the troposphere, the light wavelengths of interest range from approximately 280 to 730 nm. This is because (i) ozone in the stratosphere blocks out most UV light shorter than 280 nm, and (ii) light with a wavelength greater than 730 nm does not participate in the photochemical reactions of interest. In the U.S., the maximum noon-time light intensity does not vary significantly with latitude during summer months. In the 300 to 400 nm range, the maximum light intensity is ~2×10^{12} photons/m²/s, which remains near this value for 4 to 6 hours. In contrast, winter varies from 0.7×10^{12} to 1.5×10^{12} photons/m²/s, depending on the latitude, and the duration near this maximum is reduced to 2 to 4 hours. Clouds or aerosols will reduce these maximum values (USEPA, 2002).

DOI: 10.1201/9781003532705-12

12.1.1 Basic Photochemical Cycle

Figure 12.1 illustrates concentration profiles of NO, NO_2, hydrocarbons (HC), O_3, and reaction products in ambient air during morning rush hours resulting from automobile emissions. As shown, the times of the concentration peaks are in the following increasing order: NO, NO_2, O_3, and reaction products. Below is a simplified explanation of the chemistry of this smog formation (USEPA, 2012).

Nitric oxide (NO) emitted from mobile (and stationary) combustion sources is rapidly converted to NO_2 as follows:

$$NO + \frac{1}{2}O_2 \rightarrow NO_2 \tag{12.3}$$

Nitrogen dioxide (NO_2), those newly formed and those directly from the tailpipes, can dissociate by sunlight photolysis to form NO and atomic oxygen (O) as follows:

$$NO_2 + hv \rightarrow NO + O \tag{12.4}$$

The atomic oxygen produced can then react rapidly with oxygen to form ozone, with various hydrocarbons working as a *catalyst* (i.e., a substance that speeds up the chemical reaction rate without being consumed in the process), as follows:

$$O + O_2 \xrightarrow{\text{Hydrocarbons}} O_3 \tag{12.5}$$

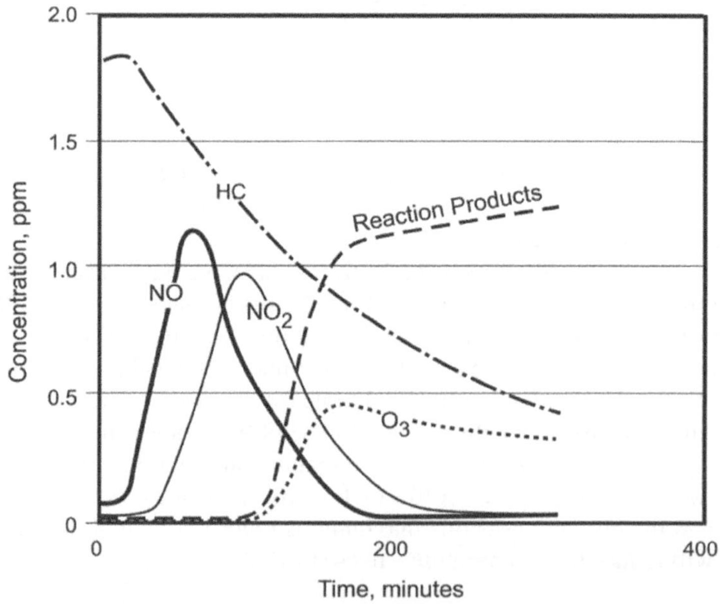

FIGURE 12.1 A typical photochemical profile (USEPA, 2014).

The primary source of these hydrocarbons is the VOCs present in the atmosphere. The hydrocarbons participate in the reaction by absorbing excess vibrational energy and stabilizing the ozone molecule.

The ozone formed from Eq. 12.5 can react with NO to regenerate NO_2 as follows:

$$O_3 + NO \rightarrow NO_2 + O_2 \tag{12.6}$$

Using the reactions shown in Eqs. 12.4 to 12.6, the steady-state ozone concentration can be calculated as a function of the initial NO_2 concentration. At a typical maximum NO_2 concentration of 0.1 ppm in urban areas, the calculated O_3 concentration is 0.027 ppm. Even using a NO_2 concentration ten times higher (i.e., one ppm), the calculated O_3 concentration is only 0.095 ppm, much smaller than the 0.5 ppm measured in urban areas. It implies that other reaction mechanisms should also exist to regenerate NO_2 from NO without the consumption of ozone, as shown in Eq. 12.6. As will be displayed in the subsequent section, that is the role that VOCs play in photochemical reactions (USEPA, 2002).

Example 12.1 Energy of a photon

Light with a wavelength of 500 nm is green. Determine the frequency and energy of this green light photon.

Solution:

a. From Eq. 12.2,

$$\nu = \left(\frac{C}{\lambda}\right) = \left(\frac{3.00 \times 10^8 \, \text{m/s}}{500 \times 10^{-9} \, \text{m}}\right) = 6.0 \times 10^{14} \, \frac{1}{\text{s}}$$

b. From Eq. 12.2 again,

$$h\nu = \left(6.626 \times 10^{-34} \, \text{J} \cdot \text{s}\right)\left(6.0 \times 10^{14} \, 1/1\right) = 3.97 \times 10^{-19} \, \text{J/photon}$$

12.1.2 Roles of VOCs in Photochemical Smog

Under radiation of light shorter than ~315 nm, ozone can dissociate into an excited oxygen atom (O^*) and an oxygen molecule:

$$O_3 + h\nu \rightarrow O^* + O_2 \tag{12.7}$$

Most of the time, the excited O* releases excess energy and returns to its ground state by colliding with N_2 or O_2. However, it occasionally will collide with water molecules to produce two hydroxyl radicals (OH):

$$O^* + 2H_2O \rightarrow 2 \cdot OH \tag{12.8}$$

Hydroxyl radicals play a crucial role in atmospheric chemistry. Unlike many radicals formed from organic molecules, they do not react with oxygen. Instead, they react with most other atmospheric compounds, including CO and hydrocarbons (RH). In the notation of RH for hydrocarbons, R represents the alkyl group. The hydroxyl radical will react with hydrocarbons to yield an alkyl radical (R) as follows:

$$\bullet OH + RH \rightarrow R\bullet + H_2O \tag{12.9}$$

The alkyl radical will then react with oxygen to yield peroxyalkyl radicals ($RO_2\bullet$):

$$R\bullet + O_2 \rightarrow RO_2\bullet \tag{12.10}$$

The peroxyalkyl radical can react rapidly with NO to form NO_2 and alkoxy radical (RO) or yield alkyl nitrates ($RONO_2$) as follows:

$$RO_2\bullet + NO \rightarrow NO_2 + RO\bullet \tag{12.11}$$

$$RO_2\bullet + NO \rightarrow RONO_2 \tag{12.12}$$

One of the alkyl nitrates potentially formed in the above equation is peroxyacetyl nitrate ($CH_3COOONO_2$), the most common one in the *peroxyacyl nitrates* (PAN) family, which has a general formula of $RC(O)OONO_2$, in which R is an alkyl group, C(O) is a carbonyl group, and $OONO_2$ is a peroxy nitrate group. Alkoxy radicals can react with oxygen to produce hydroperoxyl radicals (HO_2) and carbonyl compounds (RCHO):

$$RO\bullet + O_2 \rightarrow HO_2\bullet + RCHO \tag{12.13}$$

Finally, the hydroperoxyl radicals can react with NO to regenerate hydroxyl radicals and complete the cycle (USEPA, 2002):

$$HO_2\bullet + NO \rightarrow NO_2 + \bullet OH \tag{12.14}$$

The participation of VOCs in the reactions described in Eq. 12.14 consumes NO. It makes it less available for ozone destruction, as shown in Eq. 12.6. Consequently, this series of responses causes the build-up of ground-level ozone and harmful reaction products.

The hydroxyl radical can also react with NO_2 to form nitric acid as follows:

$$\bullet OH + NO_2 \rightarrow HNO_3 \tag{12.15}$$

In summary, this series of reactions includes the conversion of NO to NO_2, forming a variety of nitrogen-containing species (e.g., nitric acid and PANs), and accumulating ozone (USEPA, 2002).

12.2 OZONE

Ozone (O_3) molecules occur in the Earth's upper atmosphere and at the ground level. Those in the stratosphere are considered good, while those at the ground level are unfavorable.

12.2.1 Tropospheric Ozone

As discussed earlier, tropospheric ozone is not a primary air pollutant but is created by photochemical reactions among VOCs, NO_x, and sunlight. It is the main component of the photochemical smog. Ozone is most likely to reach unhealthy levels on hot sunny days in urban environments, but high-level ozone concentrations have also been observed in cold months. Ozone can also travel long distances by wind to elevate ozone levels in rural areas.

Ozone is one of the six criteria pollutants under CAA. The ozone pollution control actions include vehicle and transportation standards, regional haze and visibility rules, and routine reviews of NAAQS at the federal level, SIPs, and other local regulations.

12.2.2 Stratospheric Ozone

The stratospheric zone (the "good ozone") occurs naturally in the upper atmosphere, 6 to 30 miles (10 to 50 km) above the Earth's surface. It forms a protective layer that shields us from harmful ultraviolet (UV) rays from the sun. As a natural process, ozone molecules are constantly formed and destroyed in the stratosphere; the total amount has remained relatively constant recently.

However, scientific evidence has shown that the ozone layer has been depleted beyond the natural process since the early 1970s. Manmade chemicals are gradually destroying this beneficial ozone. At locations where the protective ozone layer has been significantly depleted (e.g., the North or South Pole), it is called a "hole" in the ozone layer. The problem is often referred to as *ozone depletion*. This allows more UV radiation to reach the Earth's surface and increases skin cancer, cataracts and other health problems. Elevated levels of UV radiation can also cause detrimental ecological effects, such as stressing the productivity of marine phytoplankton, an essential component of the oceanic food web.

Ozone can be destroyed by radicals such as hydroxyl (OH), nitric oxide ($NO^•$), chlorine ($Cl^•$), and bromine ($Br^•$). The hydroxyl and nitric oxide radicals naturally occur in the stratosphere, while chlorine and bromine radicals have significantly increased due to human activities. *Ozone-depleting substances* (ODS) can travel to the stratosphere due to their relative inertness in the troposphere and release chlorine or bromine radicals by solar UV radiation to destroy ozone. ODS that may release chlorine radicals include chlorofluorocarbons (CFCs), hydrochlorofluorocarbons (HCFCs), carbon tetrachloride (CCl_4), and 1,1,1-trichloroethane ($C_2H_3Cl_3$). Those that can release bromine radicals include bromofluorocarbons and bromomethane (CH_3Br). The discussion below uses trichlorofluoromethane ($CFCl_3$, CFC-11) and simplified reactions to illustrate ozone depletion by ODS. A variety of catalytic reaction cycles are occurring. $CFCl_3$ can be broken down by solar UV radiation to release chlorine radicals as follows:

$$CFCl_3 + \text{solar UV radiation} \rightarrow Cl\bullet + \bullet CFCl_2 \tag{12.16}$$

The formed chlorine radical can destroy an ozone molecule and form a chlorine monoxide radical (ClO):

$$Cl\bullet + O_3 \rightarrow ClO\bullet + O_2 \tag{12.17}$$

ClO radical can then react with atomic oxygen (O) to form a chlorine radical and an oxygen molecule, as follows:

$$ClO\bullet + O \rightarrow Cl\bullet + O_2 \tag{12.18}$$

The net reaction between the two reactions shown immediately above is:

$$O_3 + O \rightarrow 2O_2 \tag{12.19}$$

The chlorine radical works as a catalyst for ozone destruction because it is reformed each time the reaction cycle is completed, resulting in destruction of an ozone molecule. In this way, one chlorine atom/radical can participate in many cycles to destroy many ozone molecules before it is removed from the stratosphere. With the presence of these substances in the stratosphere, ozone can be destroyed more quickly than it is naturally created.

ODS are regulated as Class I or II controlled substances in the U.S. *Class I ODS*, including CFCs, halons, 111-TCA, CCl_4, and CH_3Br, have ozone depletion potentials ≥ 0.2 and have been completely phased out in the U.S, with a few exceptions [Note: *halons* are a group of chemicals containing, C, Br, Cl, and/or F, and they were once used in fire suppression systems]. *Class II ODS* are all HCFCs with an ozone depletion potential of <0.2. They are transitional substitutes for many Class I ODS. Most of them will be phased out soon. The most common HCFC currently in use is HCFC-22 (or R-22), a refrigerant in existing air conditioners and refrigeration equipment.

12.3 ACID RAIN

Acid rain, or *acid deposition*, includes any form of precipitation with acidic components (i.e., sulfuric or nitric acid) that falls from the atmosphere to the ground in wet or dry form. *Wet deposition* is when sulfuric and nitric acids formed in the atmosphere fall to the ground as rain, snow, fog, or hail. Acidic particles and gases in the atmosphere can also deposit on surfaces (such as water bodies, vegetation, and buildings) without moisture, that is, *dry deposition*.

The acidic components in the acid rain come from SO_x and NO_x in the atmosphere. They react with water and oxygen to form sulfuric and nitric acids, then mix with water and other materials before depositing onto the ground. Since wind can blow them long distances, acid rain can be a regional and even international problem.

12.3.1 Acid Rain Basics

Formation of nitric acid (HNO_3) from NO_x, oxygen, and water can be illustrated as:

$$NO_{(g)} + \frac{1}{2}O_{2(g)} \rightarrow NO_{2(g)} \tag{12.20}$$

$$4NO_{2(g)} + 2H_2O + O_2 \rightarrow 2HNO_{3(aq)} \tag{12.21}$$

Formation of sulfuric acid (H_2SO_4) from SO_x, oxygen, and water can be illustrated as:

$$SO_{2(g)} + \frac{1}{2}O_{2(g)} \rightarrow SO_{3(g)} \tag{12.22}$$

$$SO_{3(g)} + H_2O \rightarrow H_2SO_{4(aq)} \tag{12.23}$$

Average rain is slightly acidic and has a pH of about 5.6 because of the dissolution of carbon dioxide into it to form carbonic acid (H_2CO_3), as follows:

$$CO_{2(g)} + H_2O \leftrightarrow CO_{2(aq)} + H_2O \leftrightarrow H_2CO_3 \tag{12.24}$$

Acid rain usually has a pH between 4.2 and 4.4, and it can be lower.

12.3.2 Effects of Acid Rain

Due to its acidic nature, acid rain can adversely impact aquatic ecosystems, soil, forests, and the built environment. The same pollutants (SO_x and NO_x) that cause acid rain contribute to respiratory problems if inhaled as fine particulates.

Both marble and limestone consist primarily of calcium carbonate ($CaCO_3$) and are frequently used as construction materials. The reaction between calcium carbonate and the sulfuric acid components in acid rain results in the dissolution of $CaCO_3$ into aqueous ions, which are then washed away:

$$CaCO_{3(s)} + H_2SO_{4(aq)} \rightarrow Ca^{+2}_{(aq)} + SO^{-2}_{4(aq)} + H_2O + CO_2 \tag{12.25}$$

12.3.3 Measures to Control Acid Rain

The component of acid rain responsible for the adverse effects is NO_x or SO_x in the atmosphere. Only a tiny portion of SO_2 and NO_x is from natural sources (e.g., volcanoes and lightning), mainly from fossil fuel combustion. The measure to control acid rain is to reduce NO_x and SO_x anthropogenic emissions, mainly from fossil fuel combustion. The Acid Rain Program (ARP), established under Title IV of the 1990 CAAA, requires significant emission reductions of SO_x and NO_x from the energy sector.

12.4 GREENHOUSE GASES

12.4.1 Greenhouse Gases of Concern

The climate on Earth is determined by the balance between the energy received from the sun and the energy emitted back to outer space from the Earth and its atmosphere. Gases in the atmosphere that trap some of the outgoing energy, retaining heat in our atmosphere, are called *greenhouse gases* (GHGs). Water vapor is a GHG. However, the GHGs emitted by human activities are of concern. They are (i) carbon dioxide (CO_2), (ii) methane (CH_4), (iii) nitrous oxide (N_2O), and (iv) fluorinated gases.

12.4.2 Global Warming and Global Warming Potential

Global warming refers to the recent and ongoing rise in global average temperature near the Earth's surface. It has risen by 1.5°F (0.83°C) over the past century and is projected to rise even further. Now, it is a common belief that increasing concentrations of GHGs mainly cause global warming in the atmosphere. However, climate change may be caused by factors other than global warming. Small changes in the Earth's average temperature can translate to significant and potentially adverse shifts in weather and climate. Climate change can affect human health, cause changes to forests and other ecosystems, impact our energy supply, and cause severe weather events (USEPA, 2025).

Each GHG can stay in the atmosphere for different durations, ranging from a few years to thousands of years. Due to their extended stay, they are well mixed in the atmosphere, so their concentrations are relatively the same worldwide. Different GHGs can have different effects on global warming. Two critical factors for the differences are their ability to absorb energy ("*radiative efficiency*") and their "lifetime" in the atmosphere. Starting in 1990, the Intergovernmental Panel on Climate Change (IPCC) has used the *global warming potential* (GWP) to allow comparisons of the global warming impacts of different GHGs. A gas with a higher GWP would absorb more energy per unit mass than those with lower GWPs, thus contributing more to global warming on a per-unit-mass basis. The GWP of CO_2 is unity, which serves as the yardstick.

Table 12.1 shows the values of lifetime and GWP of GHGs (Source: IPCC, 2013; USEPA, 2024). The USEPA develops an annual report called the Inventory of U.S. Greenhouse Gas Emissions and Sinks (Inventory), which tracks U.S. Greenhouse Gas Emissions and Sinks by source, economic sector, and greenhouse gas starting in 1990. The Inventory complies with international GHG reporting standards under the United Nations Framework Convention on Climate Change (UNFCCC). UNFCCC guides currently require using the GWP values from the IPCC's Fifth Assessment Report (AR5), published in 2013.

TABLE 12.1
Lifetime and GWP of GHGs

	Lifetime in Atmosphere (year)	100-year GWP	20-year GWP
CO_2	Varies	1	1
CH_4	12.4	28	84
N_2O	121	265	264
Fluorinated Gases			
PFCs	Up to 50,000	Up to 11,100	Up to 8,210
HFCs	Up to 242	Up to 12,400	Up to 10,800
NF_3	500	16,100	12,800
SF_6	3,200	23,500	17,500

Source: USEPA (2024)

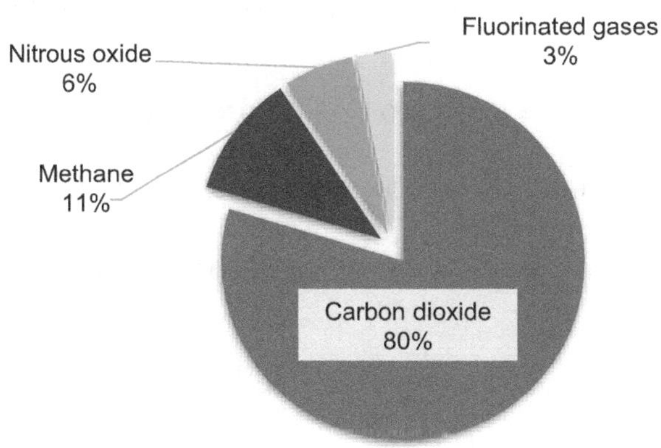

FIGURE 12.2 U.S. GHG emissions in 2022 by gas type.

The Inventory uses the 100-year GWP, based on the energy by a gas absorbed over 100 years, to measure the relative impact of different GHGs. However, some metrics used by the scientific community can vary.

Using a 20-year GWP would prioritize gases with shorter lifetimes than CO_2 in the atmosphere. For example, because of its shorter lifetime, the GWP values for CH_4 are 84 and 28 for the 20-year and 100-year cycles. On the other hand, the 100-year GWPs of fluorinated gases would be more significant than their 20-year GWPs (see Table 12.1).

The total U.S. GHG emissions in 2022 were 6,343 million metric tons of CO_2 equivalent (MMT CO_2 Eq.). CO_2 is the dominant one at 5,053 MMT CO_2 Eq. (79.7% of total), followed by CH_4, 702 MMT CO_2 Eq. (11.1%); N_2O, 390 MMT CO_2 Eq. (6.1%); and fluorinated gases, 198 MMT CO_2 Eq. (3.1%) (also see Figure 12.2). These gross emissions exclude emissions from the Land Use, Land-Use Change, and Forestry (LULUCF) sector (USEPA, 2024). To calculate CO_2-equivalent emissions,

$$\text{kg } CO_{2\ \text{Eq.}} = (\text{kg emission of gas}) \times (\text{GWP}) \tag{12.26}$$

$$\text{MMT CO}_{2\text{ Eq.}} = \{(\text{metric ton emission of gas}) \times (\text{GWP})\} \div 10^6 \quad (12.27)$$

Note: One U.S. "short" ton = 2,000 lb; one metric ton = 1,000 kg ≈ 2,200 lb (1 lb = 454 g). One metric ton is approximately 10% more than one U.S. "short" ton.

Example 12.2 Greenhouse gas emissions

Calculate the annual GHG emissions as MMT CO_2 Eq. for the following two cases:

a. An industrial boiler emits 1,000 kg CO_2/h. It operates continuously for 11 months per year (1 month for routine maintenance).
b. A semiconductor factory uses 1,000 kg/yr of nitrogen trifluoride for the etching and cleaning process, and 15% of it is emitted to the atmosphere.

Solution:

c. The amount of CO_2 emitted per year = (1,000 kg/h)(24 h/day)(365 d/yr)(11/12)

$$= 8,030,000 \text{ kg/yr} = 8,030 \text{ ton/yr}$$

$$\text{MMT CO}_{2\text{ Eq.}} = \{(8,030) \times (1.0)\} \div 10^6 = 8.03 \times 10^{-3}/\text{yr}$$

d. The rate of NF_3 emission = (1,000 kg/yr) × (15%) = 150 kg/yr = 1.5×10^{-1} ton/yr.

$$\text{MMT CO}_{2\text{ Eq.}} = \{(1.5 \times 10^{-1}) \times (16,100)\} \div 10^6 = 2.415 \times 10^{-3}/\text{yr}$$

12.4.3 Sources of GHGs

Emissions from the transportation, electric power, industry, residential & commercial, and agriculture sectors are 29%, 25%, 23%, 13%, and 10% of the total U.S. GHG emissions in 2022, respectively (USEPA, 2024); see Figure 12.3.

GHG emissions from transportation primarily come from burning fossil fuels for cars, trucks, ships, trains, and planes. GHG emissions from the electric power sector include emissions from electricity production used by other end-use sectors, such as industry. In 2022, 60% of electricity came from burning fossil fuels. GHG emissions from industry primarily come from burning fossil fuels for energy and specific chemical reactions necessary to produce goods from raw materials. GHG emissions from the commercial and residential sector come from using fossil fuels for heat, refrigeration, and cooling in buildings, as well as non-building-specific emissions such as waste handling. GHG emissions from agriculture come from livestock such as cows, agricultural soils, and rice production. While not shown in the figure, land areas can act as a sink (i.e., absorbing CO_2 from the atmosphere) or a source of GHG emissions. Since 1990, managed forests and other lands have been a net sink, offsetting 13% of total GHG emissions in the U.S. (USEPA, 2025).

Figure 12.4 depicts the annual trends of GHG emissions from 1990 to 2022. It shows that the emissions peaked in 2007 and have decreased since. The subsequent subsections discuss the sources and reduction measures of these GHGs individually.

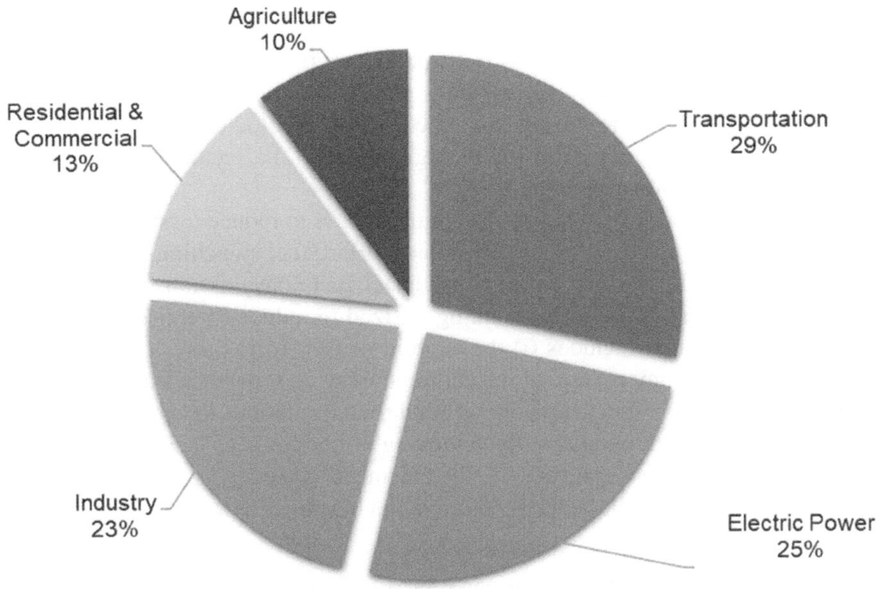

FIGURE 12.3 U.S. GHG emissions in 2022 by sectors.

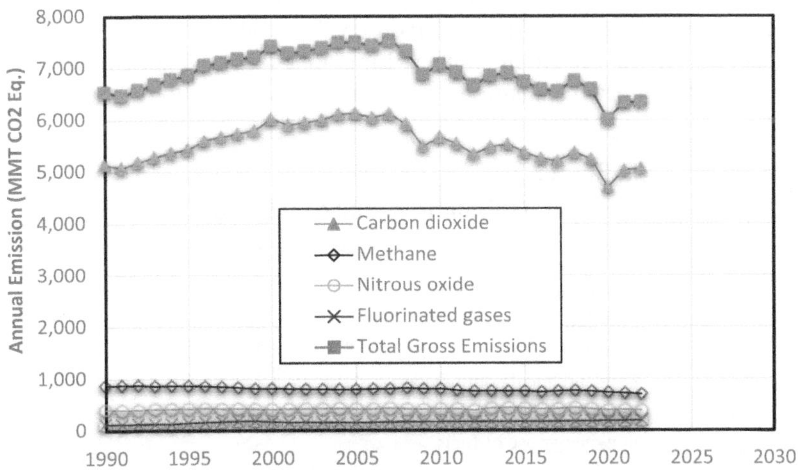

FIGURE 12.4 Trends of annual U.S. GHG emissions (1990–2022).

12.4.4 Sources and Reduction of CO_2 Emissions

Carbon dioxide is the primary GHG emitted through human activities. It accounted for 5,053 MMT CO_2 Eq. (79.7% of total) of all U.S. anthropogenic GHG emissions in 2022. Carbon dioxide is part of the Earth's carbon cycle, and CO_2 emissions come from various natural sources. However, anthropogenic emissions have been responsible for increased CO_2 concentrations in the atmosphere since the Industrial Revolution.

CO_2 enters the atmosphere mainly through burning fossil fuels (for electricity, transportation, industry, and heating) and specific chemical reactions (e.g., cement manufacturing). CO_2 is removed from the atmosphere (or termed "sequestered") when plants take it up as part of the biological cycle. The sectors' contributions of CO_2 emissions are shown in Table 12.2.

The most effective way to reduce CO_2 emissions is to reduce fossil fuel consumption, which can be achieved by energy conservation, fuel switching, and improving energy efficiency. In addition to reducing fossil fuel consumption, *carbon dioxide capture and sequestration* (CCS) is being utilized to reduce its emissions. CCS is a three-step process that includes (i) the capture of CO_2 from power plants or industrial processes, (ii) compression of the captured CO_2, and (iii) then transport through pipelines and injection into deep underground rock formations for storage and potential reuse. CO_2 capture is reportedly occurring at hundreds of facilities in the U.S., and the significant beneficial reuse is for enhanced oil recovery with other uses in metal fabrication and manufacturing of food and beverage, as well as pulp and paper.

12.4.5 Sources and Reduction of CH_4 Emissions

Methane accounted for 702 MMT CO_2 Eq. (11.1%) of all U.S. anthropogenic GHG emissions in 2022. As shown in Table 12.3, natural gas and petroleum systems, enteric fermentation, landfills, and manure management are the primary sources.

Regarding emission reduction, employing effective LDAR programs can reduce fugitive emissions from the oil and gas industry. Emission control that captures landfill gas is an excellent reduction strategy. Emissions from the agriculture sector can be reduced through better manure management and modifications to animal feeding

TABLE 12.2
U.S. CO_2 Emissions in 2022 by Sectors

Source	(%)
Transportation	35
Electricity	30
Industrial	16
Residential and commercial	12
Other	5
Non-energy use of fuels	2

Source: USEPA (2024).

practices to reduce enteric fermentation. The author of this book conducted a research project for the California Air Resources Board (CARB) in 2008, titled *Clearinghouse of Technological Options for Reducing Anthropogenic Non-CO_2 Greenhouse Gases Emissions from All Sectors*, which compiled and evaluated technologies for the reduction of methane, nitrous oxide, and high GWP gases (CARB, 2008; Kuo, 2012).

12.4.6 Sources and Reduction of N_2O Emissions

Nitrous oxide accounted for 390 MMT CO_2 Eq. (6.1%) of all U.S. anthropogenic GHG emissions in 2022. As shown in Table 12.4, agricultural soil management is the primary source, accounting for 75% of the total emission.

N_2O emissions can be reduced by applying nitrogen-based fertilizers more efficiently and by modifying manure management practices in the agricultural sector. Other reductions can be achieved by lowering fuel combustion, switching to alternative fuels, and technological upgrades of adipic acid production.

TABLE 12.3
U.S. CH_4 Emissions in 2022 by Sectors

Source	(%)
Natural gas & petroleum systems	28
Enteric fermentation	25
Landfills	16
Manure management	9
Other	9
Flooded Land	6
Coal mining	6
LULUCF emissions	2

Source: UDEPA (2024).

TABLE 12.4
U.S. N_2O Emissions in 2022 by Sectors

Source	(%)
Agricultural soil management	75
Wastewater treatment	6
Stationary combustion	6
Manure management	4
Transportation	4
LULUCF emissions	2
Other	2

Source: USEPA (2024).

TABLE 12.5
U.S. Fluorinated Gas Emissions in 2022 by Sectors

Source	(%)
Substitution of ozone-depleting substances	90
Electrical equipment	3
Electronics industry	2
Fluorochemical production	4
Production & processing of aluminum and magnesium	2

Source: USEPA (2024).

12.4.7 Sources and Reduction of Fluorinated Gases Emissions

Different from other GHGs, emissions of fluorinated gases only come from human activities. Their GWPs are much higher, and they are often called *high-GWP gases.* Their lifetimes are also long, up to thousands of years. There are four main categories of fluorinated gases: (i) hydrofluorocarbons (HFCs), (ii) perfluorocarbons (PFCs), (iii) sulfur hexafluoride (SF_6), and (iv) nitrogen trifluoride (NF_3). Fluorinated gases accounted for 198 MMT CO_2 Eq. (3.1%) of all U.S. anthropogenic GHG emissions in 2022. As shown in Table 12.5, the substitution of ODS is the primary source, accounting for 90% of the total emission [Note: *substitutes for ozone-depleting substances* are HFCs used as refrigerants, aerosol propellants, foam blowing agents, solvents, and fire retardants in products. The primary emission source of these compounds is their use as refrigerants – for example, in air-conditioning systems in vehicles and buildings. These chemicals were developed to replace CFCs and HCFCs because they do not deplete the stratospheric ozone layer].

Overall, fluorinated gas emissions in the U.S. have increased by 105% between 1990 and 2022. This increase has been driven by a 349% increase in HFC emissions since 1990, as they have been widely used as a substitute for ODS. Emissions of PFCs and sulfur hexafluoride (SF_6) have declined during this time due to emission-reduction efforts of PFCs in the aluminum production industry and SF_6 in the electrical transmission and distribution industry (USEPA, 2025).

Reductions of fluorinated gas emissions come from better handling of the uses of these gases, minimizing leaks from the systems using them (e.g., air-conditioning, refrigerators, electricity transmission, and distribution systems). Substitutes with lower-GWP compounds can also reduce emissions.

EXERCISE QUESTIONS

1. Ultraviolet (UV) disinfection is popular for water/wastewater disinfection. There are low-pressure UV lamps and medium-pressure UV lamps. The light spectrum of medium-pressure lamps is broad, while that of the low-pressure lamps is very narrow, around 254 nm, which is most biocidal. The wavelength of red light is the longest of the visible light. Compare the

frequency and energy of UV light with a wavelength of 254 nm with those of red light with a wavelength of 700 nm.

2. Calculate annual greenhouse gas emissions as MMT CO_2 Eq. for the following two cases:
   a. A dairy cow emits ~100 kg of methane/yr from enteric fermentation, and the annual methane emission is ~ 20 kg/head. A dairy farm raises 500 dairy cows.
   b. An industrial facility has annual nitrous oxide emissions of 5 metric tons from its operation.

BIBLIOGRAPHY

CARB. (2008). *Clearinghouse of Technological Options for Reducing Anthropogenic Non-CO_2 Greenhouse Gases Emissions from All Sectors*, Contract No. CARB 05-328, prepared by J. Kuo for California Air Resources Board, State of California, 384 pages.

IPCC. (2013). *Climate Change 2013: The Physical Science Basis. Contribution of Working Group I to the Fifth Assessment Report of the Intergovernmental Panel on Climate Change* (ed. By T.F. Stocker, D. Qin, G.-K. Plattner, M. Tignor, S.K. Allen, J. Boschung, A. Nauels, Y. Xia, V. Bex and P. M. Midgley), Cambridge University Press, Cambridge, United Kingdom, and New York, NY.

Kuo, J. (2012). *Technological Options for Reducing Non-CO_2 Greenhouse Gases Emissions in Handbook on Climate Change Mitigation* (ed. by W. Y. Chen et al.), Springer Science, New York, NY.

USEPA. (2002). *APTI 482: Sources and Control of Volatile Organic Air Pollutants - Student Manual*, 3rd edition, prepared by J.W. Crowder for Air Pollution Training Institute, United States Environmental Protection Agency, Research Triangle Park, NC.

USEPA. (2014). *APTI 418: NO_x Emission Control from Stationary Sources - Student Manual*, prepared by NESCAUM in 2009 (revised and reorganized by Doyle B.W. & Slot S. in 2014) for the United States Environmental Protection Agency.

USEPA. (2024). *Inventory of U.S. Greenhouse Gas Emissions and Sinks*: *1990–2022*, EPA-430R-24004, https://www.epa.gov/ghgemissions/inventory-us-greenhouse-gas-emissions-and-sinks-1990-2022.

USEPA. (2025). *Overview of Greenhouse Gases*, USEPA, https://www.epa.gov/ghgemissions/overview-greenhouse-gases, last updated on 01/16/2025.

13 Other Topics

13.1 VISIBILITY, HAZE, AND PLUME OPACITY

Visual range measures the distance over which a person can see and distinguish objects in the atmosphere, and it is a key factor of air clarity. Fine particles, smoke, smog, fog, rain/snow, and moisture affect the visual range at a specific location. The typical visual range in the western U.S. is 60 to 90 miles (100 to 150 km), or about one-half of what it would be without haze-causing air pollution. For the rest of the U.S., the typical range is 15 to 30 miles (25 to 50 km), or about one-third of the visual range under natural conditions (USEPA, 2006)

Haze is caused when sunlight encounters tiny particles in the air; some light will be absorbed or scattered. Most of these particles originated from anthropogenic sources. More fine particles will result in more absorption and scattering, which reduces the clarity and color of what we see. Some particles (e.g., sulfates) would scatter more light, especially during humid conditions. The federal regulations toward visibility include the 2005 Clean Air Interstate Rule, the 1999 regulations to reduce haze and protect visibility, and the 1992 Acid Rain Rule (USEPA, 2006).

Plume opacity refers to how transparent or opaque a visible emission plume (usually from a smokestack or exhaust) appears. It is a visual indicator of the amount of light blocked by particulates and gases in the air. It is often used to assess air pollution levels from industrial or combustion sources. USEPA Method 9 is a standardized visual observation method to determine the opacity of visible emissions. A trained observer stands at a specified distance away and at an appropriate angle to record observations (i.e., opacity in 5% increments) at a 15-second interval over 6 minutes (24 readings). The average of the highest six consecutive readings is typically used as the reported opacity value. Opacity monitors are often installed on stacks for real-time data as part of the continuous emission monitoring (CEM) systems. The opacity is then calculated as:

$$\text{Opacity }(\%) = \left(1 - I/I_o\right) \times 100\% \tag{13.1}$$

where I is the light intensity at the detector and I_o is the light intensity at the source. An opacity value of 0% means the medium is entirely transparent (i.e., no visible plume), while 100% means completely opaque (i.e., no light passes through).

Example 13.1: Visual Plaque Measurement

An inspector recorded the following 24 consecutive opacity readings (in %) during the inspection of a visual plume: 5, 5, 10, 15, 20, 25, 30, 35, 30, 25, 25, 15, 5, 5, 10, 20, 20, 25, 15, 20, 5, 10, 5, and 10. What should the reported value be?

DOI: 10.1201/9781003532705-13

Solution:

a. The six consecutive readings with the highest average: 25, 30, 35, 30, 25, 25.
b. The average value of the six = (25 + 30 + 35 + 30 + 25 + 25) ÷ 6 = 170 ÷ 6 = 28.3.
c. The reported opacity = 28.3%.

13.2 *ODOR*

Odor is a property of a substance that stimulates the olfactory organ. Odors from a single compound can be regulated like a criteria pollutant. However, odors generally result from a combination of compounds in large numbers. In addition, concerning air pollution control, an odor is not an air pollutant but rather a property of air pollutants that can only be detected or measured through its effects on human organs. Common symptoms include headaches, nasal congestion, and eye, nose, and throat irritation. Common *environmental odors* (i.e., odors in the outdoor air) can come from sources including (i) animals (e.g., confined animal feeding operations), (ii) agriculture (e.g., manure and chemical fertilizer or pesticides), (iii) human activities (e.g., compost, sewage, garbage, fires, household cleaning agents), (iv) industries or processes (e.g., oil refineries, landfills, paper mills, wastewater treatment, food processing, coal-fired power generation, wood treatment, semi-conductors, pharmaceuticals, textiles, and rubber), (v) nature (e.g., moist soil, gardens, fires), and (vi) vehicles (e.g., diesel exhaust). Odor-producing substances related to industries or processes include ammonia (NH_3), hydrogen sulfide (H_2S), sulfur oxides (SO_x), formaldehyde (HCHO), creosote, benzene (C_6H_6), toluene ($C_6H_5CH_3$), ethylbenzene ($C_6H_5C_2H_5$), xylenes ($C_6H_4(CH_3)_2$), and pentachlorophenol (C_6Cl_5OH) (ATSDR, 2024).

A compound having an odor is called an *odorant*. Concentrations of odorants in air are typically low. Even though low concentrations can be measured using sophisticated analytical instruments, a panel of people must determine the exact odor strength. The odor *detection threshold* (DT) is the lowest concentration of an odorant perceivable by the human sense of smell. The thresholds cannot be accurately detected; however, they are measured through extensive tests using human subjects instead. The threshold values can differ; for example, 100 ppm for methanol and 0.0021 ppm for trimethylamine (C_3H_9N). USEPA (1992) published an article titled "*Reference Guide to Odor Thresholds for Hazardous Air Pollutants Listed in the Clean Air Amendments of 1990*".

The unit to represent the odor strength/concentration is the *odor unit* (OU), and its value is determined by diluting an odorous air sample with a measured amount of odor-free air and presenting the diluted mixture to an odor panel. One OU is the value where the odor is at the DT. If a sample were diluted 200 times to reach the DT, its order strength is 200 OU. Many environmental regulations use OU to set limits.

The elimination of odors is the most essential part of any odor management. Air pollutants responsible for an odor should be controlled so that their concentrations in ambient air do not reach their threshold concentrations. Prevention or elimination of sources of odorous compounds should always be the priority. The destruction or removal of odorants in the air is commonly accomplished by direct or catalytic incineration, GAC adsorption, condensation, wet scrubbers, or the use of chemical reactants. USEPA's article, titled *Biosolids and Residuals Management Fact Sheet – Odor Control in Biosolids Management*, provides information on the control of odors from biosolids production facilities and the prevention of odors from the storage, distribution, and application of the biosolids product (USEPA, 2000). It provides good coverage of odors and their management.

13.3 LEAD

Lead is one of the six criteria pollutants. Sources of lead emissions to the atmosphere vary from one area to another. Significant sources of lead in the air of the U.S. are ore and metals processing, and piston engine aircraft operated on leaded aviation fuel. Other sources are waste incinerators, utilities, and lead-acid battery manufacturers. As a result of USEPA's regulatory efforts, including the removal of lead from gasoline, levels in the air decreased by 98% between 1980 and 2014 (USEPA, 2025). When lead is released into the air, it may travel long distances before being deposited on the ground, where it is usually adsorbed onto soil particles.

The USEPA regulates lead emissions under the NAAQS and several other programs, including NESHAPs and new source review (NSR), and each state needs to include lead control in its SIP.

13.4 INDOOR AIR POLLUTION

According to the USEPA, *indoor air quality* (IAQ) refers to the air quality within and around buildings and structures, especially concerning the health and comfort of building occupants. We usually spend more than half of our time indoors. Therefore, IAQ is as important as, if not more important than, outdoor air quality. Health effects of indoor air pollutants may be immediate or long-term after exposure. With regard to IAQ concerns, the USEPA classifies the buildings into three types: (i) homes, (ii) schools, and (iii) offices and other large buildings. IAQ in homes is more critical to human health since we stay inside our homes for at least one-third of our lifetime.

13.4.1 Sources of Indoor Air Pollutants

The concentration of compounds of concern (COCs) in a building depends on three main factors: (i) input rate from the outdoor air, (ii) interior emission/production rate, and (iii) the air exchange rate. Mass input from outdoors and emission/generation from inside are the total pollutant mass; the actual concentrations will also be affected by the frequency of air exchange.

13.4.2 Categories of Indoor Air Pollutants

Although numerous indoor air pollutants may be present in a building, they typically fall into three basic categories: (i) biological, (ii) chemical, and (iii) non-biological particles.

Biological. Elevated levels of bacteria, viruses, fungi, mold, dust mites, animal dander, and pollen may be present due to inadequate humidity control or water spills/intrusion/flooding/condensation (OSHA, 2011).

Chemical. Chemical pollutants (gases or vapors) can come from building materials (e.g., walls, floor covering, adhesives, and paints), office equipment (e.g., photocopy machines and printers), maintenance products (e.g., pesticides, cleaning products), and human activities (e.g., the use of consumer products, the use of heating devices, and releases from the occupants themselves). Common indoor chemical air pollutants include VOCs, CO, CO_2, formaldehyde, and pesticides. Radon, a radioactive gas, is found in nearly all soils and can enter any building.

Non-biological particles. The outdoor air may carry dust, dirt, or other substances into the building. They can also be produced by activities occurring inside the building, such as construction activities and the operation of office equipment (OSHA, 2011).

13.4.3 Effects of Indoor Air Pollution

Good IAQ in a building would contribute to a favorable and productive environment for building occupants. It would give them a sense of comfort, health, and well-being. On the other hand, indoor air pollutants (e.g., VOCs) may have short- and long-term health effects.

The type and severity of health effects resulting from poor IAQ depend on the type and concentration of the pollutant(s) and the frequency and duration of the exposure(s). Short-term effects may include allergies, stress, colds, and influenza, while long-term effects may include respiratory disease, heart disease, and cancer (OSHA, 2011). *Sick building syndrome* (SBS) describes situations in which building occupants experience acute health or comfort-related effects that appear to be linked to time spent in the building. Regardless, no specific illness or cause can be identified. In contrast, building-related *illness* (BRI) is used when symptoms of a diagnosable illness are identified and can be attributed directly to airborne building contaminants (USEPA, 1991).

13.4.4 Measures to Reduce Exposure to Indoor Air Pollution

A 718-page guide, *Indoor Air Quality Guide – Best Practices for Design, Construction, and Commissioning,* was developed by the American Society of Heating, Refrigerating and Air-Conditioning Engineers (ASHRAE), the American Institute of Architects, Building Owners and Managers Association International, Sheet Metal and Air Conditioning Contractors' National Association, U.S. Building Council, and USEPA (ASHRAE, 2009). The USEPA funded the project, and the guide provides practical information covering the full range of important IAQ information for architects, engineers, and building owners. The guide can be downloaded free of charge. Access to this guide is highly recommended. The guide first describes eight objectives needed for improving IAQ and then provides detailed guidance to achieve these objectives. The names of eight objectives alone give a good idea of the control/management measures needed to improve IAQ. One should consult the guide for details. These eight objectives are:

1. Manage the design and construction process to achieve good IAQ.
2. Control moisture in building assemblies.
3. Limit the entry of outdoor contaminants.
4. Control moisture and contaminants related to mechanical systems.

5. Limit contaminants from indoor sources.
6. Capture and exhaust contaminants from building equipment and activities.
7. Reduce contaminant concentrations through ventilation, filtration, and air cleaning.
8. Apply more advanced ventilation approaches.

Typical engineering controls include the following three measures:

1. *Local exhaust.* Using local exhaust (e.g., a canopy hood) removes point sources of pollutants before they can be dispersed into the indoor air.
2. *General dilution ventilation.* A well-designed and functioning heating, ventilating, and air-conditioning (HVAC) system would control temperature and relative humidity levels to provide comfort, distribute adequate outdoor air, and dilute/remove odors and other pollutants.
3. *Air cleaning.* Air cleaning primarily involves the removal of particulates from the air as the air passes through the HVAC equipment. Other devices, such as granular activated carbon, should be considered if needed.

13.4.5 Air Exchange Rate

The *air exchange rate* (AER) measures how quickly outside (fresh) air replaces the air inside a building or space. It is often expressed as the number of *air changes per hour* (ACH). If a room has an AER of one ACH, the total volume of the indoor air is replaced by outdoor/fresh air once every hour. For example, typical ACHs for residential homes (through natural ventilation) and office/classrooms are 0.35 to 1 and 3 to 6 ACH, respectively; those for the hospitals and operating rooms are much higher, at 15 to 20 ACH.

AER is essentially the inverse of the air detention/residence time, which is the ratio of the air ventilation rate (Q) and the volume of the space/room (V), as follows:

$$\text{AER} = Q/V \tag{13.2}$$

13.4.6 Indoor Air Concentration Calculations

Indoor air concentration calculations estimate the concentration of a pollutant present in indoor air, and this concentration is often used for exposure assessments, indoor air quality standards, or ventilation design. A common approach is to use the mass balance equation for a single-zone and well-mixed space. The general mass balance equation can be expressed as:

$$\begin{bmatrix}\text{Rate of mass}\\ \text{ACCUMULATED}\end{bmatrix} = \begin{bmatrix}\text{Rate of mass}\\ \text{IN}\end{bmatrix} - \begin{bmatrix}\text{Rate of mass}\\ \text{OUT}\end{bmatrix} \pm \begin{bmatrix}\text{Rate of mass}\\ \text{GENERATED or DESTROYED}\end{bmatrix} \tag{13.3}$$

Assuming the ventilation rate is constant (i.e., $Q_{in}=Q_{out}=Q$), the change of indoor concentration (C) of a pollutant with time t (i.e., dC/dt) in a room of volume V can be expressed as:

$$V\frac{dC}{dt}=Q\times C_{in}-Q\times C_{out}\pm\begin{bmatrix}\text{Rate of mass}\\ \text{GENERATED or DESTROYED}\end{bmatrix} \quad (13.4)$$

where C_{in} is the pollutant concentration in the incoming air and C_{out} is the pollutant concentration in the effluent air.

Assuming there is a chemical reaction occurring and there is an emission source of the pollutant in the room, then

$$V\frac{dC}{dt}=Q\times C_{\text{in}}-Q\times C_{\text{out}}+\left[\pm\left(\text{Rate of Reaction}\right)\times V+\text{Rate of Emission}\right] \quad (13.5)$$

If the pollutant has a first-order decay rate with a rate constant k (hr^{-1}) and there is an emission source at a rate of S, Eq. 13.5 becomes

$$V\frac{dC}{dt}=Q\times C_{\text{in}}-Q\times C_{\text{out}}+\left[-(kC)V+S\right] \quad (13.6)$$

With $C_{\text{in}}=C_{\text{outdoor}}$ and assuming the room is well mixed (i.e., $C_{\text{out}}=C$), Eq. 13.6 becomes

$$V\frac{dC}{dt}=Q\times C_{\text{outdoor}}-Q\times C+\left[-(kC)V+S\right] \quad (13.7a)$$

$$V\frac{dC}{dt}=Q\times C_{\text{outdoor}}+S-Q+kV\times C \quad (13.7b)$$

The steady-state concentration of the indoor pollutant (C_{ss}) can be found by setting the left-hand side of Eq. 13.7 equal to zero (i.e., $dC/dt=0$).

$$C_{\text{ss}}=\left(S+QC_{\text{outdoor}}\right)/\left(Q+kV\right)=\left(\frac{S}{V}+\frac{Q}{V}\times C_{\text{outdoor}}\right)\Big/\left(Q/V+k\right) \quad (13.8)$$

The $(Q/V+k)$ is essentially $(AER+k)$, and the inverse of it is often referred to as the *time constant* (τ), as in NCEES (2025). That is

$$\tau=\frac{1}{\left(Q/V+k\right)}=\frac{1}{\left(AER+k\right)}=\frac{V}{\left(Q+kV\right)} \quad (13.9)$$

Eq. 13.8 can be rewritten as:

$$C_{ss} = \left[S/V + (Q/V) \times C_{outdoor}\right] \times \tau \tag{13.10}$$

The steady-state concentrations for four special cases can be found by using the equations below:

i. With chemical decay, clean outdoor air (i.e., $C_{outdoor}=0$), and indoor emission $= S$:

$$C_{ss} = S/(Q + kV) \tag{13.11}$$

ii. Without an indoor emission ($S=0$) and $C_{outdoor}=$ constant:

$$C_{ss} = \left[\frac{1}{1 + k(V/Q)}\right] \times C_{outdoor} \tag{13.12}$$

iii. Without chemical decay ($k=0$), indoor emission $= S$, and $C_{outdoor}=$ constant:

$$C_{ss} = C_{outdoor} + S/Q \tag{13.13}$$

iv. Without chemical decay, no indoor emission ($S=0$), and $C_{outdoor}=$ constant:

$$C_{ss} = C_{outdoor} \tag{13.14}$$

Example 13.2: Steady-State Indoor Air Pollutant Concentration

An office building has a volume of 10,000 m^3, and the ventilation rate is 50,000 m^3/h The outdoor air carries a pollutant with a 10 mg/m^3 concentration. The building has an indoor pollutant emission rate of 1 kg/h. The pollutant decays following the first-order reaction with a rate constant of 0.1/h. Determine:

a. the air exchange rate,
b. the steady-state indoor pollutant concentration,
c. the steady-state indoor pollutant concentration, if the outdoor air is clean,
d. the steady-state indoor pollutant concentration, if the indoor emission source were absent, and
e. the steady-state indoor pollutant concentration, if no decay of the indoor pollutant.

Solution:

a. Using Eq. 13.2, AER = Q/V = 50,000 m³/h ÷ 10,000 m³ = 5/h (i.e., ACH = 5).
b. Using Eq. 13.8 for $C_{outdoor}$ = 10 mg/m³, S = 1 kg/h, and k = 0.1/h,

$$C_{ss} = (S + QC_{outdoor})/(Q + kV) =$$
$$\left(\frac{10^6\,\text{mg}}{\text{h}} + \frac{50{,}000\,\text{m}^3}{\text{h}} \times \frac{10\ \text{mg}}{\text{m}^3}\right)\Big/\left(\frac{50{,}000\,\text{m}^3}{\text{h}} + \frac{0.1}{\text{h}} \times 10{,}000\ \text{m}^3\right) = 29.41\ \text{mg/m}^3$$

c. Using Eq. 13.11 for $C_{outdoor}$ = 0 mg/m³, S = 1 kg/h, and k = 0.1/h,

$$C_{ss} = (S)/(Q + kV) = \left(\frac{10^6\,\text{mg}}{\text{h}}\right)\Big/\left(\frac{50{,}000\,\text{m}^3}{\text{h}} + \frac{0.1}{\text{h}} \times 10{,}000\ \text{m}^3\right) = 19.61\,\text{mg/m}^3$$

d. Using Eq. 13.12 for $C_{outdoor}$ = 10 mg/m³, S = 0 kg/h, and k = 0.1/h,

$$C_{ss} = \left[\frac{1}{1 + k(V/Q)}\right] \times C_{outdoor} = \left[\frac{1}{1 + 0.1(10{,}000\ \text{m}^3/50{,}000\ \text{m}^3/\text{h})}\right] \times 10\frac{\text{mg}}{\text{m}^3} = 9.8\ \text{mg/m}^3$$

e. **Using** Eq. 13.13 for $C_{outdoor}$ = 10 mg/m³, S = 10 kg/h, and k = 0/h,

$$C_{ss} = C_{outdoor} + S/Q = 10\frac{\text{mg}}{\text{m}^3} + \frac{10^6\text{mg}}{\text{h}} \div \frac{50{,}000\,\text{m}^3}{\text{h}} = 30\ \text{mg/m}^3$$

Discussion:

1. Out of the steady-state concentration of 29.41 mg/m³ (part (b)), 19.61 and 9.8 mg/m³ are contributed from the indoor emission (part (c)) and outdoor air (part (d)), respectively.
2. Without decay of the indoor pollutants, C_{ss} = 30 mg/m³ (part (e)), which is slightly larger than that with decay, 29.41 mg/m³ (part (b)).
3. If the indoor emission source were absent and there is no decay, C_{ss} would be the same as $C_{outdoor}$ = 10 mg/m³, slightly larger than that of part (d), 9.8 mg/m³.

The transient indoor pollutant concentration, $C(t)$, can be estimated from its steady-state concentration (C_{ss}) as follows:

$$C(t) = \left[1 - e^{-\left(\frac{Q}{V}+k\right)t}\right] \times C_{ss} \tag{13.15}$$

With a measured steady-state concentration (C_{ss}), the emission rate (S) can be back-calculated:

$$S = \left[\left(\frac{Q}{V} + k\right) \times C_{ss} - (Q/V) \times C_{outdoor}\right] \times V \tag{13.16}$$

Example 13.3: Transient Indoor Air Pollutant Concentration

An office building has a volume of 10,000 m³, and the ventilation rate is 50,000 m³/h. The outdoor air carries a pollutant with a 10 mg/m³ concentration. The building has a constant pollutant emission rate of 1 kg/h. The pollutant decays following the first-order reaction with a rate constant of 0.1/h. Assuming the initial indoor concentration (C_o) = 0, determine the indoor concentration at 10, 30, and 60 minutes.

Solution:

a. Using Eq. 13.15 for t = 10 minutes = 0.167 hour, and C_{ss} = 29.41 mg/m³ (from part (b) of Example 13.2),

$$C(t = 0.167\ \text{hour}) = \left[1 - e^{-\left(\frac{Q}{V}+k\right)t}\right] \times C_{ss} = \left[1 - e^{-\left(\frac{50{,}000}{10{,}000}+0.1\right)(0.167)}\right] \times 29.41 = 16.8\ \text{mg/m}^3$$

b. Using Eq. 13.15 for t = 30 minutes = 0.5 hour, and C_{ss} = 29.41 mg/m³,

$$C(t = 0.5\ \text{hour}) = \left[1 - e^{-\left(\frac{50{,}000}{10{,}000}+0.1\right)(0.5)}\right] \times 29.41 = 27.1\text{mg/m}^3$$

c. Using Eq. 13.15 for t = 60 minutes = 1.0 hour, and C_{ss} = 29.41 mg/m³,

$$C(t = 1\ \text{hour}) = [1 - e^{-\left(\frac{50{,}000}{10{,}000}+0.1\right)(1.0)}] \times 29.41 = 29.21\ \text{mg/m}^3$$

Discussion:

1. The indoor air pollutant concentrations change exponentially. The concentrations are 16.8, 27.1, and 29.41 mg/m³ at 10, 30, and 60 minutes, which are 57%, 92%, and >99% of the steady-state concentration (29.41 mg/m³), respectively.
2. The time (t) should be in hours to match the units of the other parameters in Eq. 13.15.

Example 13.4: Back-Calculate the Strength of the Indoor Emission Source

An office building has a volume of 10,000 m³, and the ventilation rate is 50,000 m³/h. The outdoor air carries a pollutant with a 10 mg/m³ concentration. There is a constant pollutant emission inside the building. The pollutant decays following the first-order reaction with a rate constant of 0.1/h. A recent indoor monitoring showed the pollutant concentration is steady at 29.41 mg/m³. What would the indoor pollutant emission rate be?

Solution:

Using Eq. 13.16 with $C_{ss} = 29.41$ mg/m^3,

$$S = \left[\left(\frac{Q}{V} + k\right) \times C_{ss} - \left(\frac{Q}{V}\right) \times C_{\text{outdoor}}\right] \times V =$$

$$\left[\left(\frac{50,000}{10,000} + 0.1\right) \times 29.41 - \left(\frac{50,000}{10,000}\right) \times 10\right] \times 10,000 = 10\,\frac{\text{kg}}{\text{h}}$$

Discussion: The estimated emissions rate, 10 kg/h, is the same as the given value in Example 13.3.

The transient indoor concentration can be determined by integrating the general mass balance equation (i.e., Eq. 13.7b):

$$\frac{dC(t)}{dt} = \left[\frac{Q}{V} \times C_{\text{outdoor}} + \frac{S(t)}{V}\right] - \left[\frac{Q}{V} + k\right] \times C(t) \quad (13.17)$$

The transient concentrations for two special cases can be found by using the equations below:

i. With a constant indoor emission S and chemical decay, clean outdoor air (i.e., $C_{\text{outdoor}} = 0$), and initial indoor concentration (C_o),

$$C(t) = C_o \times e^{-\left(\frac{Q}{V}+k\right)t} + \left[S/(Q + kV)\right] \times \left[1 - e^{-\left(\frac{Q}{V}+k\right)t}\right] \quad (13.18)$$

ii. With an instantaneous pulse pollutant emission at $t = 0$ (M_o) and clean outdoor air,

$$C(t) = e^{-\left(\frac{Q}{V}+k\right)t} \times \left(C_o + \frac{M_o}{V}\right) \quad (13.19)$$

Example 13.5: Transient Indoor Air Pollutant Concentration with a Constant Emission Source

An office building has a volume of 10,000 m^3, and the ventilation rate is 50,000 m^3/h. The building has a constant pollutant emission rate of 1 kg/h. The pollutant decays following the first-order reaction with a rate constant of 0.1/h. Assuming the initial indoor concentration (C_o) = 0 and the outside air is clean, determine the indoor concentration at 10, 30, and 60 minutes.

Solution:

a. Using Eq. 13.18 for $t = 10$ minutes $= 0.167$ hour and $C_o = 0$,

$$C(0.167) = [S/(Q + kV)] \times \left[1 - e^{-\left(\frac{Q}{V}+k\right)t}\right] = \left[\frac{1}{(50{,}000 + (0.1)(10{,}000))}\right] \times \left[1 - e^{-\left(\frac{50{,}000}{10{,}000}+0.1\right)(0.167)}\right] = 11.24 \text{ mg/m}^3$$

b. Using Eq. 13.18 for $t = 30$ minutes $= 0.5$ hour, and $C_o = 0$,

$$C(0.5) = \left[\frac{1}{(50{,}000 + (0.1)(10{,}000))}\right] \times \left[1 - e^{-\left(\frac{50{,}000}{10{,}000}+0.1\right)(0.5)}\right] = 18.08 \text{ mg/m}^3$$

c. Using Eq. 13.18 for $t = 60$ minutes $= 1.0$ hour, and $C_o = 0$,

$$C(1.0) = \left[\frac{1}{(50{,}000 + (0.1)(10{,}000))}\right] \times \left[1 - e^{-\left(\frac{50{,}000}{10{,}000}+0.1\right)(1.0)}\right] = 19.49 \text{ mg/m}^3$$

Discussion: The indoor air pollutant concentrations change exponentially. The concentrations are 11.24, 18.08, and 19.49 mg/m^3 at 10, 30, and 60 minutes, which are 57%, 92%, and >99% of the steady-state concentration contribution of the indoor emission source (19.61 mg/m^3), as shown in Example 13.2, respectively. The percentages of the steady-state concentration are the same as the corresponding ones in Example 13.3.

Example 13.6: Transient Indoor Air Pollutant Concentration with an Instantaneous Emission

An office building has a volume of 10,000 m^3, and the ventilation rate is 50,000 m^3/h. There is an instantaneous pollutant release of 1 kg inside the building. The pollutant decays following the first-order reaction with a rate constant of 0.1/h. Assuming the initial indoor concentration (C_o) = 10 mg/m^3 and the outside air is clean, determine the indoor concentration at 10, 30, and 60 minutes.

Solution:

a. Assuming the released pollutants are thoroughly mixed in the building, it will cause an instantaneous increase of the indoor concentration by:

$$M_o/V = (1 \text{ kg}) \div 10{,}000 \text{ m}^3 = 100 \text{ mg/m}^3$$

b. Using Eq. 13.19 for $t=10$ minutes $=0.167$ hour, $C_o=10$ mg/m^3, and $M/V=100$ mg/m^3,

$$C(0.167)=e^{-\left(\frac{Q}{V}+k\right)t}\times\left(C_0+\frac{M_o}{V}\right)=e^{-\left(\frac{50{,}000}{10{,}000}+0.1\right)(0.167)}\times(10+100)=46.9\text{ mg/m}^3$$

c. Using Eq. 13.19 for $t=30$ minutes $=0.5$ hour, $C_o=10$ mg/m^3, and $M/V=100$ mg/m^3,

$$C(0.5)=e^{-\left(\frac{Q}{V}+k\right)t}\times\left(C_0+\frac{M_o}{V}\right)=e^{-\left(\frac{50{,}000}{10{,}000}+0.1\right)(0.5)}\times(10+100)=8.6\frac{\text{mg}}{\text{m}^3}$$

d. Using Eq. 13.19 for $t=60$ minutes $=1.0$ hour, $C_o=10$ mg/m^3, and $M/V=100$ mg/m^3,

$$C(1.0)=e^{-\left(\frac{Q}{V}+k\right)t}\times\left(C_0+\frac{M_o}{V}\right)=e^{-\left(\frac{50{,}000}{10{,}000}+0.1\right)(1.0)}\times(10+100)=6.7\frac{\text{mg}}{\text{m}^3}$$

Discussion:

1. The concentrations dropped exponentially from 110 mg/m^3 initially to 49.6, 8.6, and 6.7 mg/m^3 after 10, 30, and 60 minutes, respectively. As shown, the decrease slows down with time.
2. If the room were clean, just set $C_o=0$ in Eq. 13.19.
3. If there is no chemical decay of the pollutant in the room, just set $k=0$ in Eq. 13.19.

Without indoor pollutant emissions, the transient indoor concentrations can be found by:

$$C(t)=\left(1-e^{-\left(\frac{Q}{V}+k\right)t}\right)\times\left[\frac{1}{1+k(V/Q)}\right]\times C_{\text{outdoor}}+e^{-\left(\frac{Q}{V}+k\right)t}\times C_o \qquad (13.20)$$

In this case, the transient concentrations for two special cases can be found by using the equations below:

i. Constant outdoor air concentration:

$$C(t)=C_{\text{outdoor}}+\left(C_o-C_{\text{outdoor}}\right)\times e^{-\left(\frac{Q}{V}+k\right)t} \qquad (13.21)$$

ii. Clean outdoor air:

$$C(t)=e^{-\left(\frac{Q}{V}+k\right)t}\times C_o \qquad (13.22)$$

Example 13.7: Transient Indoor Air Pollutant Concentration without Indoor Emission

An office building has a volume of 10,000 m^3, and the ventilation rate is 50,000 m^3/h. The pollutant decays following the first-order reaction with a rate constant of 0.1/h.

a. Assuming the initial indoor concentration (C_o) = 10 mg/m^3 and the outside air is clean, determine the indoor concentration at 10, 30, and 60 minutes.
b. What would be the corresponding concentrations if $C_{outdoor}$ = 4 mg/m^3?

Solution:

a. For the case of clean outdoor air, use Eq. 13.22:

$$C(0.167) = e^{-\left(\frac{Q}{V}+k\right)t} \times C_o = e^{-\left(\frac{50,000}{10,000}+0.1\right)(0.167)} \times 10 = 4.27 \text{ mg/m}^3$$

$$C(0.5) = e^{-\left(\frac{50,000}{10,000}+0.1\right)(0.5)} \times 10 = 0.78 \text{ mg/m}^3$$

$$C(1.0) = e^{-\left(\frac{50,000}{10,000}+0.1\right)(1)} \times 10 = 0.61 \text{ mg/m}^3$$

b. For the case of $C_{outdoor}$ = 4 mg/m^3, use Eq. 13.21:

$$C(0.167) = C_{outdoor} + (C_o - C_{outdoor}) \times e^{-\left(\frac{Q}{V}+k\right)t} = 4 + (10-4) \times e^{-\left(\frac{50,000}{10,000}+0.1\right)(0.167)} = 6.56 \text{ mg/m}^3$$

$$C(0.5) = 4 + (10-4) \times e^{-\left(\frac{50,000}{10,000}+0.1\right)(0.5)} = 4.47 \text{ mg/m}^3$$

$$C(1.0) = 4 + (10-4) \times e^{-\left(\frac{50,000}{10,000}+0.1\right)(1.0)} = 4.04 \text{ mg/m}^3$$

Discussion:

1. If there is no chemical decay, just set k = 0 in Eqs. 13.21 and 13.22.
2. As shown, the concentrations of both cases would drop exponentially, and both would approach the outdoor air concentrations (i.e., 0 and 4 mg/m^3, respectively).

13.5 WORKER EXPOSURE LIMITS

Occupational Safety and Health Administration (OSHA) is a federal agency within the U.S. Department of Labor. It assures workers safe and healthy working conditions by setting and enforcing standards and providing training, outreach, education, and assistance. Many air toxics have their concentration limits in the workplace.

Regarding air quality for the workspace, OSHA establishes *permissible exposure limits* (*PELs*), which are legal limits in the U.S. for exposure of an employee to a

TABLE 13.1
Examples of Exposure Limits of OSHA, NIOSH, and ACGIH

	OSHA		NIOSH		ACGIH	
Substance	**PEL (8-hour TWA)**	**STEL**	**REL (10-hour TWA)**	**Ceiling**	**TLV (TWA)**	**Ceiling**
Benzene (ppm)	1	5	0.1	0.1	0.02	—
Formaldehyde (ppm)	0.75	2	0.016	0.1	0.1	0.3
Lead (μg/m^3)	50	—	50	—	50	—

chemical substance (or physical agent such as loud noise). A PEL is usually given as a *time-weighted average* (TWA) over an 8-hour work shift, which balances short peaks and lower exposures into one overall daily average. However, there are *short-term exposure limits* (STEL) or *ceiling limits.* A STEL is the maximum concentration workers can be exposed to for a short period (usually 15 minutes) without harmful effects, and it is used to protect against acute effects, such as dizziness, irritation, and narcosis.

The National Institute for Occupational Safety and Health (NIOSH) is a research agency, part of the Centers for Disease Control and Prevention (CDC), and it develops recommendations for safe exposure levels, called *Recommended Exposure Limits* (RELs). RELs are based on the best available science and often are more stringent (i.e., more protective) than OSHA's PELs because some PELs have not been updated since the 1970s. However, NIOSH's RELs are not legal limits, so they are not enforceable.

The American Conference of Governmental Industrial Hygienists (ACGIH) is a scientific organization that studies occupational and environmental health issues. The ACGIH publishes annual guides for "*Evaluation and Control of Workplace Exposures to Chemical Substances and Physical Agents*", which contain *Threshold Limit Values* (TLVs) and *Biological Exposure Indices* (BEIs). Like NIOSH's RELs, ACGIH's TLVs are often more protective than OSHA's PELs and are not legal limits. Table 13.1 shows the exposure limits of OSHA, NIOSH, and ACGIH of benzene, formaldehyde, and lead as examples and for comparison.

EXERCISE QUESTIONS

1. An office building has a volume of 10,000 m^3, and the ventilation rate is 100,000 m^3/h. The outdoor air carries a pollutant with a 10 mg/m^3 concentration. The building has a constant pollutant emission rate of 1 kg/h. The pollutant decays following the first-order reaction with a rate constant of 0.1/h. Determine:
   a. the air exchange rate,
   b. the steady-state indoor pollutant concentration,
   c. the steady-state indoor pollutant concentration, if the outdoor air is clean,
   d. the steady-state indoor pollutant concentration, if the indoor emission source were absent, and
   e. the steady-state indoor pollutant concentration, if no decay of the indoor pollutant.

   Note: Compare your results with those of Example 13.2.

2. An office building has a volume of 10,000 m^3, and the ventilation rate is 100,000 m^3/h. The outdoor air carries a pollutant with a 10 mg/m^3 concentration. The building has a constant pollutant emission rate of 1 kg/h. The pollutant decays following the first-order reaction with a rate constant of 0.5/h. Determine:
   a. the air exchange rate,
   b. the steady-state indoor pollutant concentration,
   c. the steady-state indoor pollutant concentration, if the outdoor air is clean,
   d. the steady-state indoor pollutant concentration, if the indoor emission source were absent, and
   e. the steady-state indoor pollutant concentration, if no decay of the indoor pollutant.

   Note: Compare your results with those of Question #1.
3. An office building has a volume of 10,000 m^3, and the ventilation rate is 100,000 m^3/h. The outdoor air carries a pollutant with a 10 mg/m^3 concentration. The building has a constant pollutant emission rate of 1 kg/h. The pollutant decays following the first-order reaction with a rate constant of 0.1/h. Assuming the initial indoor concentration $(C_o)=0$, determine the indoor concentration at 10, 30, and 60 minutes.
4. An office building has a volume of 10,000 m^3, and the ventilation rate is 100,000 m^3/h. The outdoor air carries a pollutant with a 10 mg/m^3 concentration. The building has a constant pollutant emission rate of 1 kg/h. The pollutant decays following the first-order reaction with a rate constant of 0.1/h. Assuming the initial indoor concentration $(C_o)=0$, use the steady-state concentration you found for part (b) of Question #1 to determine the indoor concentration at 10, 30, and 60 minutes.
5. An office building has a volume of 10,000 m^3, and the ventilation rate is 100,000 m^3/h. The building has a constant pollutant emission rate of 1 kg/h. The pollutant decays following the first-order reaction with a rate constant of 0.1/h. Assuming the initial indoor concentration $(C_o)=0$ and the outside air is clean, determine the indoor concentration at 10, 30, and 60 minutes. Note: Compare your results with those of Example 13.5.
6. An office building has a volume of 10,000 m^3, and the ventilation rate is 100,000 m^3/h. There is an instantaneous pollutant release of 1 kg inside the building. The pollutant decays following the first-order reaction with a rate constant of 0.1/h. Assuming the initial indoor concentration $(C_o)=10$ mg/m^3 and the outside air is clean, determine the indoor concentration at 10, 30, and 60 minutes. Note: Compare your results with those of Example 13.6.
7. An office building has a volume of 10,000 m^3, and the ventilation rate is 100,000 m^3/h. The pollutant decays following the first-order reaction with a rate constant of 0.1/h.
   a. Assuming the initial indoor concentration $(C_o)=10$ mg/m^3 and the outside air is clean, determine the indoor concentration at 10, 30, and 60 minutes.
   b. What would be the corresponding concentrations if $C_{outdoor}=4$ mg/m^3?

   Note: Compare your results with those of Example 13.7.

BIBLIOGRAPHY

ASHRAE. (2009). *Indoor Air Quality Guide – Best Practices for Design, Construction, and Commissioning*, U.S. EPA Cooperative Agreement XA-83311201 funded the project, Developed by the American Society of Heating, Refrigerating and Air-Conditioning Engineers (ASHRAE), the American Institutes of Architects, Building Owners and Managers Association International, Sheet Metal and Air Conditioning Contractors' National Association, the USEPA, and the U.S. Building Council.

ATSDR. (2024). *Environmental Odors*, Agency for Toxic Substances and Disease Registry (ATSDR), U.S. Department of Health and Human Services, https://www.atsdr.cdc.gov/odors/about/index.html, https://www.atsdr.cdc.gov/odors/about/index.html.

NCEES. (2025). *PE/Environmental Reference Handbook (Version 1.6)*, National Council of Examiners for Engineering and Surveying (NCEES), Greenville, SC.

OSHA. (2011). *Indoor Air Quality in Commercial and Institutional Buildings*, OSHA 3430-04, Occupational Safety and Health Administration (OSHA), U.S. Department of Labor.

USEPA. (1991). *Indoor Air Facts No. 4- Sick Building Syndrome (Revised)*, Office and Air Radiation, U.S. Environmental Protection Agency.

USEPA. (1992). *Reference Guide to Odor Thresholds for Hazardous Air Pollutants Listed in the Clean Air Amendments of 1990*, EPA/600/R-92/047, March 1992, Office of Research and Development, U.S. Environmental Protection Agency.

USEPA. (2000). *Biosolids and Residuals Management Fact Sheet – Odor Control in Biosolids Management*, EPA 832-F-00-067, September 2000, Office of Water, United States Environmental Protection Agency, Washington, DC.

USEPA. (2006). *How Air Pollution Affects the View*, EPA-456/F-06-001, Office of Air Quality Planning and Standards, U.S. Environmental Protection Agency, Research Triangle Park, NC.

USEPA. (2025). *Basic Information about Lead Air Pollution*, U.S. Environmental Protection Agency, https://www.epa.gov/lead-air-pollution, last updated on 07/30/2025.

Index

Note: **Bold** page numbers refer to tables and *italic* page numbers refer to figures.

E

F

G

H

I

J

K

L